Systems Analysis and Design THIRD EDITION

Systems Analysis and Design THIRD EDITION

James C. Wetherbe

Professor and Director
Management Information Systems Research Center
School of Management
University of Minnesota

West Publishing Company

St. Paul - New York - Los Angeles - San Francisco

Copy Editor: Maggie Jarpey
Text Design: David Corona Design
Composition: Carlisle Graphics
Artwork: Carlisle Graphics

Printed in the United States of America

Library of Congress Cataloging-in-Publication Data

Wetherbe, James C.
  Systems analysis and design.

  Includes index.
  1. Management information systems.   I. Title.
T58.6.W47   1988   658.4′032   88-102
ISBN 0-314-73098-2

To Jamie and Jessie

Contents

Preface

While revising this book for a third edition, I came across a cartoon in which a son looked in amazement at his father and wondered, "Are you trying to tell me Paul McCartney was in a band before he was in Wings?" Somehow that cartoon captured my feelings as I tried to incorporate all the new techniques, technologies, and applications of information systems into the text while deciding which ones to eliminate. Do I dare cover batch processing, flowcharting or structured programming for fear I will date the book, or do I dare leave them out for fear an employer will be shocked that a recent graduate does not know what they are? To make matters worse, what if the graduate is trained only in techniques and technologies that the employer does not have available or has not yet heard of?

There is no question that in recent years new techniques and technologies have proliferated. Fourth-generation languages (4GLs), computer-aided software engineering (CASE), prototyping, end-user computing, data modeling, and information architecture have dramatically changed the way information systems are being developed and used. One of the biggest problems is that many of these techniques are used individually, without consideration for how they can be integrated with other new techniques and technologies. In the worst cases, some systems professionals are viewing complementary techniques as competitive. For example, articles have been written comparing data-flow diagrams with prototyping, and prototyping with data modeling, arguing that one is better than another. However, these techniques are not competitive; they are complementary when properly understood and integrated into the systems development process.

This book attempts to integrate the new techniques into the system development process. The older techniques and technologies that are still useful for functional, or at least historical, purposes remain, while those that are deemed outdated have been eliminated.

Sections and Chapters

SECTION I of the book, "Conceptual Foundations" (Chapters 1–4), establishes the concepts, theories, and frameworks useful for understanding and applying the techniques and technologies covered in the remainder of the book. Theories and frameworks remain fairly stable over time, and consequently provide a useful anchor with which rapid technique and technological advances can be interpreted. For example, an important *theory* is that managers do not know their own information requirements. This unawareness causes extensive revisions after systems development. In the 1960s the user sign-off was a *technique* to deal with this problem. The user sign-off contractually obligated users to like the system. The technique was politically useful but usually resulted in dissatisfied users.

In the 1970s the techniques of prototyping and heuristic design came along, and in the recent decade, data modeling and CASE were added. These new techniques and technologies helped us come closer to solving the same conceptual problem—that managers don't know what they need.

SECTION II, "Techniques and Technologies for Systems Analysis and Design" (Chapters 5–11), was extensively revised and features a new chapter on project management. Chapter 5, "Problem and Opportunity Recognition," is stronger in its emphasis on information systems for competitive advantage and justification systems. Chapter 6, "System and Information Analysis," has been enhanced with information requirements determination tools, CASE technology, and more extensive treatment of data-flow diagrams.

Chapter 7, "Information Systems Design," has been totally reworked to incorporate data modeling and includes new concepts for prototyping/heuristic design, and CASE technology and tools. Chapter 9, "Selecting Technology and Personnel," was revised to reflect new developments. Chapter 8, "Systems Development," was upgraded with CASE technology, with an emphasis on the programmer's workbench. Chapter 10, "Implementation and Evaluation," includes more material on behavioral factors. Chapter 11, "Project Management for Systems Development," is a new chapter provided in response to requests for in-depth coverage of project management techniques. A special feature of Chapter 11 is a section on project leadership and team building.

SECTION III, "Strategic Administration, and Higher-Level Concepts and Techniques," was extensively revised, and a new chapter on end-user computing has been added. Chapter 12, "Systems Administration," includes new material on centralization versus decentralization of information system activities. Chapter 13, "Strategic Planning for MIS," has been updated with the inclusion of the latest techniques. Chapter 13, "Analysis and Design of Decision-Support Systems," now contains a section on expert systems and knowledge engineering. Chapter 15, "End-User Computing," is a new chapter designed to integrate this proliferating phenomena into systems analysis and design thinking activities. Chapter 16, "Future Considerations of Systems Analysis," provides a crystal ball for consideration of the future.

Case Book

The text has a companion case book that is useful for experiential learning in preparing systems design specifications. The case book may be used in conjunction with this text or by students taking other courses, as it provides a beneficial, pragmatic exposure to systems design basics. In its new form, the case book can be used with CASE products. Its full title is *Case Studies for Systems Analysis and Design*, and it is available from West Publishing.

Acknowledgements

In conclusion, I wish to acknowledge the contributions of several individuals. I am especially indebted to V. Thomas Dock, University of Southern California, who originally encouraged me to write this book and served as consulting editor. The reviewers whose helpful suggestions are reflected throughout the text are: Marilyn Bohl, IBM; Thomas I.M. Ho, Purdue University; Roger Hayen, University of Nebraska; Maryam Alavi and Richard Scamell, University of Houston; Charles Paddock, Arizona State University; Everald E. Mills, Wichita State University; Robert T. Keim, Arizona State University; Kenneth W. Veatch, San Antonio College; Dennis Guster, St. Louis Community College; Albert L. Lederer, University of Pittsburgh; Gary W. Strong, Drexel University; Warren Briggs, Suffolk University; Sandra Fabyan, Columbus Tech; Rod Conner, University of Richmond; Carl Hossler, Kent State University; Rani Mehta, Computer Applications; M. Bond Wetherbe, Navy Automation Division; and Sal March, Dave Naumann, Detmar Straub, Cynthia Beath, Craig Rogers and Patricia Carlson, University of Minnesota.

I offer sincere thanks to my patient, tolerant, and forgiving support staff: Katherine A. Cooper, Leslie Maggi, Susan M. Scanlan, Sara Thurin, and Mark Saari for their editing and word-processing efforts.

Finally, I acknowledge the efforts and support of my family—Smoky, Jamie, and Jessie, who are so supportive in my professional efforts.

As I acknowledge the efforts of all who have contributed, I also assume full responsibility for any inadequacies or discrepancies in the text.

JAMES C. WETHERBE
Minneapolis, Minnesota
1988

Conceptual Foundations

The first section of this book covers the concepts that set the foundation for studying, understanding, and applying the systems analysis and design material covered in Section II, and the strategic and administrative material covered in Section 3.

Chapter 1 provides an overview of systems development, Chapter 2 focuses on concepts of systems, Chapter 3 on information and organizations, and Chapter 4 provides a review of basic data processing. Chapter 4 should be considered optional for readers familiar with the basics of information technology. ■

Overview of Systems Development

Introduction

Organizations are under greater pressure than ever before to increase productivity. Increasing productivity has become a key to remaining competitive and viable in today's post-industrialized, information-oriented society. Yet the very resources that organizations need to remain competitive cost more. People, equipment, facilities, capital, and energy all cost more and therefore increase the cost of producing goods and services. The only exception (and a glaring exception it is) is computer technology. Computers continue to drop in cost at a phenomenal rate. An analogy is often made between the computer industry and the automotive industry. If the automotive industry had had the same breakthroughs that the computer industry has had during the past thirty years, then you could buy a new luxury car today for about a penny. No other industry in our society is offering that much increase in cost performance.

Accordingly, computer technology and information systems are the best game in town with which to increase organizational productivity. Those people who know how to apply this powerful technology hold the keys to the future.

Organizations are just now realizing that we are truly moving into the information society. Information is now viewed as a strategic corporate resource. Most Fortune 500 companies have identified computers and information systems as one of their top strategic issues for remaining competitive in the future.

As a resource, information has characteristics that organizations have never really dealt with before. For one thing, unlike other resources such as fuel and real estate, information is not a scarce resource. It also is a sharable resource. It is a resource that you can have, give to someone else, and yet still hold. You can't do that with a barrel of oil or with investment money.

Indeed, it is this ability to be shared that makes information such a powerful resource. It is the glue that holds organizations together. It is the resource that allows the left hand to know what the right hand is doing. It is the resource that can avoid the incredible waste that occurs when an organization is poorly coordinated due to poorly informed managers.

These characteristics of computers and information, and the need for organizations to increase organizational productivity, are what make the study of systems analysis and design such a worthwhile endeavor. This is true whether you aspire to be a systems analyst or to hold any other staff or management position within an organization.

Irrespective of where you ultimately end up working, computers and information systems are going to be a key dimension of what you do. No matter what management or staff position you hold, the most important thing you will be doing is making decisions. Other than your intellect, the most important resource you will need to do your job well will be information.

To set the stage for studying material in this book, certain key foundations must be established. These will be discussed next.

Theory versus Technique

First, it is very important to understand the difference between theory and technique. Theory is really nothing more than the notions or ideas about the way things work or the way things are. Techniques are just different methods or approaches to performing different tasks. Ideally, techniques should be based upon certain theories.

Some people argue that the key to success is to have good technique, while others argue that the key is to have good theory. The two concepts, however, complement each other and are both necessary. Someone who knows technique without theory is potentially a dangerous person. Such a person uses various techniques or methods without really understanding the reasons behind them. Anytime someone uses a particular technique without knowing the reason or the theory behind it, that person is apt to use the technique when it is inappropriate. For example, consider the young man who would always cut a roast in half before he would cook it. His wife, observing him do this, asked him, "Why do you cut the roast in half before you cook it?" The husband replied, "I don't know. My mother always did, so I always have." After awhile the young man became curious as to why it had been a family tradition to cut the roast, so he asked his mother about it. His mother explained that she had always done it because *her* mother had always cut the roast in half before cooking it. Their curiosity led them to question the grandmother as to why the roast was always cut in half before it was cooked. The grandmother had a simple explanation—the reason she cut her roast in half was because her pan was so small it was the only way that she could fit the roast into the pan. Though the grandmother had a reason for using the roast-cutting technique, no one else did.

We often find people in different professions, including systems analysts, using techniques that are outdated or no longer appropriate. This generally happens when they don't understand the theory behind the technique.

It is not enough, however, just to know theory. Someone who knows theory and no technique has little to contribute. What is needed is an understanding of both the theory behind something, and the techniques that can be used to solve problems based upon that theory.

Ignorance, Knowledge, and Counterknowledge

The importance of knowing both theory and technique is best explained by first discussing ignorance, knowledge, and counterknowledge. Ignorance is when someone does not know how to do something—perhaps a technique for doing something. Knowledge is when someone does know how to do something or a technique for doing something. Counterknowledge is when someone thinks he or she knows how to do something or the best technique for doing it, but is wrong. This person might be wrong because he or she is using a technique in the wrong situation or because the technique is no longer the best technique for solving a particular problem. You will generally find people guilty of counterknowledge when they really do not understand the theory behind the technique they are using.

A common criticism made of the educational process is that it tends to teach students how to do exercises rather than how to solve problems. For example, we teach students how to write a computer program, how to solve a mathematical equation, or how to do a financial analysis. The students, however, often do not know when it is appropriate to *use* any of these techniques. This is similar to the story of the student who is taught how to use a screwdriver—the screwdriver being symbolic of some technique such as computer programming or linear programming or net-present-value analysis. Upon being employed, the student is excited about applying the screwdriver to screws that need to be tightened. The new employee goes about the organization looking for screws to tighten and tightens them. Eventually, however, there are no screws left to tighten. Looking for new opportunities to apply the skill of using a screwdriver, the new employee gets out a hack saw and starts filing slots into the heads of nails. He then gets out the screwdriver and uses it upon the nails. As silly as this story may seem, it is unfortunately representative of the application of an inappropriate technique to the solution of a problem.

It is very important when solving problems to understand what the real problem is, what theory is relevant to the problem, and what technique, or techniques best fit the situation.

Methodology Lags behind Technology

A key aspect of selecting the right technique or methodology to solve a problem is being aware of the most current techniques or methodologies. In all fields, techniques and methodologies tend to lag behind technology. In other words, a set of techniques or methodologies is developed based upon technology that is available at a specific point in time. As time passes, new technologies are introduced, but people tend to use the old methodologies that were developed for a previous generation of technology. This often results in considerable loss in problem-solving potential and productivity.

For example, let's move out of a computer and information systems context, and consider the Revolutionary War. During the Revolutionary War, the British used a methodology of warfare in which soldiers in festively colored uniforms lined up shoulder to shoulder several deep, and marched through open fields. The Revolutionaries, on the other hand, wore drab clothing that tended to blend in with the rocks and the trees that they hid behind. Thus they were able to have a strong competitive advantage in the warfare. Now it might seem a little peculiar that the British were willing to wear bright red jackets with white Xs across the front and march into open fields and open gunfire. Why, in combat, would they use a methodology that clearly was not in their best interest? The problem was that their methodology lagged behind their technology. The method of lining up in a formation of thousands and marching at the enemy was born during the sword-and-shield technological era. The idea of creating an awesome-looking army and marching at the enemy evidently was quite effective during that era. The introduction of rifle technology, however, made such a methodology fatally obsolete. The British literally offered target practice to the revolutionaries and suffered as a result of it.

Similarly, many aspects of what we are now doing in information systems development are quite obsolete, given the technologies now available. In other words, methodology is lagging behind technology. One way to become sensitive to this phenomena, and therefore have the opportunity to minimize it, is to be aware of both the theories and techniques of systems development. Theories do not tend to change as rapidly as do methodologies or techniques. For example, one theory pointed out in this book is that managers do not know what information they need. Though that theory has been around since the 1960s, the techniques used to deal with it have changed significantly. Consequently, a good understanding of theory can assist one in reviewing and revising methodologies to catch up with technologies.

Throughout the book we will emphasize what is theory and what is technique, and try to point out how we are catching up with technology.

What Makes for a Good Systems Analyst

Information systems analysts are people who are involved in analyzing, designing, implementing, and evaluating computer-based information systems to support the decision making and operations of an organization.[1] They are ostensibly a boundary spanner between computers and management as illustrated in the following circle graph:

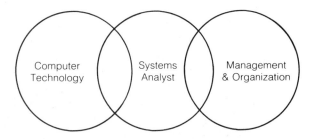

As the graph shows, systems analysts must understand the technology, the organization, and the skills of their trade. To refine this further, the key ingredients that make for a good systems analyst are:

1. Understanding of, and commitment to, the organization
2. People skills
3. Conceptual skills
4. Technical skills

Each ingredient is discussed in the following text.

Understanding of, and Commitment to, the Organization

One of the major complaints made about systems analysts is that they don't understand, and are not committed to, the organizations in which they work. The typical complaint is that systems people are more interested in the technology they work with than in the organization that uses it.

This problem can be illustrated by an example. A medium-sized department store chain was installing its first computer. In the process this chain was switching from electromechanical cash registers to electronic, optical-wand retail terminals. They had hired a young man with an M.B.A. and a few years of experience in

---

1. This discussion is based on a chapter from James C. Wetherbe, *Executive Guide to Computer-Based Information Systems* (Englewood Cliffs, NJ: Prentice-Hall, 1983), pp. 156–60.

systems design. The president of the company planned to groom the young man into an information systems manager. Within two weeks, however, the president was ready to fire his new employee.

As it turned out the president had dropped by the new employee's office and asked if he was busy. When the young man replied that he was not, the president told him that the manager of the shoe department was on vacation and the assistant manager had taken ill; there was no one to run the department. The president asked the systems person if he would like to find out a little bit about retailing by running the department for the day.

The president received a flat refusal. (What would you have done?)

Here's some more background. The president had worked his way up from a salesclerk to become president. He was often seen in different stores doing salesclerk tasks to keep a feel for the business. Now he was being told by an insubordinate subordinate, "I don't do shoes."

Of course the young man had a point. He had just gotten his M.B.A.; what if one of his classmates came into the shoe department and wanted to be fitted with new wingtips?

The real problem for the president was that he had a systems analyst with no real interest in retailing; the systems person was not interested in being exposed to the front lines, though he could have used his technical expertise to evaluate the support systems in use. In other words, the young man had no interest in gaining new understanding, and he was not committed to the organization. What he had considered as an insult, could have been viewed as an opportunity. His course of action ended his career with the retail organization.

People Skills

Another complaint made about systems people is that they don't interact well with other people in the organization. Again, they are generally more interested in the technology they work with than with the user who needs better information or computer support for operations. They can also be arrogant toward users, feeling that the technical expertise of a systems analyst gives them intellectual superiority over other people in the organization. (Many of them like to use Mr. Spock from Star Trek as a role model.)

An anecdote will illustrate this phenomena and its associated problem. I had a particularly bright computer scientist working for me. He was continually alienating users by ridiculing their lack of expertise in the use of computers. Hard as they would try, the users never won a technical argument with my staff member. I tried to point out to him that he was gaining no friends, gaining no advantage, and hurting himself professionally by his actions. He acknowledged that he knew it was bad form to do what he was doing. I asked him why he did it. He paused, thought for awhile, and said, "because it feels so good."

A great deal of insight into systems professionals and their people-orientation comes from research conducted by Couger and Zawacki.[2] They found that, when compared to other professionals, systems people on the average have the highest

---

2. Daniel Couger and Robert A. Zawacki, "What Motivates DP Professionals?" *Datamation*, September 1978, pp. 116–23.

desire for acquiring new technical skills and expertise and the lowest desire for social interaction.

If a systems analyst is a boundary spanner, he or she must be willing and able to communicate effectively.

Conceptual Skills

A study conducted by Nick Vitalari while a Ph.D. student at the University of Minnesota evaluated the problem-solving strategies used by systems analysts. The subjects used in the studies were practicing systems analysts who were either regarded as outstanding systems analysts or poor systems analysts by the organizations for which they worked.

The systems analysts who were not rated well by their respective organizations approached problem solving differently from those who were rated well. The poorly rated systems analysts, when faced with a problem, tended to go into specific detail issues and lose sight of the overall problems. The highly rated systems analysts, on the other hand, tended to approach problem solving by first going to a high conceptual perspective of the problem and then factoring down to specific detail issues that needed to be addressed to solve the problem.

In other words, poor systems analysts tended to put themselves in a position where they couldn't "see the forest for the trees" while good systems analysts always kept the big picture in mind.

Technical Skills

It goes without saying that a good systems analyst must understand information systems technology. However, this technical expertise does not need to be "nuts and bolts (e.g., computer circuitry, compiler processing)." Rather, the expertise needs to be applications-oriented. The systems analyst must know the technical alternatives used to solve different information problems and how the technology should be applied.

The Systems Development Cycle

As a way of introduction, it is helpful to have a preliminary overview of the systems development process. The process described will be greatly expanded in Chapters 5 through 11, but only after the conceptual foundation and theories have been covered in Chapters 2 through 4.

All organizations have several types of information systems. At any point in time, varying degrees of these information systems may be computer-based. For example, an organization's payroll information system may be computer-based, but its strategic planning information system may be processed manually. As time passes, organizational processes change. The organization's information systems must change in response to the changing needs of the organization. When this occurs, management usually initiates some form of a systems development life cycle (SDLC) to address the problem.

Figure 1.1   **Overview of the Systems-Development Life Cycle (SDLC)**

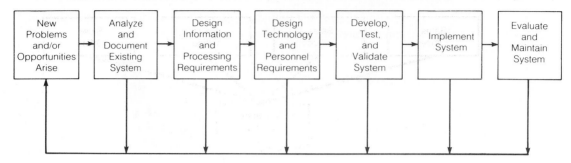

Figure 1.1[3] shows a conceptualization of the basic components of the SDLC. The steps in the cycle are not completely discrete, one-time processes. Rather, the systems development cycle is iterative and evolutionary. The steps in the cycle, however, retain a basic sequential flow from the point of origin of the development cycle—"new problems and/or opportunities arise." Each step represents a major checkpoint, or milestone, and has identifiable deliverables that symbolize completion. At any step in the cycle, previously unidentified problems and/or opportunities may be discovered. In such cases, it is important to ensure proper integration of a solution to a new problem and/or a new opportunity. Making even a minor change to a system without proper consideration of what has been established previously can cause unanticipated and undesirable rippling effects. These rippling effects can seriously damage the operation of the system. Therefore, as Figure 1.1 shows, all steps of the systems development cycle allow return to the point of origin at any time.

The time required to complete the cycle for a given problem or opportunity may range from a few hours to a few years, depending on the complexity and difficulty of the task.

Identify Problems and Opportunities

As depicted in Figure 1.1, the impetus for initiating a systems development cycle is the identification of new problems and/or opportunities. For example, a wholesale distributor receives customer complaints concerning late deliveries of orders. These late deliveries result in lost orders and loss of customer goodwill. Action is necessary.

An overview of an order-processing system for a wholesale distributor is given in Figure 1.2. The delays in delivery of orders may be caused by several factors (e.g., inefficiencies in processing customers' orders, inventory stock-outs, and programming errors).[4] To clearly understand the exact nature of the problem requires a thorough analysis and understanding of the existing system.

---

3. The components of the systems development cycle are described in varying degrees of detail and categorization by different writers. However, there is general agreement on the flow of the development steps.

4. The delays in delivery of orders can pertain to problems other than those of the information system (e.g., employee negligence or inadequate warehouse capacity). For the sake of brevity, the example used in this introductory chapter assumes the problems are primarily related to the information system.

Figure 1.2   Overview of an Order-Processing System for a Wholesale Distributor

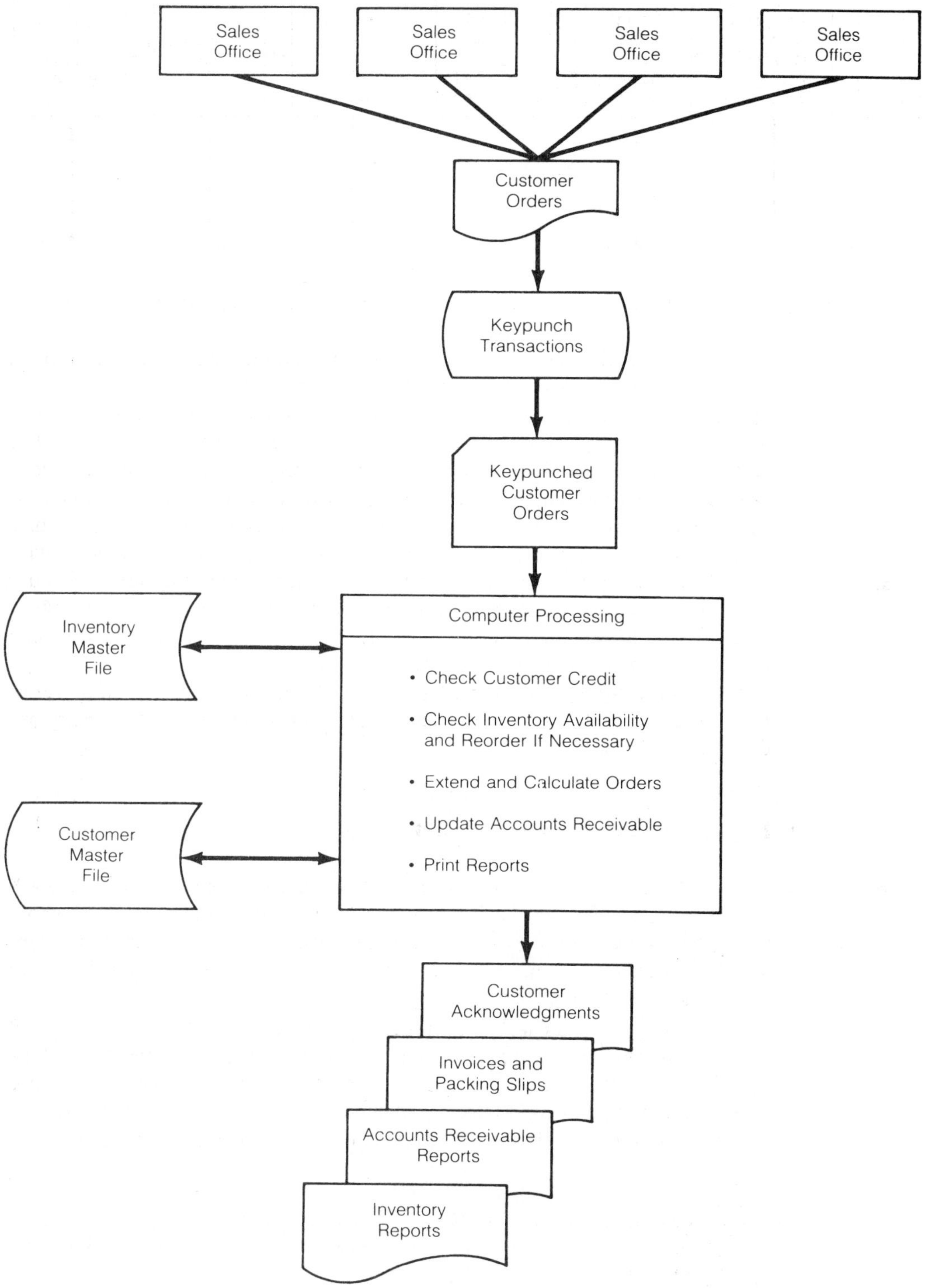

Analyze and Document Existing System

At this point, management may initiate a project team (or task force) to analyze the existing order-processing system. This project team can be headed by an individual who is a systems analyst or by someone who has systems analysis skills.

Analysis of the existing system consists of the following activities:

- Review work flow
- Define decision making associated with work flow
- Review current information available to support decision making (e.g., transactions and reports)
- Isolate deficiencies in the information system

Deficiencies in information systems can be categorized as *inclusion* and/or *structure deficiencies*. Inclusion deficiencies pertain to *what* information, technology, and personnel are included or lacking in the system. Structure deficiencies pertain to *how* the information, technology, and personnel are organized and interrelated throughout the system. This concept is illustrated in Table 1.1.

Structure and inclusion deficiencies relate to a subtle but nevertheless key concept: that an *appropriate* information system must be designed, and that it must be designed *appropriately*. An appropriate information system includes the necessary information and also the technology and personnel necessary to properly process the information. An information system is designed appropriately when the information, technology, and personnel included in the system are structured and operated correctly. In other words, an appropriate information system is

Table 1.1		Information System Deficiencies	
		Inclusion Deficiency	Structure Deficiency
Information	Definition	Appropriate information is lacking, and/or inappropriate information is included in the system.	The manner in which information is collected, stored, and/or reported in the system is difficult or inconvenient to utilize.
	Example	Information that is required for approving a loan application is missing, and/or information that is not required for approving the application is included.	Detail data are not properly summarized.
Technology	Definition	Appropriate technology is lacking, and/or inappropriate technology is included in the system.	The hardware and software included in the system are not organized to most effectively and efficiently process the available information.
	Example	Batch processing is used when an on-line system is required.	Computer programs are inefficiently written or contain programming errors.
Personnel	Definition	Appropriate personnel are lacking, and/or inappropriate personnel are included in the system.	The personnel associated with the system are poorly organized and/or managed.
	Example	Personnel do not have the aptitude, training, or motivation to work with the system or make decisions from information that it generates.	Personnel responsibilities are not clearly defined, or work load is not evenly distributed among personnel.

effective (i.e., produces the desired information). If an information system is structured appropriately, it is *efficient* (i.e., produces information with a minimum of effort, expense, or waste).

Design Information Requirements

Once the deficiencies in the information system have been determined, solutions to those deficiencies can be designed. The end result of an information system is the provision of information necessary for decision making. Therefore, solutions to information problems can be defined in terms of the inclusion or structure of information.

In the case of the wholesale distributor, it may be determined that historical information about customers' orders would facilitate better inventory forecasting. This solution would necessitate the inclusion of new information in the system. A structure-oriented solution could consist of providing a summarized average-demand-by-month report. Since this information is available, all that is required is a restructuring of the existing information into a new form.

In defining information solutions, it is important that the solutions relate to improving work flow and/or the decisions being made. In other words, solutions should improve productivity. Any temptation to include additional information that is not needed should be avoided. Otherwise, the expense of the information system may be increased needlessly, and managers may be overloaded with voluminous reports irrelevant to their decision making. For example, historical information about customer orders that is more than three years old may not be useful for forecasting purposes. To include such information may increase processing costs in order to produce useless reports.

Design Technology and Personnel Requirements

The design of information and processing requirements establishes the criteria for identifying alternative means for solution achievement. That is, the previous step defines *what* is desired. This step defines *how* to do it. Viable technologies and personnel are identified that, if included in the system, can be structured to support the solution defined in the previous step. Generally, several alternatives offering varying degrees of solution achievement are available. For example, the wholesale distributor currently posts customer orders to the inventory file on a weekly basis. Transactions are delivered to keypunch personnel on a daily basis and are keypunched during the week. The inventory file is updated, and inventory reports are printed and distributed once a week. The problem with this system is that inventory information is up to a week old. Consequently, sales personnel often overcommit existing inventory. Also, the need to reorder out-of-stock items is not detectable until the weekly processing has been completed.

One means for improving the timeliness of inventory reporting is to restructure the existing system to expedite the current processing procedure. Within this alternative, varying degrees of expediting are possible. Transaction posting and report generating could be done twice a week or even daily. Processing twice a week would require an additional computer run. Daily processing would require daily computer runs and would also very likely require either overtime or additional personnel in keypunch.

Inclusion of a new technology is another possibility. For example, on-line technology could be used for entry of transactions through computer terminals. A less expensive version of this alternative consists of logging transactions onto a microcomputer at each sales office. At the end of the day, transactions could be transmitted from the microcomputer to the central computer for further processing and preparation of daily reports. Another version of this alternative is to have terminals connected directly to a computer system. As each transaction occurs, inventory information is updated immediately. This way, the quantity-on-hand information would always be up to date.

It is important to identify all viable means of resolving the problem under consideration. In the technologically dynamic field of computers, particular care must be taken not to simply resort to previously used approaches that may already have become obsolete due to the development of more effective and/or more efficient technologies.

Evaluation and selection of technology and personnel consists of thoughtfully considering the possible approaches to problem resolution in a cost-versus-benefit framework. Sometimes costs can be accurately estimated, however, it is generally difficult to accurately assess the benefits of improving an information system. For example, how much benefit will be derived from improved goodwill due to expediting customer orders? Consequently, selection of one system over another is often based on subjective evaluations.

Management involvement in the selection decision is particularly important. Management has the best vantage point from which to make subjective-type decisions within the organization.

Develop, Test, and Validate System

At this point, the desired solutions and the means of achieving them have been identified. Specifically, inclusion and structure decisions are established for the information and the technology of the new system. The actual structuring, or development and testing, of the system is now possible. This step consists of installing any additional hardware or software required and coding and testing computer programs where necessary. The end product of this step is an operable and reliable system that actualizes the original design objectives.

Implement System

After the new system has been developed and tested, conversion from the old system to the new system is required. The actual implementation process can be expressed in terms of a continuum that ranges from parallel to discrete.

In a parallel implementation, the old and new systems are concurrently processed until the new system stabilizes. This allows the reliability of the new system to be confirmed prior to abandoning the old one. In a discrete implementation, the old system is terminated as the new one is begun. In many cases, some parts of a new system are implemented in parallel, whereas others are implemented in a discrete manner.

Parallel implementation reduces risk when implementing a new system. It does, however, have disadvantages. First, parallel implementation is more expensive than discrete implementation. Often, additional personnel (or overtime) are required and/or additional equipment is needed. For example, assume the wholesale distributor is converting from batch to on-line processing of customer orders. To parallel process during the implementation will require preparing the transaction documents for keypunching (the old system) in addition to entering the transactions through terminals (the new system). Keypunch machines and keypunch personnel and the computer-processing and report-generating functions of the old system will still be required. If the computer that supports the on-line processing is to replace a computer in the old system, both the new and the old computers will be required until conversion is completed and parallel processing is terminated.

Even if warranted, parallel processing is occasionally not practical. For example, it is not plausible to parallel process two on-line airline reservation systems. In such cases, rigorous testing of the new system is critical prior to attempting a discrete implementation.

Evaluate and Maintain System

After a new system has been implemented, it is important to review how effectively and efficiently the solutions to the new problems and/or opportunities have been achieved. Evaluation, therefore, consists of assessing the degree of variation between planned and actual systems performance.

This evaluation relates to the concepts of inclusion and structure deficiencies. If the availability and the form of the information are consistent with design specifications, then the appropriate information is included, and it is appropriately structured. This can be determined by surveying users of the information system to determine if they have the information necessary for decision making. The information included in the system can also be evaluated by assessing decision-making performance. For example, does having information available on average-demand-by-month reduce inventory costs by 10 percent?

If the timeliness and accuracy of the information are consistent with design specifications, then the appropriate technology and personnel are being used, and they are structured appropriately. For example, are computer terminal operators consistently able to achieve response times of less than two seconds for queries made to the inventory file?

If the new system fails to achieve the design objectives or presents new problems or opportunities, a new SDLC may have to be initiated. If the new system performs satisfactorily, then the system can be maintained at the current operating level until new problems and/or opportunities arise.

Summary

The need for more timely, accurate, and complete information is continually increasing as organizations increase in complexity and size. Accordingly, there is an increasing need for systems analysts who can assist in bridging the gap between computer technology and organization needs. The role of the systems analyst in designing information systems is analogous to the role of the architect in designing a building. He or she must develop an understanding of, and an intimacy with, the organization's problems and work with the organization in order to design appropriate solutions.

The SDLC consists of seven steps. It evolves and recycles with time. The steps in the cycle are:

- Identify new problems and/or opportunities as they arise.
- Analyze and document the existing system.
- Design information and processing requirements.
- Design technology and personnel requirements.
- Develop, test, and validate the system.
- Implement the system.
- Evaluate and maintain the system.

Throughout the SDLC, the inclusion and structuring aspects of information, technology, and personnel must be carefully considered to achieve an effective and efficient information system.

Exercises

1. Discuss the difference between theory and technique.
2. Define and discuss counterknowledge and its causes.
3. Define and discuss the steps of the systems development cycle.
4. Differentiate between inclusion and structure as they pertain to information, technology, and personnel.
5. Visit with several managers, preferably in different organizations, and question them as to the deficiencies in the information available to them. Categorize these deficiencies according to the matrix in Table 1.1.

Chapter 2

Concepts of Systems

Introduction

In Chapter 1 the importance of two necessary skills was discussed. The first is to understand theory as well as technique. The second is the ability to conceptualize, that is, being able to see the forest as well as the trees.

This chapter aims to provide both a strong theoretical framework and a conceptual framework within which powerful insight into information systems problems can be gained and techniques can be identified to solve them. The discussions in this chapter will at times be abstract, but that is necessary in order to achieve a conceptual understanding or perspective of complex problems. This chapter is as theoretical and abstract as this book ever gets. An understanding of the material presented in this chapter allows for very effective and efficient discussion of a variety of techniques throughout the book.

The word *system* is widely used. It has become fashionable to attach the word *system* to add a contemporary flair when referring to things or processes. People speak of exercise systems, investment systems, delivery systems, and information systems. This chapter introduces general systems concepts, with emphasis placed on concepts used later in the text in discussions of analyzing and designing information systems.

Definition of a System

In the most general sense, a system is a collection or arrangement of *entities*, or things, related or connected such that they form a unity or whole. In systems designed and controlled by people, the entities are generally arranged so that they can interact to accomplish one or more objectives. For example, the entities of an educational system consist of students, faculty, administrators, textbooks, buildings, equipment, and so forth. The entities of the system interact (e.g., faculty teach students, students read texts, administrators schedule classroom facilities), and ultimately the primary objective of the system—learning—is accomplished.

Usually, some entities are retained in (internal to) the system and other entites are transient to it. Transient entities are inputted to the system and subsequently outputted from the system—generally, after some transformation process has occurred. During the time that transient entities are in the system, they are part of the system. However, prior to being inputted or after being outputted, transient entities are external to the system. Anything external to a system is referred to as the *environment* of the system.

In an educational system, the entities retained in the system include facility, administrators, textbooks, buildings, equipment, and the like. Students are transient entities, inputted to and outputted from the system. The primary transformation processes consist of transforming students with one level of knowledge to students with a higher level of knowledge.

General Model of a System

It is helpful to view systems in terms of models. A basic model of a system, consisting of inputs, outputs, and the system itself, is shown in Figure 2.1*a*. As

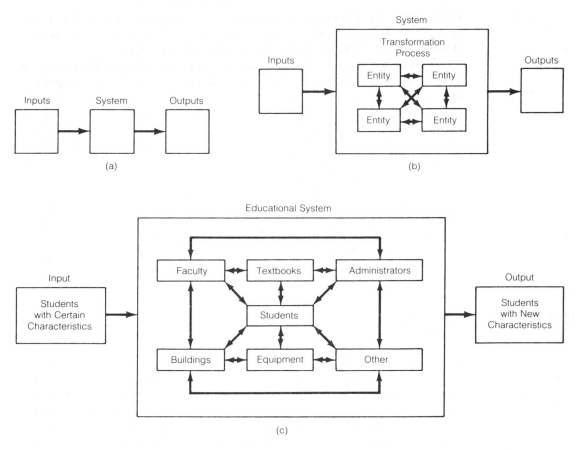

previously defined, a system is composed of entities that interact to achieve a transformation process. An expanded version of the basic model depicting system entities is given in Figure 2.1*b*. The expanded model of a system can be applied to an educational system as shown in Figure 2.1*c*.

Classification of Systems

Systems can be viewed or classified in several ways. For the purposes of this text, the most useful classification scheme consists of the categories of closed/stable/ mechanistic systems and open/adaptive/organic systems. Few systems are exclusively categorized one way or the other. Rather, any given system can be defined as existing on a continuum between the two categories. This concept, as applicable to three systems (A, B, and C), is illustrated by the following sketch:

System A   System B   System C

Closed/Stable/Mechanistic ⟵　◆　　　◆　　　◆　⟶ Open/Adaptive/Organic

Closed/Stable/Mechanistic Systems

A closed/stable/mechanistic system tends to be self-contained. It seldom interacts with its environment to receive input or generate output. Therefore, the entities of primary concern in a closed/stable/mechanistic system are those retained in the system. The interactions of the entities are stable and predictable. Consequently, the operation of the system tends to be highly structured and routine. A model of a closed/stable/mechanistic system is given in Figure 2.2.

An example of a closed/stable/mechanistic system is a terrarium (a glass container enclosing a garden of small plants). The primary entities in this system include plants, moisture, soil, oxygen, and carbon dioxide. These entities interact with each other in a very stable and predictable fashion.

No known system can continue to operate for prolonged periods of time without interacting with its environment. Systems are subject to deterioration or slow decay. They eventually need to input new entities of material, energy, or information in order to survive. In the case of the terrarium, water occasionally needs to be added so that the system can continue to operate. Another example of a closed/stable/mechanistic system is an electric motor powered by a battery. This system can continue to operate without interacting with its environment until the battery runs down or the motor needs repair.

A closed/stable/mechanistic system does not necessarily have to interact with its environment to continue to exist. However, in the long run, it will have to interact to continue its normal operation.

Open/Adaptive/Organic Systems

An open/adaptive/organic system continually interacts with its environment for replenishment of material, energy, and information. Therefore, in an open/adaptive/organic system, entities internal and external to the system are of concern. The interactions of these entities are probabilistic and changing in nature and are consequently less predictable than those in a closed/stable/mechanistic system. The operation of an open/adaptive/organic system tends to be less structured and

---

Figure 2.2   A Closed/Stable/Mechanistic System (no inputs or outputs)

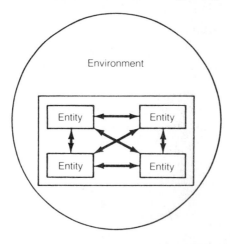

---

routine than that of a closed/stable/mechanistic system. An excellent example of an open/adaptive/organic system is a business organization.

A distinctive feature of an open/adaptive/organic system is that it can adapt to changing internal and environmental conditions. It can change the entities retained in or transient to the system. It can also change the way that entities interact. An open/adaptive/organic system is self-organizing, and it can change its organization in response to changing conditions. For example, a business organization can change its personnel, equipment, or products; it can also change the way in which they are organized and what they do.

Effect of Inputs and Outputs

Though all systems that continue to operate have to interact with their environment, the *frequency* of this interaction is a variable that plays a key role in defining the position of a system on the continuum between closed/stable/mechanistic and open/adaptive/organic. The other key variable associated with interaction is the *characteristics* of the inputs and outputs to the system. If the inputs and outputs are known, defined, and predictable, even though the system is not closed, it is still relatively stable and mechanistic. Alternatively, if some of the inputs and outputs are unknown, undefined, and unpredictable, then the system has to be adaptive and organic to cope with these unknown variables. Models of relatively closed/stable/mechanistic and open/adaptive/organic systems are given in Figure 2.3.

Figure 2.3 **Illustrations of (a) Closed/Stable/Mechanistic and (b) Open/Adaptive/Organic Systems**

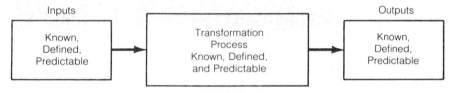

Relatively Closed/Stable/Mechanistic System

(a)

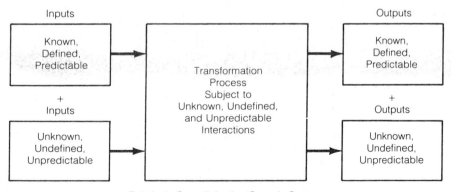

Relatively Open/Adaptive/Organic System

(b)

An example of a relatively closed/stable/mechanistic system is an internal combustion engine. The inputs are gasoline and air, and the outputs are power and exhaust. The input, output, and transformation processes are stable and mechanistic.

An open/adaptive/organic system can be illustrated by the educational system discussed earlier. Students, which are the primary input to the system, are variable and often unpredictable. The best way to achieve the transformation process (learning) is not highly defined, and, of course, the characteristics of the outputs are variable.

System Control

To ensure the proper operation of a system, some form of control is required. To accomplish this, a feedback process must be added to the basic system model. A system without feedback for regulatory purposes cannot intelligently adapt to changes in internal or external entities to the system and is subject to entropy or slow decay. The addition of feedback to the basic system model (Figure 2.1a) is shown in Figure 2.4.

This model illustrates that controlled characteristics or conditions are sensed or measured by a sensor. A comparator is utilized to compare actual conditions to desired conditions or objectives. Finally, an activator invokes action to bring about change (if necessary) in the operation of the system.

An example of a system with feedback is a thermostat-controlled heating system. Once the desired temperature has been set, the system continually senses the room temperature and compares it to the thermostat setting. When necessary, the processes that either turn on or turn off the heating system are activated.

Organizations require a multiplicity of control systems to monitor performance. For example, operating costs need to be compared to budgets. Sales and marketing research provide feedback on how well products are being received. Employees are given feedback through annual performance reviews.

Feedback provided within an organization is commonly provided through information systems. In fact, providing feedback is a key function of information systems usually in the form of reports.

Figure 2.4   The Addition of Feedback to a Basic System Model

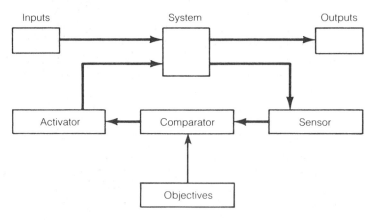

Feedback is a very powerful phenomena. We as human processors of information enjoy feedback and respond to it. Carnegie, the great industrialist, used to go into plants if they were having a productivity problem and cleverly used feedback to solve their problem. For example, suppose a large industrial manufacturing concern produced a certain type of machinery. Carnegie would visit the manager of the day shift and ask how many units were produced during that shift. The manager might respond that production was 120 units. Carnegie would write 120 in a visible spot, perhaps on a wall, where all workers could see it as they left for the day. He then would visit the evening, or second shift as they completed work and ask that manager how many units they produced. The manager might respond with 140 units. Carnegie would record that information where everyone could see it as he did with the day shift, and then leave. He would return at the *end* of the night, or third shift and find that they had produced, perhaps, 165 units during their eight-hour work period. Recording this information where everyone could see it, Carnegie would again leave and return again at the end of the day shift. Their productivity might have increased to 180 units. After this information was recorded as before, Carnegie would leave and return at the end of the evening shift to find they had produced, say, 210 units. This information would be recorded, and at the end of the night shift productivity might have gone up to 220 units. The shifts would finally stabilize at, say, 240 units. By doing nothing more than providing feedback on performance, Carnegie would have doubled productivity in the plant.

You may have observed the same phenomena yourself. For example, if you have ever gone jogging you may have noticed the difference that carrying a stopwatch can make in the effort you expend in jogging. People generally note that when they have a stopwatch on they tend to push themselves a little harder, as they can compare one day's performance with another. Without a stopwatch they get no feedback and tend not to expend as much effort.

A system operating without feedback is analogous to a person hitting golf balls into the dark. Without visual feedback on performance, any necessary adjustments for improvement cannot be determined. Indeed, few people would be interested in going out to hit golf balls in the dark, simply because the lack of feedback takes a lot of the pleasure out of the process.

The good news about feedback is that there is a theory associated with it that says feedback tends to increase effort. People playing ping-pong or tennis exert more effort when they start keeping score than when they are simply warming up. There is however, *bad news associated with feedback*. If you provide the wrong type of feedback, you may increase efforts in areas that are not in the best interest of the system under consideration. For example, students who notice that an instructor tends to count the number of pages in a term paper as a criteria in grading will focus on length rather than quality of the content in the paper. As another example, police officers who are evaluated based on the number of citations they give to motorists may focus on easy ways to get citations such as staking out behind poorly marked stop signs instead of focusing on the prevention of accidents. To prevent accidents, a police officer should report a poorly marked stop sign if people are failing to see it.

It is important to understand the power of feedback. When the theory of feedback is used in designing information systems, it is important to focus on designing feedback that will increase effort in those areas most helpful to the organization. A helpful hint about feedback: Praise is good feedback; criticism is good feedback if applied sensitively, but is bad feedback if applied insensitively.

Subsystems

The preceding discussion of systems considered only one system at a time. In actuality, the world is full of systems. Every system except the suprasystem is part of a larger system. The *suprasystem* contains all other systems. This suprasystem is beyond our comprehension, but it at least contains the universe and whatever else there might be.

Since every system is part of a larger system, it is also a *subsystem* of a larger system. For example, a sales department is a subsystem of an organization. The organization also contains other subsystems such as production, engineering, and accounting departments. The organization can be viewed as a subsystem of a particular industry that includes other organizations in the same business. The industry is a subsystem of all industries. This process of defining larger systems can be continued until the suprasystem is reached. In other words, all systems (with the exception of the suprasystem) *are* subsystems *and themselves can have* subsystems until the last or smallest subsystems (e.g., neutrons, electrons) are reached.

In order to study or design a system, it is important to define what system is under consideration. Though the primary idea of the systems concept is to work with the whole system, this is often not practical. Consequently, it may be helpful to simplify the study or design task by factoring the system into subsystems. The interactions among subsystems can then be defined in the same way that the interactions among entities are defined. This concept is illustrated in Figure 2.5.

Figure 2.5   System Composed of Subsystems

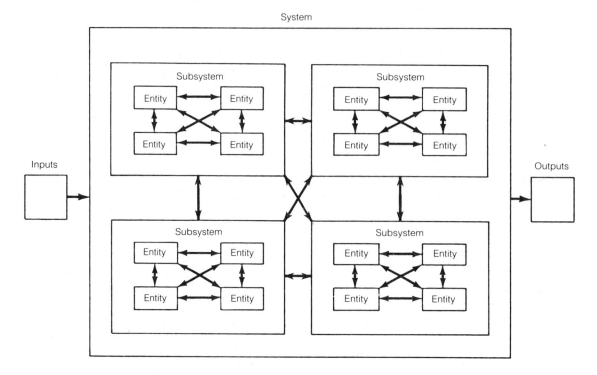

It can now be pointed out that an entity is actually a subsystem that cannot usefully be further simplified. For example, one of the subsystems of a sales department is personnel. Personnel also has subsystems such as circulatory and respiratory subsystems, which have further subsystems, all the way down to molecular subsystems. However, for most analyses, considering a particular subsystem as an entity (without additional factoring) is quite adequate.

Interface

The interaction among subsystems consists of the exchange of inputs and outputs of material, energy, and information. With knowledge of what inputs go into and what outputs result from the operations of a particular subsystem, the subsystem can be *interfaced*, or connected to other subsystems which either generate inputs needed by the subsystem or consume outputs generated by the subsystem. This concept is illustrated in Figure 2.6.

In order for the output of one subsystem to be acceptable to another subsystem, the interface between the two subsystems must be operational. That is, the output from one subsystem must adhere to certain standards to ensure that it is acceptable as input to the next subsystem. For example, the interface between electrical power and an electrical appliance (i.e., the wall socket and the plug) is standardized so that the two subsystems can be connected.

The more rigidly standards are adhered to, the more conveniently subsystems can be interfaced. In the case of the electrical appliance, if variously shaped plugs and wall sockets are used, achieving interface requires intervention in the form of changing plugs or sockets.

It is usually difficult to maintain rigid standards for interfacing many systems (or subsystems). Consequently, various *adapting techniques* are used to allow more flexibility in interfacing one system with another. The two general categories of adapting interfaces are *translation* and *slack resources*. Translation consists of converting an output of one system into an input form suitable for the next system. For example, a plug adapter can be used to interface an incompatible wall socket with a plug. As another example, consider the issues associated with interfacing accounts payable and accounts receivable within one company. Now consider the same issues for interfacing systems between two companies.

A common problem when interfacing systems results from the inability of one system to consistently generate output at a rate that optimizes the performance of the next system. For example, the manufacturing department cannot produce

Figure 2.6   Interfaces between Subsystems

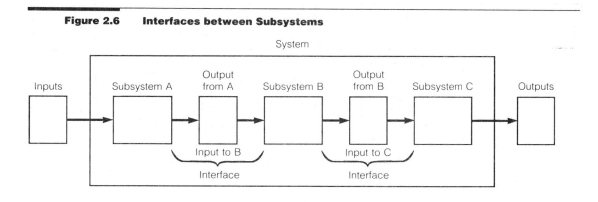

Figure 2.7 **(a) Interface by Adding a Function to One or Both Systems (b) Interface by Creating Additional Subsystem**

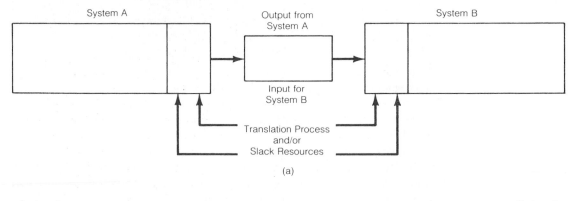

(a)

(b)

just the right amount of output to accommodate the sales department. Sales demands are usually cyclical, and, therefore, slack resources are required to maintain the desired interface between the systems. Slack resources are achieved by maintaining inventories and/or maintaining excess capacity. The use of inventories ensures the availability of sufficient output from one system to be used as input to another. Inventories absorb variations in demand. Excess capacity allows a system to increase output to accommodate increased demand from another system.

The more slack resources available, the more expensive (i.e., inventory carrying costs and/or underutilized capacity) the interfacing between systems becomes. However, poor interfacing between systems can result in a system not being able to operate due to the lack of inputs. There are costs associated with such losses in system productivity. Therefore, the costs of interfacing must be balanced against the costs of curtailing or stopping a system. For example, the costs of having excess capacity in a factory are offset by the revenues generated from being able to accommodate fluctuations in customer demand.

The interfacing processes of translation and slack resources are achieved either by adding them to the existing systems or by creating an additional subsystem to handle the interface. These concepts are illustrated in Figure 2.7.

The Systems Approach

The term *systems approach* emerged in the 1950s as a label for what was becoming a more analytical orientation to management and problem solving. During that era, Kenneth Boulding provided a perspective of the systems approach that is still commonly used today:

> The systems approach is the way of thinking about the job of managing. It provides a framework for visualizing internal and external environmental factors as an integrated whole. It allows recognition of the function of subsystems, as well as the complex suprasystems within which organizations must operate. Systems concepts foster a way of thinking which, on the one hand, helps the manager to recognize the nature of complex problems and thereby to operate within the perceived environment. It is important to recognize the integrated nature of specific systems, including the fact that each system has both inputs and outputs and can be viewed as a self-contained unit. But it is also important to recognize that business systems are a part of larger systems—possibly industrywide, or including several, maybe many, companies and/or industries, or even society as a whole. Further, business systems are in a constant state of change—they are created, operated, revised, and often eliminated.[1]

The systems approach is based upon general systems theory, which is concerned with the development of a systematic theoretical framework for the empirical world. The ultimate (and currently unachievable) goal of systems theory is to define a framework that ties all disciplines together in a meaningful relationship.

Implementation of the systems approach draws upon (1) systems philosophy, (2) systems analysis, and (3) systems management.[2] *Systems philosophy* is involved with thinking about or conceptualizing phenomena in terms of wholes consisting of entities or subsystems with emphasis placed on their interrelationships. *Systems analysis* involves methods or techniques of problem solving or decision making. Systems analysis is further concerned with becoming aware of the problem, identifying the relevant variables, analyzing and synthesizing the various factors, and determining an optimum or at least satisfactory solution or program of action. *Systems management* is concerned with the application of systems theory to managing organizational systems and/or subsystems.

The systems approach can thus concurrently be considered (1) a way of thinking, (2) a method or technique of analysis, and (3) a managerial style. The integration of these concepts is illustrated in Figure 2.8.

The traditional scientific approach to problem solving involves isolating and simplifying a system so that its various elements can be analyzed and improved. This process, a process of reductionism, plays a critical role in problem solving. Once a system has been reduced to its simplest form, problems with individual system elements can be more readily identified and resolved. However, reductionism alone is not adequate when studying many systems of major concern today.

A vast, complex system has many elements with significant interactions that are often not understood. Isolating the elements for analysis may lead the systems analyst to overlook the interactions that occur when the elements are combined into a system. (For example, understanding hydrogen (H) and oxygen (O) does not ensure understanding of the combination of H_2O). In particular, a system

---

1. Kenneth Boulding, "General Systems Theory—The Skeleton of Science," *Management Science*, April 1956, pp. 197–208.
2. Richard A. Johnson, Freemont E. Kast, and James E. Rosenzweig, *The Theory and Management of Systems*, (N.Y.: McGraw-Hill, 1973).

Figure 2.8   Integration of Systems Concepts within a Systems Approach

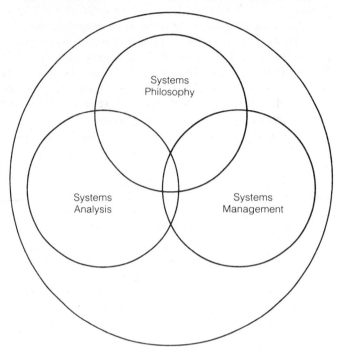

that involves human behavior (e.g., an organization) involves varying and changing variables that interact with, and/or compromise, the system. One of the key realizations contributed by the systems approach is that the interactions among elements in a system can be as, or more, important than the elements themselves. A system is more than just the sum of its parts and must be considered as such. Being able to use the systems approach to conceptualize when solving problems greatly increases the ability of the systems analyst as discussed in Chapter 1.

Designing Systems

One of the most useful frameworks for designing systems was developed by Christopher E. Nugent and Thomas E. Vollman.[3] Within their framework, the first step in designing a system involves clearly establishing the criteria and objectives of the system. The second step involves discovering which system or systems to use to satisfy the systems objectives. The search for the desired system constitutes the core of the design process. It involves:

1. Deciding what entities to include in (or exclude from) the system. This is referred to as the *inclusion process*.

2. Deciding how to structure the attributes of the set of chosen entities. This is referred to as the *structuring process*.

---

3. Christopher E. Nugent and Thomas E. Vollman, "A Framework for the Systems Design Process," *Decision Sciences*, 1972, pp. 84–109.

The inclusion process logically precedes the structuring process. However, in practice the systems analyst alternates between the inclusion and structuring processes. Inclusion is a highly creative process wherein the systems analyst must look beyond the obvious entities and search for new and significant entities possessing attributes that can better contribute to the performance of the system. For example, a soccer player's (an entity) kicking ability (an attribute) has been found to be most effective for field-goal kicking (an objective) in football (a system). Interestingly, it took over 100 years of football before anyone thought to use soccer players for field goal kickers.

Exclusion is a necessary part of systems design. It is frequently a difficult and frustrating process—for example, when some relevant entities have to be excluded from a system to constrain complexity. A systems analyst may recognize the potential contribution an entity offers to a system but not be able to deal with a system that includes the entity. For example, inclusion of solar energy in transportation systems is desirable but not yet practical.

The structuring process involves organizing the attributes of the entities included in the system. This involves determining the key relationships among the attributes of the chosen entities. The structuring process is important and yields payoffs. However, often too much effort is directed toward improvements in structure (e.g., queuing theory, mathematical programming, or game theory) when inclusion changes might yield higher payoffs. Recognizing when to emphasize inclusion and when to emphasize structure is key to good systems design.

Illustrations of Goal Setting, Inclusion, and Structuring

To illustrate the difference between goal setting, inclusion, and structuring, consider professional sports. First of all, what is the objective of a professional sports team? Is it to win? To become national champions? To entertain? To make money? Or to fulfill an ego need of the owner? The answers to these questions would vary depending on the ownership of the sports team. For example, living in Minnesota, I have found that in the 1970s and early 1980s the objective of the Minnesota Twins' owner was primarily to make money. There was less concern about winning, becoming the national champions, or fulfilling an ego need of the owner. Accordingly, the Twins' management traded away any player whose salary became excessive. As a result, the salary for the entire Twins team was less than many individual salaries for ball players on other teams. This resulted in not having a highly competitive team to watch. Nevertheless, enough fans came out to watch baseball for the love of the game, to see Minnesota's new rookies, or to watch other visiting teams with their superstars. Given the low cost of running the Twins ball club, as long as enough revenue came in, the team could be very profitable.

On the other hand, a team like the New York Yankees over this same time period focused on winning and becoming national champions (many critics claimed the primary purpose was to fulfill the ego needs of the owner). Accordingly, they have had one of the highest total salaries in professional baseball. Consequently, if the Yankees do not have an outstanding season, they may not have enough revenue to cover their high salaries. The point is that the objectives of a system are greatly determined by those who are responsible for it. It is important to understand what the objectives are before you design the system. In the case of the Minnesota Twins, the focus was on finding good rookies. For the Yankees, the focus is on finding the best talent that money can buy.

It is interesting to note that after being purchased by new management in 1985, the Twins adopted more of a Yankees' philosophy. Their salaries have skyrocketed. They have lost money for the past three years, however, in 1987 they won the World Series!

To understand goals and objectives, you must take the time to communicate with people, and understand what they want. Explaining the inclusion process is a little more difficult. Inclusion is perhaps the most creative process associated with designing a system. Good inclusion requires focusing on the attributes needed in the system and then determining where the attributes can be obtained. Continuing with our professional sports example, let's say that we are now dealing with a football team, and our objective is to have a profitable, winning team. This means that we would like to get good talent at a good price. One of the subsystems of a football team is the offense; within the offense we can factor into the subsystems of the passing game and the running game. Within the passing game we can factor into the receivers and, further, to the wide receivers. One question at this point would be: What are the attributes of a good wide receiver? The two most important of these would be blazing speed and good hands. Where might we find the attribute of blazing speed other than from the college football draft, where most teams automatically look for players? The answer is, of course, from track teams. In fact, if one were to look for potential wide receivers among world-class sprinters and find among them the additional attribute of good hands, one could find potentially good wide receivers. A few teams in the National Football League have done that.

Once the various players on any team are included within the system, a great deal of care must be taken in the structuring process. However, if serious errors are made in the inclusion area (e.g., the wrong attributes are introduced into the structure), no amount of structuring is going to result in a highly effective system. Therefore, it is critical to consider inclusion prior to structure.

As a final illustration, consider the following riddle: A father takes his son for a drive in a car. They have an accident. The father is killed, and the son is seriously injured. The son is rushed to a nearby hospital for emergency treatment. The physician who is to perform surgery walks up to the boy, looks at him, and says, "I cannot operate on this boy. He is my son." Now your objective, without reading any further, is to solve the riddle. How can the surgeon claim the boy to be a son?

Let's use the systems approach to illustrate how to expand our thinking. Again, the objective is to solve the riddle. During the inclusion process it is important to identify the attributes necessary to solve the problem. Then we can identify the attributes. In this case, we want to identify those attributes the surgeon would need in order to claim the patient as a son. To claim someone as a son, the necessary attribute would be that that person would have to be his parent. The attribute "parent" means mother *or* father. Since the father died in the accident, the answer to the riddle is that the surgeon is the boy's mother. If you didn't solve the riddle originally, notice how going through the attribute process of inclusion expands your thinking, enabling you to consider a broader realm of possibility before proceeding to the structure stage of problem solving.

One final element of good systems design is that one should accommodate systems that are dynamic by explicitly and implicitly considering a time dimension. Inclusion of the time dimension denotes that a system is designed to perform over time. A system should be able to evolve over time and in some cases to accelerate that evolution.

Summary

A system is a related collection of entities generally designed to accomplish multiple objectives. Systems receive inputs that are processed to produce outputs.

Systems can be classified as being either closed/stable/mechanistic or open/adaptive/organic in orientation. Closed/stable/mechanistic systems have more routine and predictable interactions among entities and subsystems internal to the system than do open/adaptive/organic systems. Closed/stable/mechanistic systems either do not interact with their environment or interact in a more predictable fashion than open/adaptive/organic systems.

System control is achieved by some form of feedback process. Feedback is compared to standards or expectations to determine adjustments needed to achieve or maintain system objectives. Systems are composed of subsystems that must be properly interfaced. If necessary, translation processes or slack resources can be introduced.

The systems approach is a way of thinking about analyzing and managing systems. It is not concerned only with the parts of a system; rather, it considers the total effect created when the parts function as a whole—the system itself. Designing systems consists of establishing systems objectives, selecting for inclusion those entities possessing attributes that can contribute to the systems objectives, and appropriately structuring the entities included in the system.

Exercises

1. Define and discuss the word *system*.

2. In the following list of entities relevant to a manufacturing concern, classify each entity as being primarily internal, transient (i.e., inputs and outputs), or external to the manufacturing system.
   - a. Equipment used in a plant
   - b. Raw materials
   - c. Assembly-line workers
   - d. Customers
   - e. Manufacturing facility
   - f. Management
   - g. Competitors
   - h. Government regulations

3. Classify the following systems as being primarily closed/stable/mechanistic or primarily open/adaptive/organic.
   - a. Clock
   - b. Human being
   - c. Fraternity
   - d. Prison
   - e. University
   - f. Mathematics

4. Discuss the use of feedback in controlling the performance of systems. What forms of feedback can be used to control the performance of the following systems or system entities?
   - a. A student
   - b. A football team
   - c. An employee
   - d. A manager
   - e. An organization

5. Discuss the relationships among systems, subsystems, and entities.

6. The interfacing of two subsystems is achieved by the use of standards and/or adapting techniques (i.e., translation or slack resources). Discuss interfacing techniques as they may be used in the following environments.

a. Interfacing the manufacture of seasonal goods (e.g., lawn mowers) with consumer demand
   b. Interfacing stereo components
   c. Interfacing high school education with college education
   d. Interfacing the manufacture of automobile replacement parts (e.g., tires and batteries) with automobiles requiring them
   e. Interfacing ambulance services with a volatile demand for ambulance service

7. Discuss the difference between the traditional approach to problem solving (i.e., reductionism) and the systems approach.

8. Determine whether the following problems are inclusion oriented or structure oriented.
   a. A college has an unusually high dropout rate due to admission standards that allow students who are not qualified to enter the college.
   b. A college has an unusually high dropout rate due to poor advising of students. Because students are not guided to take courses in the proper sequence (e.g., calculus is taken before algebra), they frequently become discouraged and leave the college.
   c. A company is losing sales of a new product because there is little demand for the product.
   d. A company is losing sales of a new product because the product is assembled poorly and is therefore of inferior quality.

Selected References

Ackoff, Russel L. "Computer Obstructed Education." *Society for Management Information Systems*, 1973, pp. 108–14.

———. "Towards a System of System Concepts." *Management Science*, July 1971, pp. 661–71.

Boulding, Kenneth. "General Systems Theory—The Skeleton of Science." *Management Science*, April 1956, pp. 197–208.

Churchman, C. W. *The Systems Approach*. New York: Dell Publishing Co., 1968.

Davis, Gordon. *Management Information Systems: Conceptual Foundations, Structure, and Development*. New York: McGraw-Hill, 1974.

Emery, F. E., ed. *Systems Thinking*. Baltimore: Penguin Books, 1969.

Johnson, Richard A.; Kast, Fremont E.; and Rosenzweig, James E. *The Theory and Management of Systems*. New York: McGraw-Hill, 1973.

Kast, Fremont E., and Rosenzweig, James E. *Organization and Management: A Systems Approach*. 2d ed. New York: McGraw-Hill, 1974.

Katz, Daniel, and Kahn, Robert L. "Organizations and the System Concept." *The Social Psychology of Organizations*. New York: Wiley, 1966, pp. 14–29.

Luchsinger, Vincent P., and Dock, V. Thomas *The Systems Approach: A Primer*. Dubuque, Iowa: Kendall/Hunt Publishing Co., 1975.

Nugent, Christopher E., and Vollman, Thomas E. "A Framework for the Systems Design Process." *Decision Sciences*, 1972, pp. 84–109.

Sayles, Leonard R., and Chandler, Margaret K. *Managing Large Systems*. New York: Harper & Row, 1971.

Von Bertalanffy, Ludwig. *General System Theory: Foundations, Development, Applications*. New York: George Braziller, 1968.

Weldon, R. J. "The Concept of a System." *The Journal of System Engineering 1*, (1969): 130–49.

Wendler, C. C. "What Are the Earmarks of Effective Total Systems." *Systems and Procedures Journal 92*, (1966): 30.

Wetherbe, James C. *Systems Analysis for Computer-Based Information Systems*. St. Paul, Minn.: West Publishing, 1979.

———. *Executive's Guide to Computer-Based Information Systems*. Englewood Cliffs, N.J.: Prentice-Hall, 1983.

3 Concepts of Information and Organizations

Introduction

This chapter has three primary thrusts. The first two provide conceptual foundations in the areas of information and organizations. The third integrates the concepts of information and organizations.

Information

Information is the result of the capture and organization of data. In an information system, data become information when they are the basis upon which efficient and effective decisions can be made. That is, information is used to increase the probability that the right decision is made. The transforming of data into information is the primary function of an information system.

Using the systems concepts discussed in Chapter 2, a simple model of an information system can be illustrated as shown in Figure 3.1.

Value of Information

Few decisions are made with perfect information. Virtually all decisions are made with either some insufficiency of required information and/or an overload of unrequired information.

The value of information is a function of the effect it has on decision making. If additional information can result in a better decision, that information has value. If additional information cannot improve a decision, it has little or no value.

Consider a marketing manager who must determine whether to market a new product. The manager has tentatively decided against introducing the product but has asked the marketing research department to survey a sample of customers to determine interest in the product. The information resulting from

Figure 3.1   Simple Model of an Information System

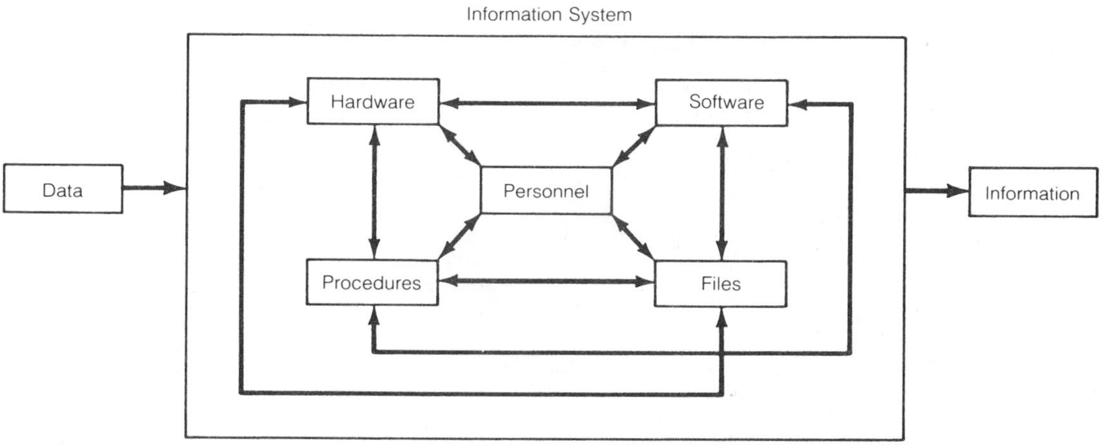

this research indicates an 80 percent probability that the new product will be successful. Based on this information, the manager decides that introducing the product is a good idea. The additional information is, therefore, valuable; it results in the manager making what appears to be a better decision.

Suppose the manager wants to be more certain as to the success of the new product. He or she requests the marketing research department to conduct additional, more comprehensive research. The results of the second study indicate a 90 percent probability that the new product will be successful. This information confirms the manager's decision to introduce the new product. However, since this course of action had been determined earlier from previous information, the additional information has no particular value, other than that of increasing the confidence in a decision made previously.

Information can improve a decision only if it leads to a change in the decision. Accordingly, obtaining additional information is of value until it no longer changes a decision. In other words, information must have a "surprise" effect to be of value. It must tell the manager something important that he or she did not already know.

Note that in the case of the marketing manager, even perfect information (i.e., such that the manager is 100 percent certain of success) would not have changed or improved the decision. The manager has already determined the correct course of action. In many situations, the optimum, or at least a reasonable, decision can be made without perfect information. This is fortunate, because obtaining perfect information for most decision making is impractical or impossible due to the factors listed in Table 3.1.

The problem with less than perfect information is that management can never be positive that additional information will have surprise value and result in a better decision. Accordingly, to the extent it is practical and possible, there

---

Table 3.1   Imperfections of Information for Decision Making

Information Imperfection	Illustrations
Unable to accurately predict or control the future	▪ Estimate customer demand for a new product. ▪ Know the future availability of natural resources.
Impractical or too costly to obtain information	▪ Develop a profile of all airline passengers. ▪ Maintain summary information on annual purchases made by customers of a supermarket chain.
Information not available	▪ Historical data on average customer balance in checking accounts. ▪ Current employment status of previous employees.
Unaware information exists	▪ A company is unaware of a new government report on the labor market. ▪ A company is unaware of a research project describing customer attitudes toward public transportation.
Information in the wrong form	▪ Sales data are summarized on an annual basis rather than monthly. ▪ Inventory turnover statistics are available by inventory item but not by customer demand.

---

is potential value in obtaining additional information. However, the question that should be asked is whether the cost of obtaining the additional information is justified.

Cost Justification of Information

There are two basic categories of cost justification for information: those cases where the benefits can be calculated, and those cases where they cannot. The benefits can be calculated when the effect on profitability of altering a decision can be determined. In this case, if the cost of additional information is less than the increase in profit, obtaining the information is cost justified. For example, consider a department store chain that has a problem with lost or stolen credit cards. These cards might illegally be used to run up large bills at several stores before a stop-credit can be enforced. If the annual losses incurred by the company amount to $250,000 per year, an information system that can resolve the problem for a cost of $75,000 is quite justifiable.

In most situations, the benefits of improved information systems are not measurable. This makes it difficult to justify the cost of obtaining the information. An example of this situation is the current move by financial institutions to implement electronic funds transfer systems. Such systems minimize the need for checks, greatly expedite the flow of money between businesses, and improve the timeliness of financial information. Though these systems are very costly, little attempt has been made to cost justify them. There seems to be a general consensus that "faster is better," "new technology is good," and "everyone (i.e., banks, stores, and customers) benefits."

The task of assessing the benefits of additional information is particularly perplexing since there is a tendency to seek or receive more information than is needed. Government agencies are often criticized for generating voluminous reports that have questionable value for decision making. Of course, since government agencies are nonprofit organizations, there is less pressure to cost justify information in terms of its contribution to profit.

Even in business settings, cost justification of information becomes obscure. Suppose a loan officer at a bank requests information on the average checking account balance of, and the number of bad checks written by, loan applicants. Certainly this information is meaningful when deciding whether to approve a loan. However, determining the number of bad loans avoided, and the consequent dollar savings, is difficult.

Information that cannot be explicitly cost justified must at least influence decisions if acquiring or processing this information is to warrant consideration. Given that information requested does influence decisions, several factors that affect the cost of information require assessment. These factors are accuracy, timeliness, and reporting interval. They are discussed and illustrated in Table 3.2.

Framework for Information Systems

Information systems can be subdivided into four subcategories, or types. Each type of system is specifically designed to perform a major function. The functions are referred to as transaction processing, information-providing, decision support, and programmed decision making.

Table 3.2   Factors That Affect the Cost of Information

Factor	Consideration	Example
Accuracy	Increased accuracy generally means increased cost. Therefore, accuracy should only be increased to the point that it still influences decisions.	At the end of each business day, bank management must decide if the books are in balance. The bank can close a couple of hours earlier by allowing its balances to be off by a few dollars. A simple adjustment entry is much less costly than spending a few hundred dollars on salaries and computer time to locate a minor mistake.
Timeliness	In general, the more quickly information is provided after the occurrence of an event, the more costly is that information. More timely information can result in earlier problem identification and resolution.	Retailers like to identify slow-moving merchandise and put it on sale to move it before it becomes even less attractive to customers. A decrease in the demand for merchandise can be determined by keeping track of inventory turnover. However, such information can usually be a few days or weeks old without serious consequences.
Reporting Interval	In general, the more frequently information is updated and reported, the higher the cost. If information that affects decisions is changing quickly, then the need is greater for shorter updating and reporting intervals.	An airline needs extremely short intervals for updating and reporting passenger reservations and changes in reservations in order to sell as many tickets as possible without overselling. However, the airline need only know the amount of passenger luggage to be carried once—just prior to take off.

Transaction-Processing Systems

Transaction-processing systems provide high-speed and accurate processing and record keeping of the basic clerical or operational processes of an organization. Examples of transaction-processing systems are the processing of sales transactions in a department store, checking accounts in a bank, or work orders in a manufacturing plant. Transaction-processing systems have traditionally been the most visible dimension of information systems, since they often displace clerical personnel. The cost savings achieved by such displacement often provides a definitive basis for cost justification of a transaction-processing system. Note that the information generated from such systems may or may not be of value. Indeed, many transaction-processing systems generate a lot of useless, bureaucratic paperwork.

Information-Providing Systems

Information-providing systems provide information necessary for decision making. This information is generally gleaned from transaction systems in the form of summary and exception reports. *Summary reports* are tabulations of detail in categories such as sales by store, by department, and by inventory item. *Exception reports* isolate detail or summary information that significantly deviates from normal. For example, an exception report may indicate large credit transactions, inventory items that are out of stock, or departmental payrolls that are exceeding the budget.

Both summary and exception reporting are powerful dimensions of information systems. They highlight and isolate important detail information. This saves decision makers considerable time.

Decision-Support Systems

Decision-support systems go beyond the simple provision of information needed for decision making. They actually assist in the decision making. Decision-support systems include such things as statistical, mathematical, and simulation models that can be used by the decision maker to experiment with information in determining a course of action. For example, a simulation model can be used to forecast revenue for a new product.

Decision-support systems are relatively new, but they are gradually increasing in use. These systems are discussed in much greater detail later, in Chapter 14 in this book.

Programmed Decision-Making Systems

Programmed decision making involves having a system make a decision rather than a person. Such decision making is preprogrammed according to very specific decision rules. For example, customer credit is refused if a customer has exceeded his or her credit. Before a system can be designed and implemented to program the computer for decision making, the decision-making process must be highly structured and defined.

The most advanced form of programmed decision making is artificial intelligence or expert systems. In these systems computers are programmed to simulate human thinking. For example, computers can be programmed to play chess or make medical diagnoses. Chapter 14 covers expert systems in greater detail.

Misconceptions about Information

Because information systems are able to process transactions, provide information for decision making, support decision making, and even make decisions, people in organizations have developed (often unrealistically) high expectations. Unfortunately, there is often a gap between what organizations expect of information systems and the performance of the systems.

This expectation gap results from six basic misconceptions about information systems and the role they play in organizations.[1] These misconceptions are identified and explained in Table 3.3.

Human Processing of Information

Understanding how humans process and store information is the key to designing information systems helpful to humans. Figure 3.2 provides a simple model of the flow of information through sensory capabilities of eyesight and hearing into short-term memory and long-term memory. Short-term memory is where we do our thinking or processing of information. Within this part of our consciousness we have very limited memory. In fact, research shows that short-term memory is capable of remembering, on the average, only seven items (chunks) of information at one time. This, incidently, is one of the primary reasons why telephone

---

1. The misconceptions are based on the author's own work as well as the following articles: Russell L. Ackoff, "Management Misinformation Systems," *Management Science,* December 1967, pp. 147–156; and Chris Argyris, "Resistance to Rational Management Systems," *Innovation,* 1970, pp. 28–35.

Table 3.3 Misconceptions about Information

Misconception	Explanation
Computers can quickly and conveniently provide managers with information they desire.	■ Considerable organizational information is not stored on computer files. It must be collected before it can be processed. Once information is available, computer programs are required to access the information.
More information means better decisions.	■ Though most managers lack a good deal of the information they *should have*, many information systems provide them with an overabundance of irrelevant information they *don't need*. Many managers receive voluminous reports that are unused because the reports are not relevant to decision-making processes.
Managers need the information they want.	■ For managers to know what information they need, they must be aware of each decision they should and do make. They also must define the process and variables required to make each decision. This situation is seldom the case. Consequently, managers tend to "play it safe" and ask for more information than they really need.
Give managers the information they need, and their decision making will improve.	■ Managers may not be aware of the best way to use information to make the best decision. For example, several managers given the same information make different decisions, and some of the decisions are better than others. Since all managers had the same information, the better decisions are a result of the decision maker's performance, not of the information.
Managers do not have to understand how information systems work.	■ Managers may lack confidence in something they do not understand and consequently not use it. Also, managers are less apt to recognize logic problems in the information generated by an information system if they do not understand how the information is processed.
Managers always welcome and encourage improved information for decision making.	■ Often, improved information reveals that an operation is poorly managed (e.g., incompetence, missed opportunities, coverups). Improved information about an operation that reveals problems can be threatening and, therefore, resisted by managers.

Figure 3.2 Human Information Processing

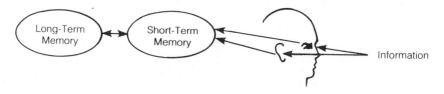

numbers are just seven digits long. Chunking information allows more of it to be stored in short-term memory. For example, in a social security number, information is chunked into three groups of information. The social security number of 585–24–7387 breaks nine digits into three chunks of information to assist recall.

Long-term memory is virtually infinite as far as human information processing goes. Evidence indicates that our long-term memory remembers everything that has ever happened to us. A good illustration of this is the ability of adults to vividly recall child experiences when they are under hypnosis. The difficulty is that we cannot always recall or retrieve into our short-term memory everything that we would like from our long-term memory.

We can process information from sight or sound into short-term memory very quickly. The speed with which we can shift things back and forth between long-term memory and short-term memory, however, is not as rapid as we would like it to be. The result is that as processors of information, humans are not very good with mental arithmetic (poor ability at performing computations and at shifting information between long-term and short-term memory). For example, try to multiply 48 times 1,324 in your head—it is a difficult task.

We are very good, however, at pattern recognition (the ability to transfer visual information to short-term memory). If four pictures of the Mona Lisa were placed in front of you, each with a slight variation such as one having brown eyes, one having blue eyes, one having a smile, and one having lighter hair than the others, you could easily point out the differences in the pictures. This illustrates the pattern-recognition ability of human information processing.

Interestingly, computer processing skills are just the opposite of human processing skills. Computers are incredibly powerful at doing arithmetic, yet have a great deal of difficulty at pattern recognition. What is child's play to a five-year-old in terms of differentiating among visual stimuli, is very difficult for a computer. Everything has to be digitized before it can be carefully interpreted. Digital patterns can quickly be determined; however, it is difficult for the computer to interpret and describe the differences in an artistic or aesthetic way. For example, a computer would have difficulty differentiating between an attractive or an unattractive work of art.

Graphics are a powerful tool to help humans chunk information and thereby understand and perceive it more readily. By converting digital information to graphic information, humans are more readily able to interpret and understand what is going on. To illustrate this point, consider whether it is more informative to get a national weather forecast on radio, or on TV. The added advantage of the graphics provided from TV allow us to understand the movements of the storm fronts better than when we just hear about them on the radio. Also it is easier to understand temperature trends when they are graphically displayed on a map than when a list of cities and their temperatures is read.

Decision-Maker Cognitive Style

Cognitive style refers to the way a decision maker approaches problem solving. It is an often overlooked but significant dimension of the decision-making process. The cognitive style of decision makers can be expressed as an existing continuum, ranging from intuitive/heuristic to systematic/analytic.

Intuitive/Heuristic

Intuitive/heuristic-oriented decision makers tend to learn from experience. They use common sense and intuition to approach decision making in what can be described as a trial-and-error mode. They tend to view a problem or an opportunity as a totality rather than as a structure consisting of from specific parts. Highly visible situational differences that vary with time are the basis from which they draw inferences about different situations.

Systematic/Analytic

Systematic/analytic-oriented decision makers tend to learn by analyzing a problem or opportunity. They do not rely heavily on previous experiences and feedback. Rather, decision making is based upon formal, rational analysis in which explicit, quantitative models are often used. These managers prefer to reduce a problem or an opportunity to a set of underlying causal functions and then look for similarities or patterns by analyzing and comparing variables.

Comparison of Cognitive Styles

If you were to ask systems analysts which category most managers tend to fall in, in terms of their cognitive style, they would say that most of them are intuitive/heuristic in their problem solving. That is, most managers use a strong trial-and-error orientation in their problem solving. Systems analysts, however, tend to view themselves as being relatively systematic/analytic in their problem-solving approach. Indeed, they often associate a certain virtue with being systematic/analytic.

However, the way most people approach problem solving involves a lot of trial and error. Most people do combine some analytical procedures in problem solving, but, nevertheless, trial and error is their dominant problem-solving mode.[2]

Consider, for example, the way you might approach a crossword puzzle. Some people might study every conceivable aspect of solving such a problem for some time, but most people quickly engage in experimenting with the problem itself. They try different approaches, commit themselves to those approaches that seem to contribute to solving the problem, and reject those that do not until the problem is ultimately solved. This process is sometimes referred to as *progressive deepening*. For instance, as you attempt to solve a crossword puzzle using a particular word, you may find that other words tend to intersect with that word successfully. You tend to get more and more committed, in a progressive manner, to using that word to solve the puzzle. However, if you reach a point where you cannot intersect another word into the crossword puzzle because of the word you were committed to, you ultimately have to reject that word and start over at an earlier stage of the problem. Then you develop a new concept for solving the problem, using a different word.

---

2. Norman L. Chervany, Gary W. Dickson, and James Senn, "Research in MIS: The Minnesota Experiments," *Management Science*, May 1977, p. 913–23.

Implications of Cognitive Style

The implications of cognitive style indicate that in designing systems to serve managers, systems analysts must be sensitive to the trial-and-error orientation that most people have in problem solving. They must allow a manager to approach a problem in different ways until a satisfactory solution is reached. This also implies that most managers are not able to analytically determine what they require from a system. Rather, they will normally go through a trial-and-error process—trying to define what it is they need, experiencing various capabilities, learning from them, and then indicating further modifications until their requirements are ultimately met.

Organizations

Organizations are human-designed and human-controlled systems. The entities of an organizational system are people, equipment, inventory, procedures, ideas, and so forth, arranged to interact to accomplish one or more objectives. For example, in a business organization, objectives typically include the production of certain goods or services, organization survival and growth, social responsibility, and, of course, profitability.

The most common way of illustrating an organization is to use an organization chart (see Figure 3.3). In this pyramidal structure, ultimate authority and responsibility reside at the top. Authority and responsibility flow down through the ever-widening succession of levels to the bottom of the organization.

An organizations have both vertical and horizontal dimensions. The vertical dimensions are defined by the organization's departments or subunits. The horizontal dimensions are defined by the organizational layers or hierarchical structure.

Vertical Dimension

The vertical dimension of the organization performs the different functions required for the organization to operate (e.g., accounting, manufacturing, marketing, and research). The number of vertical dimensions in an organization is a function of the differentiation (i.e., division of labor) and specialization required to perform the various tasks in the organization.

In a small, one-owner proprietorship, the owner may perform all of the various tasks associated with sales, accounting, purchasing, inventory, and so forth. However, the volume and complexity of a large corporation require that the various tasks be "divided and conquered." Different departments are established to manage different functions, and people specialize in performing the tasks associated with particular functions.

Horizontal Dimension

The horizontal dimension of an organization can be broadly categorized as top, middle, and operating management. In Figure 3.3, top management includes the board of directors, president, and vice-presidents. Middle management includes

Figure 3.3   An Organization Chart

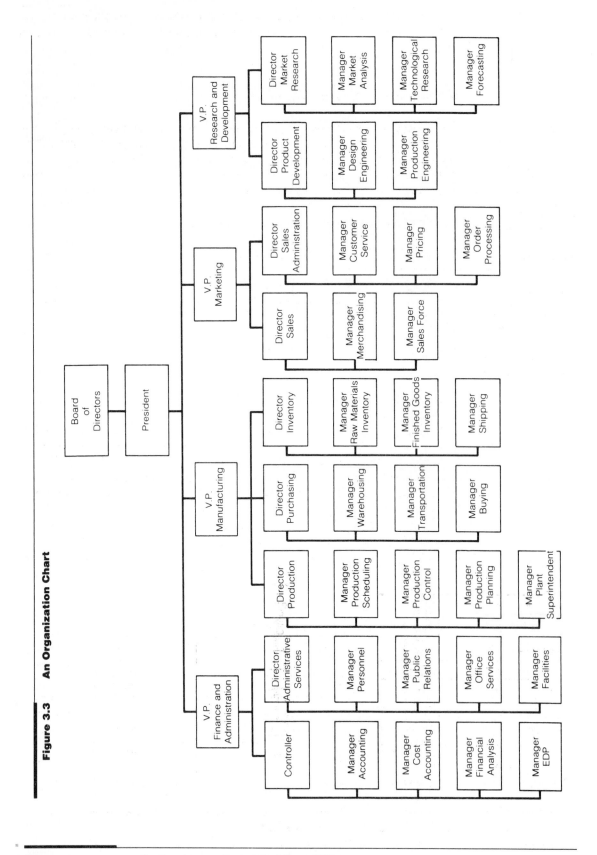

the controller and various directors. Operating management includes the remaining management positions. The major functions of these levels are given in Table 3.4.

Management of Organizations

The concept of organization implies rational, integrated, and coordinated interaction of the organizational subsystems and entities. However, inevitable conflicts and discord arise. These can be dysfunctional to the performance and even the survival of an organization. Management is required to maintain harmony of the (often diverse) dimensions of the organization to ensure a viable and workable system.

Since an organization is a system, it depends on its environment to supply it with inputs (material, energy, and information) and to consume its outputs (goods and/or services). Since the demand for goods and services changes with time, an organization must be sensitive to market conditions to ensure consumption of its output. An organization must also be concerned about the continued availability of necessary inputs (e.g., natural resources). Therefore, besides addressing issues internal to the organization, management must be concerned with coaligning organizational processes with the inputs available from, and the outputs desired by, the organization's environment.

Approaches to Management

Over the years, various theories of management have been proposed. Many people have attempted to determine the best way to manage organizations. Broadly speaking, existing management theories form a continuum between two extreme positions with respect to management: the classical approach and the human relations approach. The primary management concepts within each approach are listed in Table 3.5.

The classical and human relations approaches to management are obviously opposing in perspective. Both approaches have been used with success in some organizations and have been unsuccessful in others. This has understandably created confusion and debate as to which is the better way to manage organizations.

Table 3.4	Horizontal Dimensions of an Organization
Management Level	*Function*
Top Management	■ Management defines overall organizational objectives and formulates strategic plans and policies to achieve objectives. Management allocates the required financial and human resources among the vertical dimensions of the organization.
Middle Management	■ Management develops, directs, and controls departmental objectives, and plans policies consistent with overall organizational objectives and available resources. Management integrates and coordinates the vertical dimensions of the organization.
Operating Management	■ Management supervises the daily production of goods and/or services consistent with performance criteria established by middle management.

Table 3.5 **Contrasting Approaches to Management**

Classical Approach	Human Relations Approach
Formal	Informal
Highly Structured	Flexibly Structured
Centralized Decision Making	Decentralized Decision Making
Dictatorial Management	Participative Management
Programmed Procedures	Dynamic Procedures
Control Orientation	Autonomy Orientation
Process Orientation	Personnel Orientation

Contingency Theory of Management

Actually, there is no single, best way to manage all organizations. Rather, the best management approach is *contingent* upon the characteristics of the organization being managed. This concept is referred to as the *contingency theory* of management. It attempts to recognize organizational differences and propose management approaches that best fit those organizational differences.

For example, the organizational characteristics of an assembly line in a manufacturing plant are different than the organizational characteristics of a group of research scientists. An assembly line can be characterized as routine and repetitive, whereas research is creative and diverse. To be effective, each group should be managed differently. The management challenge becomes one of systematically identifying the relevant differentiating characteristics and then applying the appropriate management techniques to each group.

Technology and Environment

The contingency theory of management has established that the technology and/or the environment of an organization are the key variables that characterize the organization. *Technology* is defined as the human and mechanical processes by which an organization produces its goods and/or services. *Environment* is defined as all entities or factors external to the organization. An organization's environment includes such things as customers, competitors, and government regulations.

The classification of systems defined in Chapter 2 can be used to characterize an organization. That is, the technology and environment of an organization establish the organization's position on a continuum ranging from closed/stable/mechanistic to open/adaptive/organic.

A closed/stable/mechanistic organization has a predominantly routine and predictable situation. Productivity is a major objective. Technology and environmental forces are relatively stable and certain. Consequently, decision making tends to be straightforward and programmable. An example of a closed/stable/mechanistic organization is the can-manufacturing industry. It has used a relatively stable technology and encountered a relatively stable environment for over twenty years.

Alternatively, an organization that is open/adaptive/organic is involved with other than routine activities. Creativity and innovation are required. Technology and environmental forces tend to be volatile, presenting considerable uncertainty. Problems encountered are complex, presenting conflict and unclear courses of action. Decision-making processes are characterized as risky, and few optimum decisions are possible. An example of an open/adaptive/organic organization is the personal computer industry; it has had an extremely dynamic technology and environment.

However, few organizations or organizational subsystems can be strictly classified as either closed/stable/mechanistic or open/adaptive/organic. Rather, organizations generally tend, in varying degrees, to be more of one type than the other. Also, an organization's position on the continuum will occasionally vary, perhaps in an extreme way, with time. This occurs when an organization encounters changing technologies and/or environments. For example, the energy crisis of the late 1970's dramatically affected the automotive marketplace. Consequently, automobile manufacturers have had to cope with a less predictable environment as they contemplate future product offerings (e.g., luxury cars versus fuel-efficient cars).

The realization that organizations vary in form from closed/stable/mechanistic to open/adaptive/organic to accommodate their technology and environment, provides the basic framework for contingently applying managerial approaches. A management style aligned with the classical approach tends to be effective in relatively closed/stable/mechanistic organizations. Management can operate in a structured, centralized, and process-oriented mode. Conversely, the human relations management approach tends to be effective in open/adaptive/organic situations where employees are coping with greater uncertainty and complexity. Management must provide a more autonomous and participative setting for what is generally a more sophisticated level of staff. In fact, in most open/adaptive/organic settings, staff members generally have more expertise in special areas than their managers. Figure 3.4 illustrates the relationship between organizations or organizational subsystems and the appropriate management orientation.

Figure 3.4 **Relationship between Organizational Characteristics and Management Orientation**

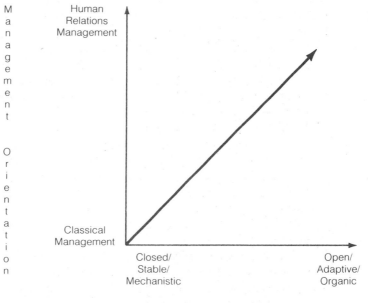

The contingency view of management and organizations can also be applied to both the horizontal and vertical dimensions within an organization. From a horizontal perspective, the lower levels of the organization are generally closed/stable/mechanistic in orientation. The orientation of the organization becomes increasingly open/adaptive/organic as one moves from the operating level to top management.

From a vertical perspective, an accounting department provides an excellent example of a process that is inherently structured and, therefore, leans toward a more closed/stable/mechanistic orientation. Conversely, a research and development department, where creativity and spontaneity are essential, represents the opposite end of the spectrum. Figure 3.5 offers a contingency view of the relationship between the horizontal and vertical dimensions of a hypothetical organization.

Organizational Considerations for Information Systems

Even those with limited experience can see that computer technology is highly structured. Computer programming requires great precision and attention to details. Accordingly, computer-based information processing is at first commonly applied in the closed/stable/mechanistic dimensions of organizations. Figure 3.5 indicates that there are both horizontal and vertical organizational dimensions that readily lend themselves to computer-based information systems.

Horizontal Dimension
In the horizontal dimension, the information needs of operating management are more easily addressed than those of middle and top management. Ostensibly, at

Figure 3.5   Contingency View of the Vertical and Horizontal Dimensions of a Hypothetical Organization

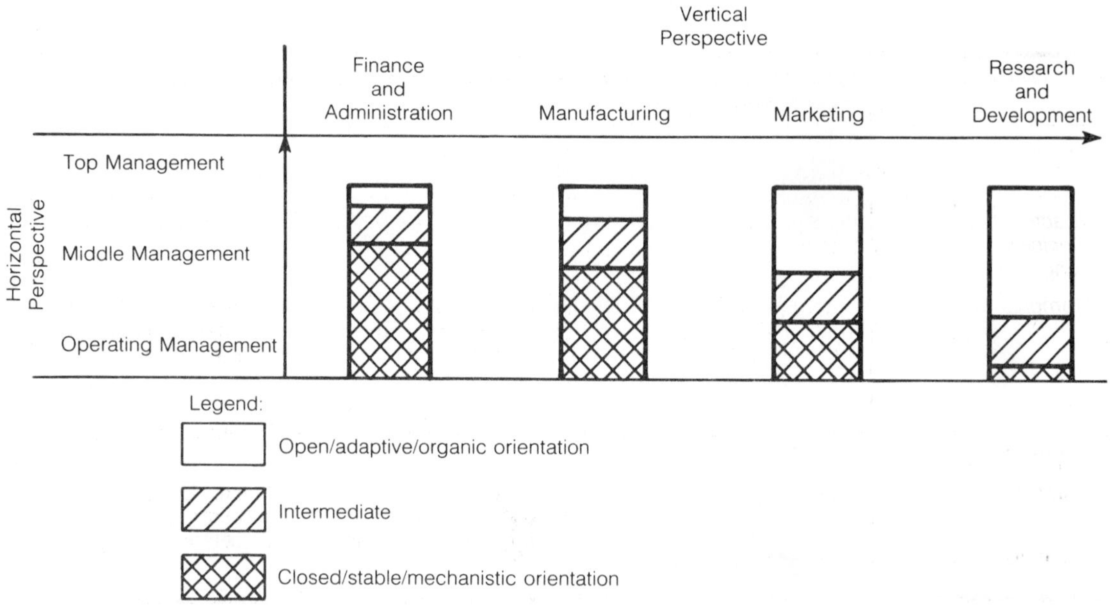

progressively higher organizational levels, it becomes increasingly difficult to accommodate the decision-making processes with the desired information.

Due to the greater structure and predictability at lower organizational levels, decision making tends to be more straightforward and computational. Most of the necessary information is internal to the organization. Usually, the information is complete. Most of the lower-level information requirements are accommodated by information generated from the transaction-processing systems.

Higher-level management is working with less definition, structure, and routine in decision making. In fact, much of decision making is involved with issues external to the organization. Accordingly, information external to the organization plays a key role in upper-management decision making. Unfortunately, external information tends to be voluminous, unstructured, volatile, and inaccessible. Consequently, it is impossible to provide to this level of management all the information necessary for decision making. The lack of definition, structure, and routineness in upper-management decision making, combined with the external and incomplete qualities of the necessary information, tends to make upper-management decisions highly judgmental.

A profile of the basic differences in both information and decision making at the various management levels is given in Table 3.6.

Vertical Dimension

The closed/stable/mechanistic and open/adaptive/organic tendencies of the vertical, or functional, dimension vary (review Figure 3.5). The finance and administrative functions, which are common to all organizations, are usually the most closed/stable/mechanistic. Consequently, accounting and administrative functions are generally the first to be included in computer-based information systems developed and/or used by an organization.

The remaining vertical dimensions vary with organizations. Their characteristics are determined by the goods or services an organization produces, and the technologies associated with producing those goods or services. For example, hospitals, banks, retailers, manufacturers, and universities differ in the goods

---

Table 3.6 **Characteristics of Decision Making and Information at Different Management Levels**

| | | Management Level | |
	Operating	Middle	Top
Characteristics of Decision Making	Computational/ Objective ———————————————→		Judgmental/Subjective
Examples	Inventory Reordering Production Scheduling Credit Approval	Short-Term Forecasting Budget Preparation Capacity Planning	New Product Planning Location of New Factory Mergers and Acquisitions
Characteristics of Information	Internal/Complete ———————————————→		External/Incomplete
Examples	Sales Orders Production Requirements Customer Credit Status	Sales Analysis Budget Analysis Production Summaries	Market Conditions Industry Forecasts Government Regulations

---

and services they deliver and consequently in their vertical structures. They all have finance and administrative functions, but their *mainstream* (core) activities vary. However, the mainstream activities do share certain commonalities. Basically, all mainstream activities involve some form of the following activities:

- Forecasting demand
- Scheduling or allocating resources
- Taking inventories
- Maintaining control

Illustrations of these four activities in the contexts of different organizations are provided in Table 3.7.

The mainstream activities of some organizations are more closed/stable/mechanistic than those of others. Consequently, certain industries have experienced earlier and more far-reaching success in developing computer-based information systems. Banking and manufacturing are examples of industries that are considered ideal for computer-based information systems. Nearly half of the computer equipment installed today is utilized by the banking and manufacturing industries. Such success has materialized more slowly in industries such as medical care and retailing, which combined, utilize less than 15 percent of currently installed computer equipment. However, generalizations should not be made about industries as to their closed/stable/mechanistic orientation. For example, in manufacturing, large-batch and mass production is more closed/stable/mechanistic than production in unit and small-batch manufacturing. The latter is involved with tailoring production to customer requirements, production of prototypes, and fabrication, which are more open/adaptive/organic processes. Therefore, in designing an information system, the particular organization and organizational processes within the organization must be considered on an individual basis. The more closed/stable/mechanistic a particular process, the more the information system can be oriented toward routine transaction processing and programmed decision making. The more open/adaptive/organic a process, the more the information system should be oriented toward providing, to the extent possible, information for decision making and decision-support capabilities.

Management Information Systems

From the discussions of vertical and horizontal information systems, it can be concluded that an organization has several information systems. For example, there are financial information systems that have subsystems consisting of general ledger, accounts receivable, accounts payable, cost accounting, and payroll/personnel. There are manufacturing information systems that have subsystems consisting of production scheduling, inventory, and material requirements. The various information systems address, to varying degrees, the information and decision-making needs of operating, middle, and top management.

The vertical and horizontal information systems can be envisioned as a federation of information subsystems that constitutes an overall organizational information system called a *management information system* (MIS). MIS is a somewhat overwhelming concept to fully comprehend. Considerable organizational and information systems knowledge and experience are required to totally appreciate

Table 3.7 Activities Common to Mainstream Processes of Different Organizations

Organization	Mainstream Process	Forecasting	Scheduling	Inventory	Control
Manufacturer	Manufacturing	Demand for existing and new products Future availability of raw materials and labor	Work orders Equipment Labor	Finished goods Raw materials Labor Equipment	Performance of product Market acceptance of product Productivity assessment
University	Instruction Research Community service	Future demand for higher education Future availability of faculty	Students Classrooms Equipment Faculty	Students Classrooms Equipment Faculty	Placement of students Students' performance on exams Quality of teaching Research activity Resource utilization
Airline	Transportation of passengers and freight	Future demand for air travel Future supply of aircraft	Passengers Freight Flights Aircraft Personnel	Repair parts Aircraft Personnel	Safety record Customer satisfaction Flight occupancy
Bank	Checking and savings accounts Installment and mortgage loans	Future supply of money in checking and savings accounts Future demand for installment and mortgage loans	Operational personnel Cash flow investments	Dollars Personnel	Market share Loan defaults Liquidity
Hospital	Health care	Future need for health care Future availability of hospital facilities and physicians	Patients Rooms Equipment Physicians	Medical supplies Patients Rooms Equipment Physicians	Patient care Facility utilization

it. To build an effective and efficient MIS, all of the various information subsystems must be designed and developed in the context of the total MIS. Where appropriate, the various subsystems are interfaced to supply information to one another. For example, the payroll/personnel subsystem requires information from, and supplies information to, every other vertical and horizontal dimension of the organization. Sales information is used to determine production schedules; production schedules are used to determine raw materials requirements, and so forth. An overall conceptualization of an organizational MIS is given in Figure 3.6.

A more comprehensive treatment of MIS is presented in the final section of this text, "Strategic, Administrative, and Higher-level Concepts and Techniques."

Figure 3.6   Conceptualization of a Management Information System

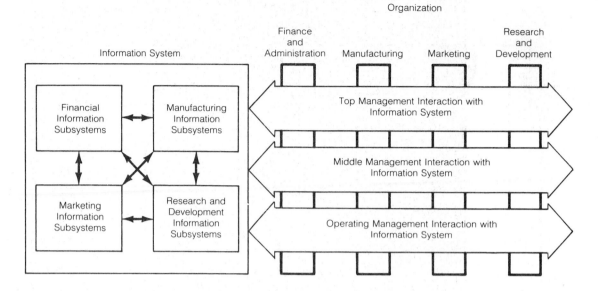

To prepare the student for that material, it is critical to first develop systems analysis skills in working with smaller, more manageable, information subsystems. Accordingly, prior to the final section, the text concentrates on systems analysis at the subsystem level. However, the student is encouraged to keep in mind that integration of these subsystems is necessary to achieve an overall organizational MIS.

Summary

Information is the primary basis for decision making. The value of information is a function of the effect it has on increasing the probability that the right decision is made. Virtually all decisions are made on the basis of imperfect information. However, additional information should be sought, only to the point that it is likely to influence a decision.

The four basic categories of information systems are transaction processing, information-providing, decision support, and programmed decision making.

There are several misconceptions about what information can do for organizations. These misconceptions are based on erroneous assumptions about the relationship between management decision making and information.

Humans have limitations as information processors, which are complemented by the power of computers. They have limits of short-term memory and have difficulty retrieving information from long-term memory. We are creative and good at pattern recognition. Our dominant problem-solving approach is trial-and-error oriented.

Organizations are human-designed and human-controlled systems. They have both vertical and horizontal dimensions, which vary between closed/stable/mechanistic and open/adaptive/organic. The more closed/stable/mechanistic a process, the more readily it can be supported by a computer-based information system. However, both improvements in technology and more advanced analytical skills are continually facilitating the development of new computer-based information systems applications in open/adaptive/organic environments.

An organization has a federation of information subsystems that constitutes a management information system (MIS). Proper interfacing of these subsystems is critical to proper integration of the collection, processing, and providing of information to the organization.

Exercises

1. Why is it often difficult to cost justify information? If information cannot be cost justified, what should be the primary criteria in assessing the utility of providing it?

2. Identify and discuss the major factors that affect the cost of information.

3. Categorize the following functions as pertaining to transaction processing, information-providing, programmed decision making, or decision support.
   a. Credit is automatically rejected because a customer has a balance past due.
   b. Management is provided with a sales analysis by department.
   c. An on-line financial terminal in a savings and loan captures data on deposits and withdrawals.
   d. A simulation model is used to forecast profitability for a company, based on projected sales.

4. Discuss the major misconceptions about information systems and the role they play in organizations.

5. What are the strengths and weaknesses of humans as information processors?

6. Discuss the differences between the vertical and horizontal dimensions of an organization.

7. Discuss the effect of an organization's technology and environment as it pertains to the organization being closed/stable/mechanistic or open/adaptive/organic.

8. Recognizing that closed/stable/mechanistic systems are easier to automate than are open/adaptive/organic, identify the following organizational processes as being relatively closed/stable/mechanistic or relatively open/adaptive/organic.
   a. Processing checking accounts in a bank
   b. Payroll processing
   c. Advertising
   d. Credit management
   e. Diagnosing medical treatment for a heart patient
   f. Determining whether to merge with another corporation
   g. Strategic planning
   h. Reordering inventory
   i. Designing an information system

9. Discuss the difference between a single information system and an MIS.

Selected References

Ackoff, Russell L. "Management Misinformation Systems." *Management Science,* December 1967, pp. 147–56.

Andrus, Roman R. "Approaches to Information Evaluation." *MSU Business Topics,* Summer 1971, pp. 40–46.

Argyris, Chris. "Resistance to Rational Management Systems." *Innovation,* no. 10 (1970), pp. 28–35.

Aron, Joel D. "Information Systems in Perspective." *Computing Surveys,* December 1969, pp. 213–36.

Bowers, David G., and Seashore, Stanley E. "Predicting Organizational Effectiveness with a Four Factor Theory of Leadership." *Administrative Science Quarterly* 2 (1966): 238–63.

Chervany, Norman L., and Dickson, Gary W. "An Experimental Evaluation of Information Overload in a Production Environment." *Management Science,* June 1974, pp. 1335–344.

Davis, Gordon B. *Management Information Systems: Conceptual Foundations, Structure, and Development.* New York: McGraw-Hill, 1974.

Dickson, Gary W., and Simmons, John K. "The Behavioral Side of MIS." *Business Horizons,* August 1970, pp. 59–71.

Etz, D. V. "The Marginal Utility of Information." *Datamation,* August 1965, pp. 41–44.

Feltham, Gerald A. "The Value of Information." *The Accounting Review,* October 1968, pp. 684–94.

Fielder, Fred E. "Validation and Extension of the Contingency Model of Leadership: A Review of Empirical Findings." *Psychology Bulletin* 76, (1971): 128–48.

Galbraith, Jay R. "Organization Design: An Information Processing View," *TIMS Interfaces,* May 1974, pp. 28–36.

Gregory, Robert H., and Atwater, Jr., Thomas V. V. "Cost and Value of Management Information as Functions of Age." *Accounting Research* 8, (January 1957): 42–70.

Hirsch, Rudolph E. "The Value of Information." *The Journal of Accountancy,* June 1968, pp. 41–45.

Hunt, Raymond G. "Technology and Organization." *Academy of Management,* September 1970, pp. 235–52.

Johnson, Richard A., Kast, Fremont E. and Rosenzweig, James E. *The Theory and Management of Systems.* New York: McGraw-Hill, 1973.

Kanter, Jerome. *Management-Oriented Management Information Systems.* 2d ed. Englewood Cliffs, N.J.: Prentice-Hall, 1977.

Kast, Fremont E., and Rosenzweig, James E. *Contingency Views of Organization and Management.* Chicago: Science Research Associates, 1973.

———. *Organization and Management: A Systems Approach.* 2d ed, pp. 510–11. New York: McGraw-Hill, 1974.

Katz, Daniel, and Kahn, Robert L. "Organizations and the System Concept." in *The Social Psychology of Organizations,* pp. 14–29. New York: Wiley, 1966.

Lawrence, Paul R., and Lorsch, Jay W. *Organization and Environment.* Homewood, Ill.: Irwin, 1967.

———. *Organizational Structure and Design.* Homewood, Ill: Irwin, 1970.

Perrow, Charles, "A Framework for the Comparative Analysis of Organizations." *American Sociological Review,* April 1967, pp. 194–208.

Pugh, D. S.; Hickson, D. J.; and Hinings, C. R. *Writers on Organization.* 2d ed. Harmondsworth, England: Penguin Books, 1971.

Terreberry, Shirley. "The Evolution of Organizational Environments." *Administrative Science Quarterly,* March 1968, pp. 490–613.

Thompson, James D. *Organizations in Action.* New York: McGraw-Hill, 1967.

Wetherbe, James C. *Systems Analysis for Computer-Based Information Systems*. St. Paul, Minn.: West Publishing, 1979.

———. *Executive's Guide to Computer-Based Information Systems*. Englewood Cliffs, NJ: Prentice-Hall, Inc., 1983.

———. "Learning to Live in the Knowledge Society," *Corporate Report*, March 1981, 42.

Wetherbe, James C. and Dickson, Gary W. *MIS Management*. New York: McGraw-Hill, forthcoming.

Wetherbe, James C., and Whitehead, Carlton J. "A Contingency View of Managing the Data Processing Organization." *MIS Quarterly*, March 1977, pp. 19–25.

Woodward, Joan. *Management and Technology*. London: Oxford University Press, 1965.

4

Concepts of Data Processing

Introduction

This chapter presents basic concepts of data processing as they pertain to computer-based information systems. The orientation of this material is fundamental and well within the grasp of students with limited data-processing exposure. For students with more extensive data-processing backgrounds, the material is a good review of data processing within a systems framework.

In Chapter 3 a model of an information system was illustrated, as shown in Figure 4.1. In this chapter the five subsystems—hardware, software, files, personnel and procedures—are defined and discussed using the systems concepts developed in Chapter 2.

Hardware

The core of data-processing hardware consists of the computer system itself. The major entities of a computer system are

- Input devices
- Output devices
- Secondary storage devices
- Central processing unit (CPU)

The historical input and output devices of computer systems are card readers and line printers, respectively. However most input and output today are done with on-line devices, such as terminals or personal computers. The most common secondary-storage devices are magnetic disk units and magnetic tape or cartridge units. Data channels and control units are used to connect these devices to the CPU. A model of a computer system is shown in Figure 4.2.

Figure 4.1   An Information System

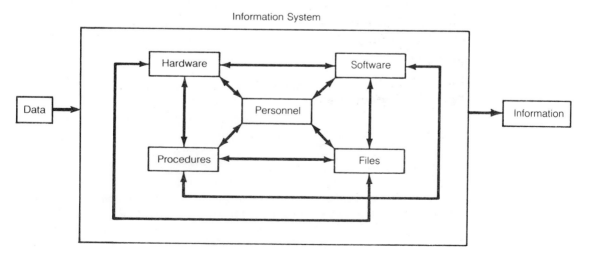

Figure 4.2   Model of a Computer System

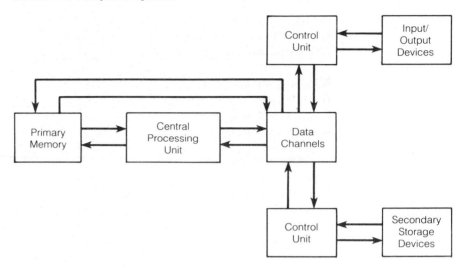

Input/Output Devices

A wide variety of input and output devices may be used on a computer system. Commonly used devices will be discussed first.

Input Devices

The card reader is the "grandfather" of data entry and is still used in many older systems. Other commonly used input devices include computer terminals, personal computers, optical scanners, magnetic ink character readers (MICRs), bar coding as used in retail establishments for point-of-sale terminals, and automatic teller machines at banks.

Output Devices

The historical and still popular form of output is a report generated on a printer. Other output devices include computer terminals, personal computers, graphic plotters, computer output microfilm (COM) units, laser printers, and audio (voice-synthesis) units.

Secondary Storage Devices

The most common data-storage media on computer systems are magnetic tapes and magnetic disks.

Magnetic Tape

A magnetic tape or cartridge unit is referred to as a *sequential-access device*. Data are sequentially stored on, and retrieved from, a magnetic tape in much the same manner as recordings are sequentially recorded on, or played back from, a tape recording. On a stereo tape, the preceding nine recordings have to be physically bypassed to get to the tenth recording. Similarly, on a magnetic tape, the preceding

1,999 payroll records have to be physically bypassed to access payroll record 2,000.

A reel of magnetic tape can generally store millions of characters (i.e., numbers, letters, or special characters). A magnetic tape unit can read or write a full reel of magnetic tape in a few minutes.

Magnetic Disk

A magnetic disk unit is referred to as a *direct-access device*. It has read/write heads that can both read and write data on the surface areas of platter-shaped magnetic disks (see Figure 4.3). As depicted in Figure 4.3, access arms are used to shift the read/write heads back and forth over the surface areas of the rotating disks.

The time associated with the shifting of a read/write head over a track of data is referred to as *seek time*. Once the read/write head is positioned over the desired track, it must wait for the desired data to rotate (or spin) under the read/write head. This is referred to as *rotational delay*. Therefore, the time required to access data on a disk is the sum of the seek time and the rotational delay, which is sixteen milliseconds or less on faster disk drives. More expensive disks eliminate seek time by having one read/write head per track.

Magnetic disk technology is superior in both storage capacity and access speed to magnetic tape technology. However, magnetic tape is less expensive. Magnetic tape units are also less costly than magnetic disk units, although disk-processing costs have dropped dramatically during recent years, and, consequently, disk processing now dominates the secondary-storage market.

Central Processing Unit (CPU)

A CPU is comprised of at least one control unit, an arithmetic-logic unit, and a primary storage unit. These components and their interaction are illustrated in

Figure 4.3   A Magnetic Disk Drive

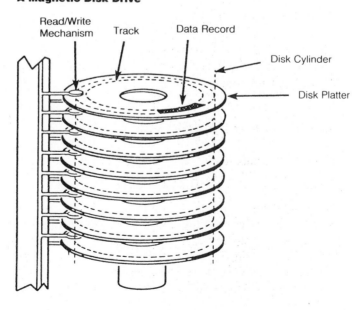

Figure 4.4. The control unit monitors and regulates the overall operation of the system and the execution of computer programs. The actual arithmetic and logic processing occur in the arithmetic-logic unit. The primary storage unit, or computer memory, is used to store data and instructions currently in use by the CPU.

The speed at which CPUs operate can generally be measured in terms of *nanoseconds*, or billionths of a second. Newer computers operate at speeds measured in *picoseconds*, or trillionths of a second. Viewed as a relative unit of measure in time perspective, there are as many nanoseconds in a second as there are seconds in thirty years. There are as many picoseconds in a second as there are seconds in 30,000 years. The CPU of a computer system is therefore operating at significantly faster speeds than any other component of the system (e.g., even disk drives are several thousand times slower than a CPU). Consequently, the average CPU can concurrently run several computer programs accessing various input, output, and secondary storage devices and still have unused (idle) capacity.

Data Entry and Reporting

Computers are quite efficient at processing data. However, the logistics associated with data entry and reporting represent two of the major difficulties and expenses of data processing. The two basic modes of data entry and reporting are batch processing and on-line processing.

Batch Processing
Batch processing is a computer processing technique whereby a number of similar input items are grouped together for processing during the same computer run (Figure 4.5a). It involves converting transactions into machine-readable form prior to inputting them to the computer system. The punched card is the historical medium for batch processing, but today, optical scanning, MICR, and terminals are more common.

Reporting in a batch-processing environment generally consists of printing and disseminating reports on a periodic basis. For example, weekly printouts are prepared to report outstanding customer orders and then distributed to the various offices that need the information.

Figure 4.4   **Central Processing Unit (CPU)**

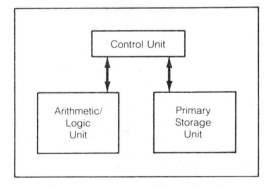

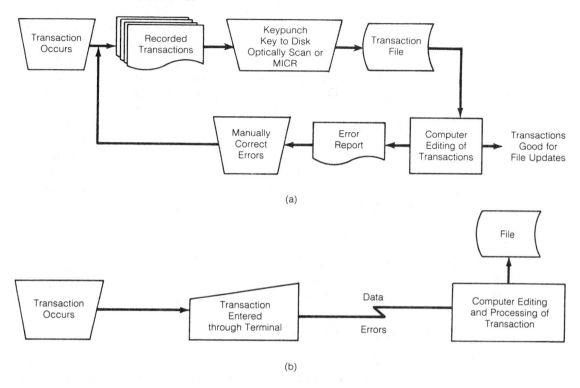

(a)

(b)

On-line Processing

On-line processing involves the direct entry of transactions into the computer system using computer terminals and is replacing most batch systems. A significant advantage to this approach is that a programmable terminal or the central computer can be programmed to edit or verify transactions as they are entered. This allows corrections to be made at a transaction's point of origin. In batch processing, a transaction may not reach the computer for editing until hours or days after the transaction has occurred. Consequently, it is more difficult to reconstruct the transaction for error correction.

Another disadvantage of batch processing is that an extra step (i.e., data conversion) is required between the occurrence of the transaction and computer processing. On-line processing allows the conversion of the transaction to a machine-readable form as a by-product of conducting the transaction. For example, use of retail or banking terminals to conduct transactions allows the data associated with each transaction to be captured and transmitted to the computer. Figure 4.5*b* illustrates on-line processing.

Types of On-line Processing

The reporting of information is greatly enhanced by on-line processing. Information stored in the computer can be accessed using computer terminals as well as printed on reports. There are two basic levels of on-line processing—one with real-time update and one without it.

In a system using on-line processing without real-time update, transactions are captured on-line, but data files are not updated at the time of capture. Rather, a transaction file is maintained (usually on disk) for subsequent updating of data files. The information displayed on a computer terminal is only as current as the last file update. This is how batch processing is done using terminals. For example, an accounts receivable department might capture customer charges and payments on-line but not update the accounts receivable file until the end of the day.

In an on-line system, data files can be updated immediately to reflect transaction activities. Such on-line systems are referred to as *real-time* systems. Each input to a real-time system is accepted and processed fast enough to affect subsequent inputs and queries. An example of an application requiring real-time processing capabilities is an airline reservation system.

Distributed Data Processing

The biggest breakthrough for on-line processing is in the area of distributed processing. Distributed processing involves distributing on-line processing from large computers to smaller, less costly processing units such as minicomputers and microcomputers.

Minicomputers are commonly used for on-line applications. For example, minicomputers can be set up to handle on-line data entry in each of several department stores. The transaction data associated with each store can be transmitted to the central computer at the end of the day for centralized processing. Note that in this example, transactions are collected *on-line* for subsequent *batch* processing. Figure 4.6 shows a distributed network.

Computer processing capabilities are also distributed to terminals using microprocessors. Such terminals are said to be *intelligent,* or *programmable.* This technology facilitates the design of industry-oriented devices such as retail terminals and banking terminals. Such devices can perform many functions, independent of a large, central computer. This not only reduces the load on the central computer, but also allows the terminal to have stand-alone capabilities that are operational even if the central computer malfunctions. For example, a retail terminal can be used independently to total sales, compute tax, and the like. The terminal can also store sales transaction data on floppy disk until the central computer is available.

The advent of the microprocessor-based personal computer has become a major factor in the use of computers. The economics of this technology indicate that virtually every employee in an organization will have a personal computer in the near future.

Communication Networks

The advent of distributed data processing has increased the need and demand for communication among computers to connect computers with other computers of different sizes in different locations. There are two general classes of communication networks—local area network (LAN) and wide area network (WAN). LANs can be used to support interconnections within a building or set of buildings that are close together. WANs (sometimes called long-haul networks) can be used for communications among remote devices as well as local devices if there is no LAN. Networks can be established in hierarchal structure, as demonstrated in Figure 4.6, or in circular structure.

Figure 4.6 **A Distributed Processing Network**

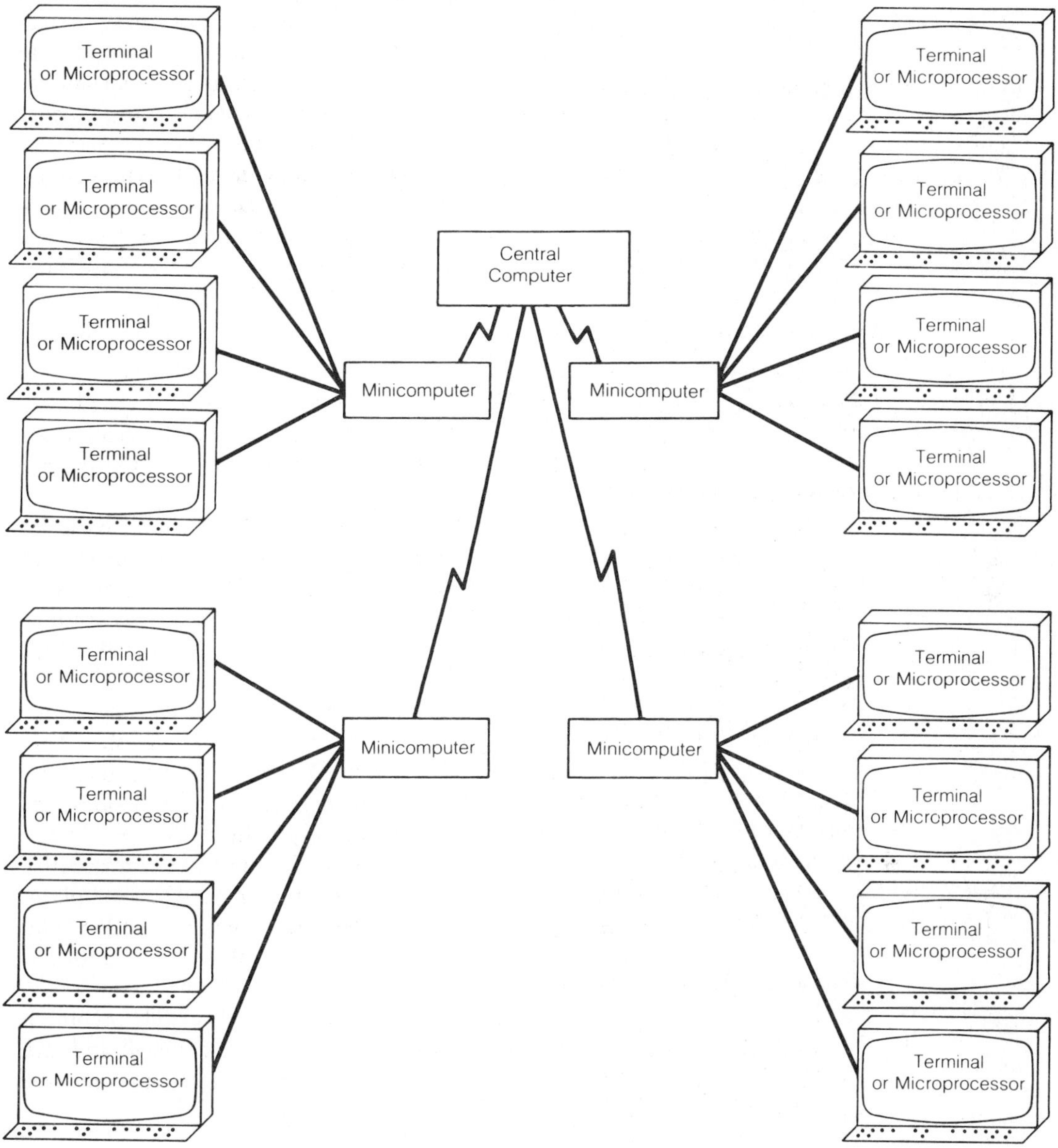

Advanced Office Systems

Advanced office systems (AOS) are another form of distributed data processing and on-line systems. Such systems offer significant capabilities for increasing the productivity of clerical and white-collar personnel. There is understandable excitement about AOS due to management's increased concern about the productivity of office workers. The white-collar segment of the labor force is growing dramatically; it currently includes approximately 50 percent of all workers nationwide. The cost of operating offices is also increasing, this at a rate of approximately 10 percent per year for large corporations. It has further been estimated that white-collar workers spend 50 percent of their time seeking information that they need for decision making. Not surprisingly, organizations are interested in a technology that will increase the productivity of the office worker.

There is considerable confusion and misunderstanding about the actual focus, capabilities, and degree of sophistication of the various AOS. Similarly, due to the evolving nature of this technology, new approaches and new systems appear constantly. Within this proliferation of technology, however, a fairly broad consensus of opinion about the basic generic functions for advanced office systems is emerging. According to this consensus of opinion, the major classifications for advanced office systems are:

Document Processing. Document processing consists of standard word-processing applications, including correspondence preparation, forms preparation, general text editing, and an electronic filing capability. Such systems primarily augment the duties of secretaries.

Electronic Mail Systems. Electronic mail systems provide for composition of managers' notes, memos, and similar communications. They also support a facility for editing compositions and transmitting the electronic messages to other persons on the automated office system. These systems also have the capability for printing and distributing messages through an automated "mailroom" facility. Transmissions may be sent to individual addresses or distribution lists may be used; they may also be broadcast to all members on specified mailing lists. Finally, filing and retrieval capability manages the mail that is received and provides for access to correspondence by author, date, address, and so forth.

Executive-Support Systems. Executive-support systems consist of a series of on-line, storage, and retrieval modules and files that directly support general executive functions with automated services. The types of functions available as executive-support systems include automated calendars (personal, departmental, etc.), directories (important telephone numbers, important officials, etc.), tickler files (for "to do" lists and to schedule future reminders), and calculator packages.

Information-System Interface. Information-system interface provides more tailored, local information systems as well as links into traditional corporate information-system capabilities. Advanced office systems extend and enhance existing information-systems technology. These systems are sets of on-line computer programs and files that provide information for decision making. Depending on their use, some programs may reside in a large-scale, centralized computer while others reside in the local system. Whatever the arrangement of hardware, software, and data, managers are provided with easy and timely access to information while at their work stations.

Software

In the most general sense, software is computer programs, procedures, and associated documentation, used in the operation of computer hardware. The two major categories of software are systems programs and application programs.

Systems Software

Systems software is a collection of computer programs used to coordinate and control the overall operation of a computer system. Systems software perform the following functions:

- It facilitates the loading and execution of application programs.

- It processes input and output commands to facilitate interaction with disk drives, tape drives, communication controllers, printers, and other devices.

- It provides a general library of utility routines for listing the contents of a tape on the printer, copying the contents of a disk to tape, and the like.

- In a multiprogramming system (i.e., where the computer is capable of concurrently running two or more application programs), it ensures that each program is serviced by the CPU in a rotating sequence or according to some other priority scheme. It allocates system resources (i.e., primary storage, secondary storage, and input and output devices) among the programs being executed and ensures that the programs do not interfere with one another's processing.

- In an on-line system, the systems programs include a data communications monitor to coordinate and control data transmissions to and from computer terminals.

Application Software

Application software are computer programs written for specific applications. These applications range from payroll processing and accounts receivable to forecasting and simulation. Application software generally depends on operating software during execution. For example, if an application program requires a record from a disk file, the operating software intervenes to retrieve the record from disk and makes it available to the application program.

Originally, most computer users wrote all of their application programs. However, it soon became apparent that some computer applications (e.g., payroll processing) were quite similar at a great number of installations. Rather than "re-invent the wheel," organizations began to buy and sell application programs to organizations with similar requirements. A large number of companies now develop and market specialized application systems for particular industries. It is usually less expensive for an organization to purchase such a system, when available, than to develop its own.

Preprogrammed application "packages" have two major disadvantages. First, they have to be highly generalized to fit a large number of customer requirements. Consequently, they are not as efficient as systems tailored to particular user needs. Second, they seldom exactly fit an organization's requirements. Therefore, modifications are necessary. In some unfortunate cases, the modifications become so extensive that packages have to be discarded.

Fourth-Generation and End-User Software

It is clear, given the insatiable demand for more and better information to support decision making and operations of organizations, that enough computer programmers cannot be graduated from our colleges and universities to handle the work load. Most organizations have a three- to five-year backlog of computer applications waiting to be programmed. This is only the tip of the iceberg of what is needed and can be done with computer-based information systems.

What the Phone Company Did

The problems ensuing from increasing productivity are similar to the ones faced by the phone company years ago. Had the phone company not automated as it did, everyone in the work force would have to be a telephone operator to handle today's telephone work loads.

What did the phone company do? To some extent, they *did* make all of us telephone operators. In the early days of phone use, we went through a telephone operator to do everything, including making a local call. Gradually, via automation, the phone company shifted more and more of the work load to us. First, we could make local calls by ourselves, then long distance, then credit card calls, etc. Today we get assistance from a telephone operator only when we can't make it on our own, and often that assistance comes from a computerized telephone operator. There are still telephone operators—the phone company is simply able to provide more service without adding substantial numbers of people. As a result, costs per transaction have gone down.

The evolution of phone use is analogous to what is happening with computer use. In the early days of computers, managers or users of computer services were totally dependent upon the computer technician for everything. The gradual evolution of "user-friendly languages" is allowing managers to do more and more on their own. The computer technician's role, like that of the telephone operator, is becoming one of "helping users to help themselves."

Evolution of Programming Language

First-generation computer languages of the early 1950s looked like this:

```
011011  0111011  0111101
011100  1100110  1010111
110000  1110000  1110010
```

By the late 1950s and early 1960s, second-generation languages looked more like this:

```
LD A   R1
AD B   R1
ST R1  C
```

In the mid- to late 1960s and 1970s, third-generation languages more commonly looked like this:

```
READ PRODUCT-FILE
    INVALID KEY PERFORM BAD-KEY ROUTINE.
MOVE PRODUCT-NUMBER TO O-PRODUCT-NUMBER.
MOVE P-DESCRIPTION TO O-DESCRIPTION.
MULTIPLY UNITS-ORDERED BY SELLING-PRICE
GIVING TOTAL-PRODUCT-LIST.
```

(Note: many additional instructions are required in order to execute the preceding statements.)

Today, fourth-generation user-friendly languages look like this:

```
SELECT PERSONNEL–RECORDS WHERE JOB–CLASSIFICATION
    IS ACCOUNTANT, SALARY
    IS LESS THAN 35,000, AND COLLEGE
    DEGREE IS MASTERS.
```

These advanced languages are easier to read and learn how to use.

Users Are Less Dependent on Technicians

If properly collected and organized data is stored on a computer, today's advanced programming languages allow managers to learn enough skills in a day or two to satisfy most of their immediate and ad hoc information requirements.

Computer technicians and information systems professionals are still needed to manage, design, and control the increasing complexity of hardware, software, and databases necessary to support this new environment. However, managers do not have to explain everything they need to have done to a technician. They have direct access to the computer; thus, overall productivity increases significantly.

In the future, systems professionals will focus more on developing systems that support the infrastructure for transaction processing or large-scale data processing. End-users increasingly will use high-level languages to directly access the various databases that are supported by these systems. In other words, users will be doing more for themselves. Systems design of the infrastructure systems will be more complicated than the types of systems developed earlier in computer-based information systems; but, fortunately, the tools available to do these systems are more powerful. Also, with the end-users being more able to help themselves, less effort will be required of the systems professional to write routine report-generating programs.

Chapter 15 of this book examines end-user computing in greater detail.

Artificial Intelligence

On the cutting edge of software is *artificial intelligence* (AI). AI refers to machines that can learn, reason, judge, and in many situations, exceed the mental capacity of humans.

The Power of Artificial Intelligence

To appreciate the speed and capability with which a computer can deal with intellectual problems, you need only purchase a microcomputer-based chess game for less than $100 and play a game of chess. The computer is kind enough to offer you several levels of difficulty. That is, it will lower its standards of play to meet your capabilities. While a human may take several minutes to ponder a move, the computer will counter in a split second with a well-calculated move. This move will reflect varying levels of sophistication, depending on how well a human has decided to let it play. Should you try to keep pace with the computer and play at its speed, you would have no chance of winning.

How Artificial Intelligence Works

AI software operates differently than traditionally programmed software. Instead of following precise "how to" instructions, programs are developed in a way that provides general guidelines for searching large data banks of knowledge. The AI program sorts through the knowledge stored in the computer and decides, on its own, the logical sequence of steps. The more advanced AI systems are able to evaluate the outcome of decisions made under a given set of circumstances. From that experience, the AI system can "learn" how to do the job more effectively in the future. Among the early applications of AI are expert systems, artificial senses, and fluent computers. Brief descriptions of each of these applications follows.

Expert Systems. The term *expert system* refers to the use of a computer as a consultant to assist humans in problem solving, such as medical diagnosis. To develop an expert system, a significant amount of research is required. It is necessary to study and structure the knowledge used by human experts in solving complex problems such as medical diagnosis, geophysical exploration, or computer design.

This process is called *knowledge engineering*. Knowledge collected in this process can be transferred to the computer. The more knowledge transferred to the computer, the better the expert system is.

Chapter 14 of this book, "Analysis and Design: Decision Support and Expert Systems," examines expert systems in greater detail.

Artificial Senses. Compared to humans, computers are quite unsophisticated in some tasks that even children take for granted. These tasks are related to sight and sound recognition. However, AI is making significant strides in these areas. AI technology can rapidly sort through signals coming from cameras and other sensors to identify images and sounds. AI capabilities in these areas will be invaluable for enhancing the capabilities of robots.

Natural-Language Interface. One of the biggest constraints to the use of computers by those unfamiliar with them is the difficulty of learning sophisticated computer programming languages. However, AI is changing that. With AI systems, it is no longer necessary for people to learn rigid computer syntax in order to instruct computers. *Fluent* (natural-language) computer systems can understand everyday English; they can also ask for more information if it is needed.

For example, an AI dialogue might go as follows:

Person:	Give me sales for product A and B in New York.
Computer:	For what time period do you want sales for A and B?
	A. last month
	B. last year
	C. year-to-date
	D. other
Person:	Last month.
Computer:	Thank you. Did you mean New York City or New York state?
Person:	New York state.
Computer:	O.K. Do you want a detail listing or summary totals?
Person:	Summry.
Computer:	Is that summary?
Person:	Yes.

Computer:	Thanks. So what you want is summary sales of products A and B for last month in New York state. Correct?
Person:	Yes.
Computer:	A = \$185,000 B = \$250,000

Implications of AI

The power and potential of AI is clear. There are, however, understandable concerns. As with all new technology, there is the issue of AI-related unemployment. Unquestionably, AI systems will replace people in many industries and offices.

However, we have to keep in mind that until the Industrial Revolution most labor was involved in producing agricultural products. In 1800, 95 percent of the United States work force was involved in agriculture; today less than 3 percent is. Few people would want to return to manual labor in agriculture and give up the standard of living we enjoy today. Similarly, in our highly educated society, fewer and fewer people are willing to do routine, assembly-line tasks or menial, clerical tasks. Accordingly, computer technology, including AI, will free people to do intellectual and service-oriented projects beyond what the computer can do for us.

Files

A file is a collection of related records. For example, a payroll file consists of a payroll record for each employee. The records in a data file are composed of data elements. Some data elements commonly included in a payroll file are employee name, social security number, address, salary, and deductions.

Accessing Files

Files are generally stored on magnetic tape or magnetic disk. Data is accessed using access techniques provided by the operating software. The three basic techniques for accessing files are

- Sequential access
- Direct access
- Indexed access

The following sections discuss common implementations of these techniques.

Sequential Access

Sequential access pertains to storing and retrieving records in a "one after the other" order. Records are generally stored in ascending or descending order by a record key. A record key is a unique, unchanging piece of information such as an account number, name, or social security number. Sequential access is the only access technique used with magnetic tape drives, which are, by design, sequential-access devices.

The storage and retrieval of records in sequential order is similar to the approach used in manual systems (e.g., of personnel folders stored alphabetically in a filing cabinet). Accordingly, sequential access has traditionally appealed to organizations converting from manual to computer-based systems. This appeal, combined with the early dominance of magnetic tape as a cost-effective storage medium, has led to the use of sequential-access files in most initial computing efforts.

Sequential access is used primarily in batch-processing environments. It is particularly effective when transaction activity is evenly distributed among the records in the file. For example, in payroll processing, *hours worked* data is captured for all hourly employees and payroll checks are printed for all employees. When a sequential file is updated, the transactions are sorted into the same sequence as the file and then sequentially matched against the file.

Direct Access

Sequential access is inefficient for batch-processing applications in which only a small proportion of the records in a file are affected by a given batch of transactions. The entire file may have to be passed to update a few records. For on-line processing, sequential access is inadequate. The time lapse of several minutes generally required to locate a record sequentially is unacceptable.

Sequential access fails to take advantage of two exceptional capabilities of computer technology: speed and direct access. Direct access is an alternative to sequential access that significantly accelerates the process of storing and retrieving records by capitalizing on both the computational speed of the CPU and the access speed of the disk drive. Disk drives are capable of directly accessing any record in a file in a matter of milliseconds. However, to access a particular record, the disk location of the record is required. The location is indicated by the address assigned to it, which is saved when the record is stored or recalculated on the basis of the record key when the record is sought.

In sequential access, the location of a record is a function of its proximity to preceding records. Since processing begins at the first record of the file, it is necessary to know only the address of the location of the first record. All subsequent records follow the first record.

Since direct accessing requires the address of the specific location of a record desired, it requires an addressing scheme that computes a unique address for each record. Generally, the record key (e.g., social security number, account number, or part number) must be transformed into a disk storage address.

The most common approach to transforming record keys into storage addresses involves an arithmetic procedure that generates "random addresses" from record keys. There are several randomizing algorithms. The most common algorithm involves the generation of addresses by dividing the record key by a positive prime number (usually, the prime number is the largest prime number that is less than the number of available addresses). The remainder of a division operation is used as the address locator. Figure 4.7 shows the generation of random addresses from record keys when 1,000 addresses are available.

A randomizing algorithm always generates the same address for a particular key. Therefore, given the key to a record, the computer can calculate the disk address and then access the record in a matter of milliseconds.

Occasionally, a random address generator generates the same address for two or more keys. The second and succeeding records with duplicate addresses are referred to as *synonyms*. When a synonym occurs, the record having the duplicate address can be stored in a location next to other synonym(s) of the record stored at the computed address, or it can be stored in a general overflow area. In either

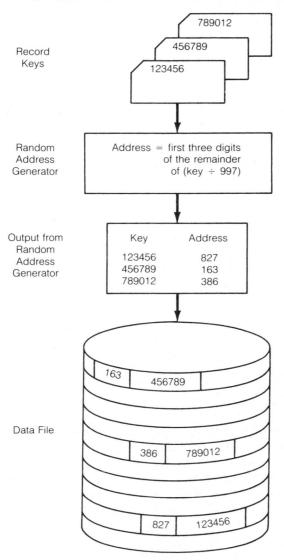

case, if a desired record is not located at the computed address (this is determined by checking the key stored in the record), a sequential search of synonyms is invoked until the desired record is located. This sequential search slows processing slightly. A good randomizing algorithm generates few synonyms. If a random address generator produces many synonyms for a particular set of keys, analysis and modification of the randomizing algorithm is needed.

Another drawback of a direct-access technique is that, by design, it usually leaves large gaps between records on a disk. This results in wasted disk space. Some of the gaps may be consumed by synonyms, but considerable wasted space still remains. An offsetting advantage is the incredible speed with which records can be accessed, regardless of the size of the file.

Perhaps the most challenging concept to grasp with respect to direct access is that the physical locations of records bear no relationship to the logical view of the data. The random generation of addresses physically scatters records throughout the disk such that without knowledge of the randomizing algorithm used to transform the keys, locating a record requires a sequential search through a nonsequentially-ordered file. The use of a randomizing algorithm is an extraordinary deviation from the way that files are maintained manually. However, it is a highly suitable technique for computer-processed files. Any one record of several thousand or even several million can be located instantaneously.

Indexed Access

Indexed access pertains to using tables of key fields that provide the disk addresses of records stored in a file. Indexed-access techniques are used with either sequential-access or direct-access files.

The simplest form of indexed access is referred to as indexed sequential. This technique involves the use of indexes to segment a sequentially stored file to facilitate quicker access. To access a specific record, a search is made of the index file to determine the address of the first record of the segment in which the record is located. A direct access is then made to that address. This positions the disk's read/write head in the proximity of the record desired. A sequential search of the file segment is then invoked until the record is located.

Since the index file is much smaller than the actual data file, indexed-sequential processing greatly accelerates the locating of a specific record; substantial sequential processing is eliminated. For large files, multiple levels of indexes (i.e., hierarchical indexes) are used to more quickly locate records. Figure 4.8 shows a hierarchical indexing scheme using three levels. The major-level index points to the intermediate-level index, which points to the minor-level index, which points to the actual data file.

Indexed-access techniques can also be used to access records by means other than record keys. This provides flexibility that is especially useful for querying a data file. For example, the personnel office can request the records of all employees who are accountants. Such access to records is accomplished by setting up indexes based on one or more fields of the records. In the same way that the index in this book points to each page that contains a particular word, a field, or *inverted* index, points back to each record that contains a certain field value. Figure 4.9 illustrates a job-classification index for a personnel file.

Indexes can be set up for several record fields. This allows even more discrimination in making a data-file query. Instead of just requesting the records of all employees who are accountants, the query can also specify that the employees have a college degree and two years of experience within the organization. One way of responding to this query requires that an index exist for job classification, education, and number of years within the organization. The query is satisfied by identifying all records pointed to by the first index, and then progressively eliminating any records that are not pointed to by the second and then the third index. After this process of elimination, only records that satisfy all the selection criteria remain. These records can then be directly accessed using the addresses provided by the indexes.

The use of indexed access on fields other than the record key is more suitable for higher-level decision-making and planning functions than for routine transaction processing. It allows convenient and less restrictive retrieval of information, but considerable programming time and computer processing are required to set up and maintain the indexes as records are added to and deleted from the data file.

Figure 4.8   **A Hierarchical Indexed Sequential Structure**

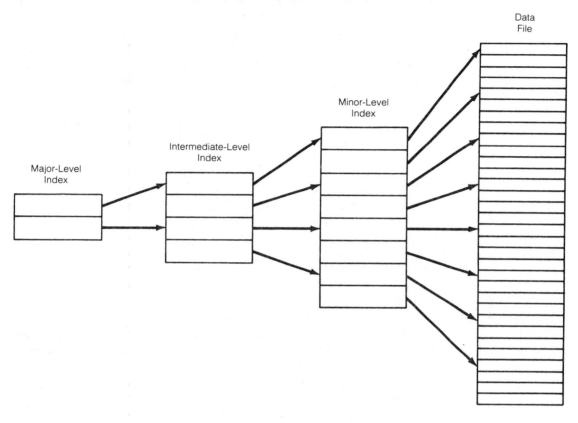

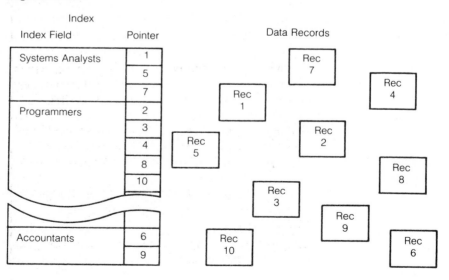

Database Management Systems

The preceding discussion of access techniques was oriented toward a single-file environment. At the end of Chapter 3, the concept of a management information system (MIS) was introduced. In that discussion, an MIS was portrayed as a federation of information systems, integrated as necessary. For example, in a university, student information in a student file needs to be linked or integrated with course information in the class schedule file (see Figure 4.10). Database technology is a step beyond file-access techniques and is a key factor in achieving an MIS. It provides software to facilitate the integration of the various data files of an organization and to provide more flexible access.

When organizational data are interrelated and stored together to serve one or more applications, the collection of data is generally referred to as a *database*. Data redundancy is seldom totally eliminated, but it is controlled and significantly reduced. Figure 4.11 is a simplified view of files integrated into a database struc-

Figure 4.10   **Linking Data between a Student File and a Class Schedule File**

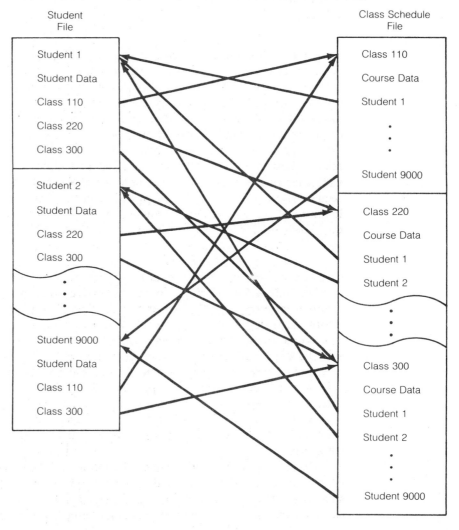

Figure 4.11   **Conceptual View of an Integrated Database Using Batch and On-line Processing**

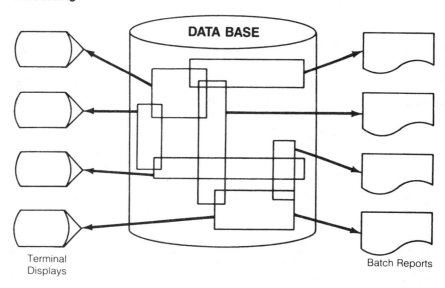

ture. An organization can have more than one database structure. For example, a university may have a database for a student information system consisting of student, course, and facilities files and a database for a financial information system consisting of general ledger, accounts receivable, accounts payable, and payroll files. When an organization decides to have two or more databases, the databases are disjoint in structure.

Without a database management system (DBMS) programming data inter-relationships to achieve integration and reduce redundancy is a complex task. Computer programs must use one or more of the file-access techniques (i.e., sequential, random, or indexed) to provide explicit instructions as to the physical locations of the data elements required to satisfy a particular on-line terminal query or to produce a particular batch report. Figure 4.12 shows the relationship among application programs, file-access software, and computer files.

For example, a program may need to access several interrelated data elements from records located in different files. The program must use indexes to compute physical addresses for each file in which data are contained, retrieve the entire record from each file, and then extract the data elements required from each record. Accordingly, programs must be coded to consider the physical locations and structures of all files, records, and data elements in the database. If a new data element is added to a record, resulting in a new record format, all programs accessing that record must be modified (even those to which the new data element is irrelevant). Ideally, application programs should be able to access the data in a database, independent of the physical structure of the data in the database. In other words, a program should be able to ask for, and get, only the data elements needed without having to get and process all the records in which the data elements are located.

During the early 1970s, generalized DBMS packages were developed by various companies to reduce the difficulties of integrating the data contained in a database. Their developers have made a significant contribution to the design and implementation of integrated, computer-based information systems.

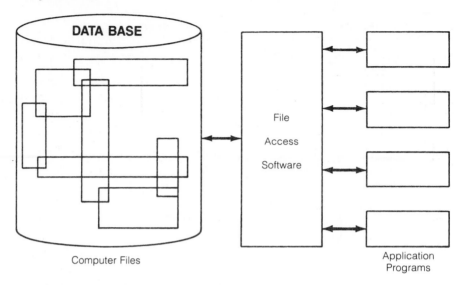

DATA BASE

File

Access

Software

Computer Files

Application
Programs

A DBMS can be envisioned as an additional layer of software between file-access software and application programs (see Figure 4.13). (Note that in systems theory terms, a DBMS performs a translation interface between programmers and access techniques.) It is beyond the scope of this text to explain the actual approaches and resulting mechanics of the DBMS's available. Their functions are described below to the extent needed for a systems analysis perspective.

A DBMS removes from the application program the function of explicitly coping with the physical locations and structures of the files and records in which data elements are stored. Therefore, the application program need only indicate the specific data elements required along with the key(s) or index(es) necessary to uniquely identify the group of data elements. With a DBMS, any data element or combination of data elements can be used to index a group of data from a record or records. (This is usually accomplished using an indexing scheme, as discussed earlier in this chapter.) The DBMS, in conjunction with the file-access software, handles the necessary logistics of retrieving and storing specified data elements in the proper physical locations within the database.

Vendors and software companies offer sophisticated advancements to their DBMS software packages in the form of query languages, which are called *non-procedural* or *fourth-generation* languages. A query language is a set of easy-to-use computer instructions. These instructions are designed to allow a person who is not trained in computer programming to retrieve, modify, add, or delete selected data elements, based upon stated conditions. These languages can usually be learned in a matter of hours.

The use of query languages is becoming increasingly popular due to the proliferation of computer terminals, ad hoc retrieval requests from all levels within an organization, and the ever-improving software packages readily available for most user applications.

Thus, a DBMS allows the users of a database to view the database in terms of the application necessary to support their departmental or functional area.

Figure 4.13   The Role of a DBMS

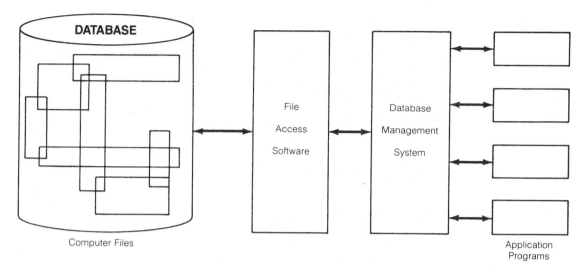

Their access to the database is not complicated by other data elements in the database. This means the data elements can be added to, or deleted from, the database, yet they affect only the application programs that use those data elements. A DBMS greatly reduces the difficulty of both establishing and modifying integrated files.

DBMSs combined with high-level query languages increase programmers productivity up to one hundred times of what was achievable using COBOL or PL/1 with traditional access techniques.

Personnel

The personnel in a data-processing or information systems department can be broadly categorized as either development personnel or production personnel. The *development personnel* include systems analysts and programmers involved with the analyzing, designing, implementing, and evaluating of information systems. Once a system has been developed and is operational, it is released to production personnel for ongoing scheduling and production processing. *Production personnel* include computer operators, data-entry operators, and production-control personnel.

Position Descriptions

Table 4.1 provides basic job descriptions for both development and production personnel. Within each job category, varying degrees of experience are generally recognized by designating positions as junior, senior, and lead levels.

Table 4.1	Position Descriptions for Basic Data-Processing or Information Systems Personnel
Development Personnel	*Position Description*
Systems Analyst	■ Analyzes, designs, implements, and evaluates computer-based processing systems to process transactions and provide information to meet organizational requirements.
Programmer	■ Designs, codes, tests, and debugs computer programs consistent with systems design specifications.
Production Personnel	*Position Description*
Computer Operator	■ Monitors and controls the operation of computer equipment.
Data Entry Operator	■ Operates microprocessor terminal to directly enter data into a computer system.
Production Controller	■ Records and maintains control information on the receipt and processing of transactions, and the subsequent distribution of generated reports.

The increasing complexity associated with computer-based information systems has resulted in the need for further specialization of the basic data-processing positions and for entirely new positions. Table 4.2 depicts the further specialization and the creation of new data-processing positions.

Basic Organizational Structure

One of the most important managerial issues regarding MIS is that of structuring the MIS organization itself. This must be done in such a way as to enhance managerial efficiency and the overall effectiveness of the unit (especially in serving the needs of the users). The most common basic structure of an information systems organization is shown in Figure 4.14. Note that the computer-operations function and the systems-development function report to the chief information officer (CIO).

The internal structure of the information systems organization must accommodate a number of situations. One involves the trade-off between managerial flexibility and efficiency versus service to the user. Another is the necessity of maintaining existing systems yet providing for evolution of these systems as user needs change and/or new external requirements appear. A third concerns the

Figure 4.14   Basic Structure of an Information Systems Organization

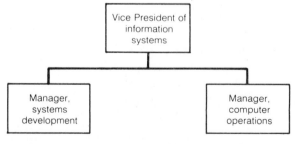

Table 4.2   Further Specialization of Basic Data-Processing Positions

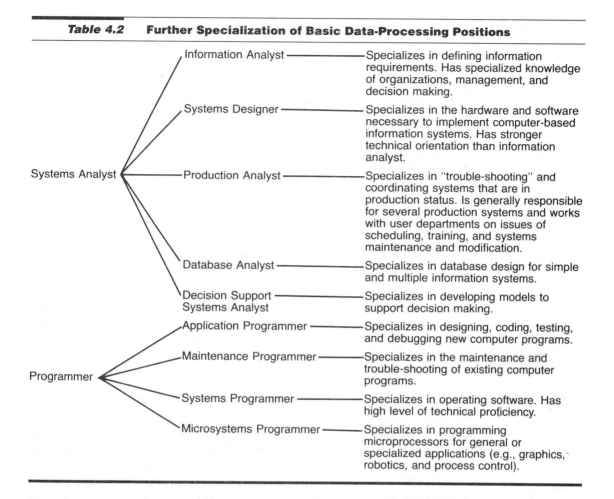

Systems Analyst

Information Analyst — Specializes in defining information requirements. Has specialized knowledge of organizations, management, and decision making.

Systems Designer — Specializes in the hardware and software necessary to implement computer-based information systems. Has stronger technical orientation than information analyst.

Production Analyst — Specializes in "trouble-shooting" and coordinating systems that are in production status. Is generally responsible for several production systems and works with user departments on issues of scheduling, training, and systems maintenance and modification.

Database Analyst — Specializes in database design for simple and multiple information systems.

Decision Support Systems Analyst — Specializes in developing models to support decision making.

Programmer

Application Programmer — Specializes in designing, coding, testing, and debugging new computer programs.

Maintenance Programmer — Specializes in the maintenance and trouble-shooting of existing computer programs.

Systems Programmer — Specializes in operating software. Has high level of technical proficiency.

Microsystems Programmer — Specializes in programming microprocessors for general or specialized applications (e.g., graphics, robotics, and process control).

monitoring of the information-systems function by the overall organization, setting the priorities for the organization, and aligning the information systems' planning with overall organizational directions.

Managerial Efficiency versus User Service

From an efficiency standpoint, the MIS manager would prefer to structure the development function as shown in Figure 4.15, assuming that one group of persons acts as analysts and another does the programming.

Figure 4.15   Information Systems Organization Structured for Managerial Flexibility

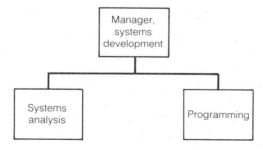

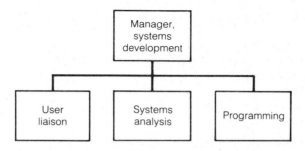

The beauty of this organization is that it is very flexible. If new systems-development requirements arise, it is easy to assign persons from the analysis and programming pool to do the job. The difficulty, especially from a user perspective, is that the analysts and programmers assigned to their project may have no experience or familiarity with the problem area or with the user.

An attempt to ease these problems has resulted in another structure, which is shown in Figure 4.16. Note that another subunit, *user liaison,* has been added. The persons in this unit, typically experienced senior analysts, are responsible for working with the user organizations to determine their information needs and with the analysts and programmers to meet these needs. Typically, one person from the user liaison gets to know the user's business and becomes well-known to persons in the user organization. In addition, the user liaison can translate business problems into a more technical language when communicating with the system developers. One problem of this organization is that there is often confusion about the responsibilities and the authority of the user liaison and the responsibilities and authority of the users or systems personnel. Another problem is that this system is very dependent on the skills of each user liaison.

Another organizational form that has been used to balance managerial needs and user requirements is the matrix organization shown in Figure 4.17. In a matrix organization, there is a pool of programmers and a pool of analysts. The people in these pools report to positions with titles such as manager of programming and manager of analysis, respectively. In addition, however, there is a subunit composed of project leaders. These persons report to a manager, perhaps having the title of manager of project development. When a new project comes along, it is assigned to a project leader, who negotiates with the manager of analysis and with the manager of programming to form a project team.

Figure 4.17   **Information Systems Organization in Matrix Form**

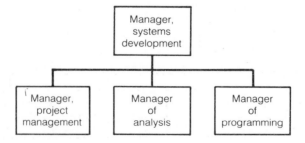

A programmer or analyst is typically involved with one project at a time, but a project leader may be involved with as many as three projects at the same time. The project leaders tend to work with particular user organizations, but the analysts and programmers do not. This organizational form is very efficient, but it has the disadvantage of each analyst and programmer having two different bosses, because each of these persons reports to a project leader for project-related matters and to the manager of their function for overall matters. Performance is evaluated by the functional manager with input from project leaders with whom an individual has worked. This system potentially leads to great difficulties when conflicts arise.

Figure 4.18 shows what is probably the best compromise structure. This is a functional organization in which analysts and programmers are grouped to service a particular user group. It is a difficult structure to manage in the sense that it is not very flexible. Slack resources may exist in one subunit while another is overloaded. This shortcoming is balanced by the fact that, over time, the systems development personnel become very familiar with the function they are serving. Furthermore, by working together over many projects, they tend to form many strong interpersonal relationships that usually help ease problems of communication and systems implementation.

Before leaving this topic, it should be noted that for very large projects it is often desirable that special project teams be formed. An organizational form not discussed in detail here is the placement of the analysis and programming function in the user organization. This is really a variant of the functional structure. Although its use has been a recent trend, it should be approached with caution. Unless good plans and standards are in place and the user organization is very familiar with managing data processing, many difficulties, not the least of which are cost escalation and violations of standard, can occur when systems development is distributed. Chapter 15 provides frameworks for facilitating more systems development into user organizations.

Analysis and Programming

In the preceding discussion, it was assumed that one set of persons does systems analysis and another group does programming. This is only one way of dividing the systems development task. Another, obviously, is to have a single group perform both analysis and programming. This approaches resolves coordination problems but reduces specialization.

Another issue to keep in mind involves career paths. Frequently it is the practice to start persons out in programming and, over time, move them into

Figure 4.18 **Information Systems Organization Structured by User Functions**

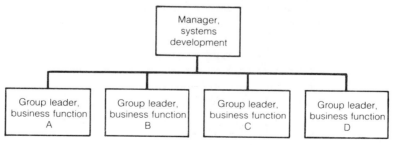

analysis. However, there is ample evidence that good systems analysts do not have to have been programmers, and some programmers do not make good systems analysts. The more technically focused programmers often do not have the inclination or ability in the interpersonal skills area.

Maintenance

The method chosen for maintenance of existing systems also affects the MIS internal organization chart. Two major alternatives exist for performing this function. The first is to have those who developed the system be responsible for correcting any errors that show up after the system is put into operation. These persons also must modify the system as changes come up as a result of new user needs or business requirements. The second approach for dealing with system change is to create a separate maintenance function for handling all system modifications of a modest nature. Sometimes such a function is placed in the systems development subunit, and sometimes the function is found in the systems operations subunit.

Each of the two approaches to performing systems maintenance has advantages and disadvantages. Having the systems developers do their own maintenance is efficient. The developers know the system well and can more easily make modifications. The difficulty is that if these persons know they will be doing their own maintenance, they tend to keep some of the documentation in their heads rather than writing everything down in the systems documentation. Further, if one of these key people should leave the company, it is very difficult for a new person to figure out what needs to be done to modify the system. Still another difficulty with this approach is that developers can spend an excessive amount of their time performing maintenance. In many cases, this allocation of time is difficult to track, and, therefore, the organization does not recognize the amount of its resources being allocated to maintenance.

A separate maintenance function has the disadvantage of requiring that the maintenance programmers go through a new learning process each time a new task is encountered. Moreover, doing nothing but systems maintenance is boring for many people over a long period of time. Some people of course, enjoy it as a long-term career and has a special aptitude for it, but in order to accommodate those who do not, some organizations use maintenance programming as a first-year training assignment for new programmers, who can later go on to other programming positions.

Advantages of a separate maintenance group are that it tends to force good systems documentation on the part of the developers, it makes for easy accounting of maintenance activity, and it buffers the organization against losing key development personnel.

The Information-Center Concept

The concept of an *information center* as an addition to the MIS organization has recently emerged. This subunit would be set up to help users get certain systems built quickly and would provide tools that could be employed by users to build their own systems. Tools such as database management systems, report generators and query languages, time-sharing, and fourth-generation software packages are provided and supported. The concepts of information centers and end-user computing are covered in depth in Chapter 15.

Procedures

Procedures are specific instructions and techniques for both the development and the operation of information systems. The accepted means of documenting and conveying procedures is through procedures manuals. Such manuals can be viewed as "cookbooks" that carefully lead personnel through the necessary and accepted steps to complete a task or resolve a problem.

Much of what this book provides in the following chapters are procedures and techniques to be used by the systems analyst in developing systems. Procedures can be broadly categorized as developmental and operational.

Systems Development Procedures

Systems development procedures provide standards for systems analysis, systems design, programming, and documentation. Adherence to development standards and techniques ensures uniformity of information systems. This uniformity is important because information systems can take months or even years to develop and then have a useful operating life of several years. The original systems analysts and programmers frequently changed responsibilities or organizations during either the development or the operation of a system. When this happened, new personnel were assigned to take over their responsibilities. The uniformity established through later development procedures greatly eased the transition of responsibilities.

Some organizations have experienced considerable hardship because of the loss of key systems analysts and/or programmers when systems development procedures did not exist or were not enforced. For example, assume a payroll system must be modified to implement new tax tables. Without procedures to explain how or which computer program computes taxes, this modification is difficult. As a result, payroll processing may be delayed.

A recent development that has greatly facilitated the management and documentation of development procedures are computer-aided systems engineering (CASE) products. They provide automated support for development procedures.

Operating Procedures

Operating procedures help to ensure that the processing of data is accomplished in an efficient, effective manner. These procedures comprise specific instructions for data collection, preparation, and processing, and for report distribution. For example, organizational departments are told how to collect and prepare payroll transactions; data-entry operators are provided instructions; and computer operators are told which computer programs and files to use.

Operating procedures are also used for control purposes. They document the actions to be taken when errors are discovered in the collection, preparation, processing, and distribution activities. For example, a payroll clerk can refer to a procedures manual to determine what action to take if payroll totals are out of balance.

Summary

The major subsystems of an information system consist of hardware, software, data files, procedures, and personnel. These subsystems interact to transform data into information suitable for decision making.

The basic hardware components are input/output devices, secondary-storage devices, and the central processing unit. The major forms of processing are batch processing and on-line processing. Distributed processing is a popular architectual concept for structuring data-processing hardware. Distributed processing involves distributing processing capabilities to minicomputers and microprocessors through communication networks.

Software is categorized as either operating software or application programs. Operating software coordinates and controls the overall operation of the computer system. Application programs are written to process specific applications.

Files are collections of related records. The records within data files can be accessed using sequential, direct, or indexed access. A database system is a collection of files and software organized to facilitate the interrelating of the files of an organization.

Procedures are specific instructions for both the development and operation of systems. Procedures manuals are set up to instruct personnel on how to perform a task or solve a problem.

The personnel in a data-processing or information systems department can be categorized as either development or production personnel. Development personnel primarily consist of systems analysts and programmers. Production personnel primarily consist of computer operators, data-entry operators, and production controllers.

Exercises

1. Identify and define the major hardware components in a computer system.
2. Discuss the major advantages and disadvantages of batch and on-line processing.
3. What technological developments have enhanced the viability of on-line processing?
4. Discuss the difference between systems programs and application programs.
5. Define and discuss the relationships among data elements, records, files, and data bases.
6. Define and discuss the three major types of file access. Which access techniques are suitable for on-line processing?
7. Discuss the role of procedures in an information system.
8. Discuss alternative ways of organizing systems analysts and programmers.

Selected References

Bateman, Barry L., and Wetherbe, James C. "Production Analyst: A New Position for an Old Problem." *Data Management*, December 1974, pp. 30–33.

Canning, R. G. "Trends in Data Management, Part 1." *EDP Analyzer* 9 (1971); "Part 2." *EDP Analyzer* 9 (1971).

CODASYL Systems Committee. *A Survey of Generalized Data Base Management Systems*. Technical Report. May 1969.

Davis, Gordon B., and Olson, Margrethe H. *Management Information Systems: Conceptual Foundations, Structure, and Development*. New York: McGraw-Hill, 1985.

Lucas, Henry C. *Introduction to Computer and Information Systems*. New York: McGraw-Hill, 1985.

―――. "Telecommuting: A Remote Possibility." *Corporate Report*, August 1981, p. 36.

―――. "Fighting Information Overlap." *Corporate Report*, November 1981, p. 56–57.

―――. "Making Distributed Data Processing Work." *Corporate Report*, January 1982, pp. 46–47.

Wetherbe, James C.; Bateman, Barry L.; and Westman, Chadwick H. "A Model of for Minicomputer Maintenance Evaluation and Penalty Compensation." *Mini and Micro Processors*, 1978.

Wetherbe, James C., and Davis, Charles K. "An Analysis of the impact of Distributed Data Processing on Organizations in the 1980s." *MIS Quarterly*, December 1979, pp. 47–56.

―――. "Planning and Controlling Distributed Data Processing." *Systems, Objectives, Solutions*, March 1981, pp. 79–87.

Wetherbe, James C.; Davis, Charles K.; and Dykman, Charlene. "Implemented Automated Office Systems." *Journal of Systems Management*, August 1981, pp. 6–12.

Wetherbe, James C., and Dickson, Gary W. *MIS Management*. New York: McGraw-Hill, 1984.

Wetherbe, James C., and Rademacher, Donald. "Computer Graphics: A Management Perspective." *Journal of Systems Management*, December 1982, pp. 6–9.

Wetherbe, James C., and Whitehead, C. J. "A Contingency View of Managing the Data Processing Organization." *MIS Quarterly*, March 1977, pp. 19–25.

Techniques and Technologies for Systems Analysis and Design

The material covered in Section 1 has provided the conceptual and theoretical foundations for a more pragmatic view of the processes and techniques of systems analysis as discussed in Section 2.

Chapter 1 provided an overview of the systems development cycle (see Figure 1.1). The seven steps involved in the systems development process are allocated among the six chapters in this section as shown in Figure 2.1.

Increasing the productivity of systems development and shortening the time between a user's request for a system and the delivery of a system that meets that request are two of the most pressing issues facing information systems management today. The chapters in this section cover the traditional, structured, and advanced techniques currently available to meet these challenges.

The concepts and techniques of information systems development have been poorly understood and poorly used throughout the short history of information systems. One problem has been the lack of knowledge and

Figure 2.1   Relationship of Chapters in Section 2 to Systems Development Life Cycle

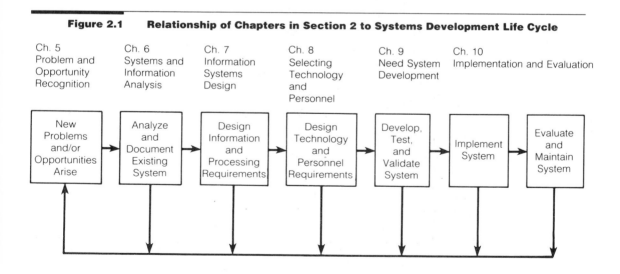

experience—systems development is an emerging discipline. The second problem is that the technology available to support information systems continues to change so swiftly that systems development techniques lag behind technology by five to ten years. No sooner are systems development deficiencies resolved for one type of technology than a new era of technology begins. For example, as soon as effective techniques were developed for batch-oriented technology, on-line technology became prevalent. Once techniques were refined for on-line systems, database and distributed data-processing technology began to emerge. The advent of high-level user languages and advanced office systems greatly complicate the development of information systems.

Fortunately, experience has resulted in a differentiation between systems development concepts or theory and systems development techniques: concepts and theory transcend technology changes, whereas techniques may need to be redesigned. For example, a major concept in systems development is determining what information a decision maker requires, regardless of the type of system proposed. The technique for making that determination when using batch technology, however, is different from the technique employed when using on-line database technology with high-level query languages.

Before proceeding directly into the chapters of this section it is helpful to review the evolution of the systems analysis and design process and techniques.

Evolution of Systems Development

When computer-based information systems were initially being developed in the 1950s, there was very little process or technique available to do the job. Systems development processes and techniques primarily resulted in what could be characterized as a "Victorian novel;" that is, it narratively described what an information system was supposed to do. When computer programmers tried to translate this narrative documentation into structured, precise software, many problems arose. The narrative approach to information systems development resulted in key steps being overlooked and left out, as well as contradictions, ambiguities, and utter confusion among those trying to develop the systems. Consequently, a great deal of dialogue was necessary to supplement the inadequacies provided by this narrative approach.

System Development Methodologies

A major development that occurred in information systems development in the late 1950s and early 1960s was the formalization of the systems development process. Using the systems development life cycle (SDLC) as a framework for developing computer-based information systems, more refined and formalized expression of the SDLC provided rigorous definition of the various steps needed to develop an information system.

This formalization of the systems development process resulted in comprehensive packages consisting of training programs and procedures manuals that could be used to lead organizations through the SDLC. These methodologies included extensive checklists and checkpoints that facilitated the management of SDLC. The checklists are analogous to the type of checklists pilots go through before they take off in an aircraft. It is hoped that by carefully evaluating each item on the checklist, one will not overlook any critical issues before proceeding.

As helpful as systems development methodologies were, there were still problems in information systems development, mainly resulting from the inadequacies caused by extensive use of narrative explanations of what an information system was supposed to do. In systems theory terms (see Chapter 2), what was needed was a more closed stable mechanistic approach to defining or articulating what an information system was to do. In this way, different analysts or programmers could look at the design specifications and clearly understand them. Without this approach, different analysts could look at the same problem and come up with totally different design specifications, and different people could look at the same design specifications and interpret them differently. The result was a great many systems that simply did not do what the user wanted.

Structured Techniques

To address this problem, a new set of techniques emerged in the mid- to late 1970s that were generally referred to as *structured techniques.* Structured techniques provided an "engineering" style of discipline that is very helpful in ensuring the correctness and communications of systems design specifications.

The structured tools commonly used for systems analysis and design are flowcharts, data flow diagrams, structured English, decision tables, decision trees, and the data dictionary. Each of these techniques is covered in the following section, as are structured techniques for software development or software engineering.

Computer-Aided Systems Engineering

The original introduction of structural techniques was based on manual, clerical products. However, during the past decade computer-assisted tools emerged to support analysis, design, and software development. Referred to as CASE (computer-assisted systems engineering), these products, though initially primitive and generally more trouble than they were worth, have recently matured into impressive, powerful tools. They are causing at least an evolution, if not a revolution, in the way systems are being developed.

CASE products include diagramming tools, syntax verifiers, automated data dictionaries, prototyping tools, code generators, project management tools, automated administrative support for the systems development and life cycle, and programmer workbenches. CASE tools are integrated into the systems development techniques throughout this section.

Information Requirements Determination

Structured techniques and CASE technology greatly enhance the ability of systems designers to develop systems that work right. It is, however, another issue to ensure that a system provides the right information. As discussed earlier, managers do not usually know what information they need. Consequently, techniques specifically designed to determine the right information are required. The task involves some structured interview techniques and other advanced techniques generally referred to as heuristic development and prototyping, which are covered in Chapters 6 and 7.

Single System versus Multiple Systems

In the following section we will be looking at developing a single information system that might involve multiple files, multiple departments, and multiple users of the system. Though for instructional purposes this is a helpful way to proceed, when systems are actually developed in organizations it is not wise to develop one or more information systems without an awareness of all the types of files and systems that will ultimately be needed by the organization. To understand how a single information system fits into the context of all organizational information systems, a strategic information systems plan is needed. That issue will be addressed in the final section of this book, which focuses on strategic, administrative, and higher-level concepts and techniques of systems development and planning. With the plan for developing systems in place, as described in the final section, you will be able to integrate a variety of information systems into your organization at any time. ■

Chapter 5

Problem and Opportunity Recognition

Introduction

This chapter discusses the opportunity- and problem-recognition dimension of systems analysis and design. The particular areas of interest are related to the methodological investigation of a problem or opportunity. The problem or opportunity is identified and separated into related components for further detailed study.

Problems and Opportunities

Before a problem or an opportunity can be analyzed, it must be clearly identified and defined. The first section of this chapter presents concepts pertinent to identifying and defining problems and opportunities.

Reaction and Opportunistic Surveillance

There are two basic approaches to management decision making that are relevant to defining problems and opportunities. These two approaches are called *reaction* and *opportunistic surveillance*.

Reaction involves addressing problems as they arise. This is a "management by exception" approach. It is commonly employed. For example, if customers begin to complain about the tardiness of order processing, management reacts by attempting to resolve the problem. Otherwise, management "leaves well enough alone."

Opportunistic surveillance involves continually seeking opportunities that might be beneficial to the organization. Rather than leave well enough alone, management seeks continual improvement. This is a much more aggressive posture, and it can pay considerable dividends. For example, management may see an opportunity to acquire more customers by implementing a new order-processing system that will provide customer deliveries superior to those provided by competitors.

Generally, the systems analyst is not called in on a major problem or opportunity until after the problem or opportunity has been identified by management. However, once the systems analyst is involved, he or she should maintain opportunistic surveillance as well as reactionary orientation. New opportunities and problems continually arise in organizations. Many of them are subsets of an original opportunity or problem identified by management. The systems analyst should be alert to such developments and call them to management's attention so that they can be properly considered and addressed.

Object System and Information System

A distinction that is sometimes subtle but nevertheless important is the distinction between an object system and an information system. An *object system* is a physical process whose purpose is to achieve one or more organizational goals. Examples of object systems include manufacturing of products and selling of merchandise.

An *information system* is a physical process that supports an object system in achieving organizational goals. The information system *runs parallel* to the object system to provide both documentation and information for decision making pertinent to the management and operation of the object system. For example, in a manufacturing plant, work orders are used to document the work flow, and production schedules are used to plan the work flow. In a retail store, sales are documented by being entered into cash registers or retail terminals, and sales information is used for merchandising decisions.

A helpful and important way to differentiate between object systems and information systems is to keep in mind that without an object system there is no need for an information system. The purpose of an information system is to facilitate the operation of an object system. Indeed, the cost of operating the information system should be offset by the savings realized because the object system can be (and is) more efficiently and effectively managed.

Generally, the systems analyst accepts the object system as given. However, there are occasions in systems analysis when potential improvements in the object system are apparent to the systems analyst. For example, rearranging work stations in a manufacturing plant might save considerable time in transporting goods-in-process through work stations. In such cases, the systems analyst should consider analyzing and designing improved methods for the operation of the object system as well as the information system.

Identifying Problems and Opportunities

There are a multiplicity of problems that can occur and opportunities that can arise in organizations. Management generally expresses problems and opportunities in terms of an "end result," or "bottom line." General management is often critical of the inability of information systems professionals to appreciate that management is ultimately interested in the end result or bottom line associated with a problem or opportunity. This point was well made by Jerry Hoffman, an information systems executive from Standard Oil speaking at the 1983 Information Systems Education Conference, held in Chicago. He stated, "More than anything else, I wish our colleges and universities could produce graduates who could tell the difference between a good idea and a bad idea."

It seems often that information systems professionals get so caught up in the technological elegance of computers that they lose sight of the contribution the information system is ultimately going to make to an organization. One of the best ways to learn how to identify problems and opportunities, and to differentiate the good opportunities from the bad, is by example. To that end, several examples of organizations that have opportunistically integrated computer-based information systems into their organizations will be discussed.

Opportunity Is a Computer

Almost all companies initially apply information systems technology to their fundamental accounting applications (general ledger, accounts receivable, accounts payable, and payroll). Such applications are defined and structured well; therefore, they are well suited for information systems technology. If done properly, the application of information systems technology to accounting applications can reduce costs.

Except for accounting firms, however, companies are not in the accounting business. Rather, they generate revenue by providing other goods or services. Accounting is a support function and does not in itself generate revenue. The more mature and sophisticated users of information systems recognize that the real pay-off comes from integrating information systems technology into the core operations of their company.

Case: Dayton's Department Stores

Dayton's, a premier department store chain located in the upper Midwest, noted a common problem with wedding gift giving. Brides and grooms historically indicate what they would like in generic terms (e.g., a blender, towels, toaster). Consequently, they generally don't get the exact item they want. And since coordination among gift givers is minimal, duplicate gift giving often occurs.

This situation results in three problems: (1) The bride and groom do not get what they want and/or may receive duplicates, resulting in gifts having to be returned or exchanged; (2) gift givers often find out later their gift was returned or exchanged; and (3) the store has to bear the expense of non revenue-generating, high-overhead transactions—returns and exchanges. To make matters worse, since wedding gifts don't usually have receipts, the store cannot even be sure that returned gifts were purchased at their store.

To solve these problems, Dayton's offered a computer-based bridal registry system. Supported by personal computers, brides and grooms can generate a computer-stored list of the exact items they would like. Once stored on the computer, gift givers can go to the bridal registry at Dayton's, key in the bride's name, and have access to the "wish list." As gifts are purchased, they are checked off the list to avoid duplicates.

The system lets the bride and groom get the gift they want, ensures the gift givers that they are giving a desired gift that will not be duplicated, and Dayton's virtually eliminates returns and almost guarantees gift givers will shop at Dayton's. As an additional benefit, gift givers tend to shop earlier so they can have a bigger selection (or buy the lower-priced items before someone else does). The early shopping improves Dayton's cash flow. ▮

Case: Fingerhut

Fingerhut, a successful direct marketer and one of the most progressive users of information systems, has recognized the need for integration as discussed here and has proceeded accordingly. Says Steve Platt, vice-president of marketing development, "The cost of everything we do is going up. Personnel, fuel, postage, printing, and merchandise all cost more. Computers are the only business expense that is going down. We made a strategic decision to invest in marketing-information systems."

How is that strategy put into effect? Platt explains: "To the extent that we use information systems technology to determine what products a customer wants, is willing to buy, and is able to pay for, we can greatly enhance our marketing effort. We maintain comprehensive customer profiles that we use for a variety of customer analyses and forecasting. Such information can be used in productive ways. For example, we can take two separate items of information and combine them to improve our marketing. If we predict that a customer will buy a product and we also know that same customer has a child who has a birthday coming up, we sent this customer a personal letter offering them the product and a free additional gift for their child's upcoming birthday." ■

Case: Investors Diversified Services (IDS)

IDS, which markets investment services to individuals and corporations out of 160 regional offices with over 3,000 salespersons, has undertaken a new project aimed at integrating information systems technology into company operations. In an effort to improve customer service, IDS is exploring the use of microprocessor-based computers to aid in the analysis of customer requirements and in the presentation of alternative investment proposals.

According to George M. Perry, vice-president of information systems, "Financial services marketing, and total financial planning in particular, is becoming an extremely complex process, often requiring several iterations—that is, modeling of alternatives—before an optimum solution can be found. We feel that sales-force productivity goals will require extensive use of information systems technology.

"To this end, we are experimenting with personal computers, portable terminals, graphic-illustration systems, and the like. We are experimenting with graphically illustrating sales proposals with multi-color graphs and bar charts using the client's home TV set as a display for the computer output. It is much easier for a customer to comprehend a proposal that is presented graphically than as a table of numbers on a computer printout."

"This is just the start," Perry maintains. "Computers are coming out of the back office and are becoming an integrated part of the product-design process and the evolution of delivery systems. In computer-dependent companies like IDS, technology will play a key role in product structure and delivery methods of the future." ■

Case: Super Valu

Super Valu, a food wholesaler for over 2,000 retail food operations across the country, has cleverly integrated information-systems technology into its goods and services. Besides providing food products to grocery stores, Super Valu provides extensive information systems services to assist in store management.

John Ferris, a senior vice-president controller, sees it this way: "Our philosophy is a total commitment to servicing our customers more effectively than anyone else could serve them. To accomplish this, we use information systems to enhance our operation as well as that of our customers. Our success is dependent upon the success of our customers. Accordingly, we provide support for electronic point-of-sale terminals and electronic order entry. We also offer specialized information systems services to our customers such as shelf-allocation studies and new-store-site location analysis." ■

Case: B. Dalton Bookseller

Besides being used to change and enhance old ways of doing business, information systems technology is often the foundation that allows a new company to compete with companies that are not taking advantage of this technology. B. Dalton Bookseller, is an example of such a company. In the mid-1960s, B. Dalton Bookseller was organized to compete in an industry in which virtually no information systems technology was being used.

R. Richard Fontaine, president of B. Dalton Bookseller, indicates that information systems were essential to the basic concept of his company: "There are over 35,000 new book titles each year and over 450,000 books in print, which means we are dealing with a lot of stock-keeping units in what has got to be the ultimate fashion industry. Inventory selection and management are critical success factors in this business. Our strategy is to offer the broadest possible selection that is appropriate for every market area we compete in, from San Juan to Anchorage, Alaska.

"This means," says Fontaine, "that the inventory of each store must be tailored to its immediate surroundings. When opening a new store or reviewing an existing one, we evaluate the demographic profile of the site location. We consider such things as: is the site downtown, near a university, or a special industry. Based upon such information, we customize the inventory to match the local market. In essence, we maintain a centralized, chain-store environment with decentralized marketing. This approach gives us the economy of scale of a large chain without becoming insensitive to the unique market characteristics of each of our stores."

B. Dalton has also placed microcomputers in stores to assist customers locate books. Books can be searched for by a variety of indexes including author, subject, even a word within the title. The system provides store location and prices. ▪

Case: National Car Rental

National Car Rental is one of the most progressive users of computers and information systems. In 1968 they became the first car rental company to introduce computerization of centralized reservations. Most industry experts agree today that if National Car Rental had not made that move they would not have be become one of the three major car rental companies. J. W. James, president of National Car Rental, recognized early the need not only to have systems people understand the business, but also to have managers well informed about computers and information systems. He has provided his management with such training.

"Today," James says, "more and more non-systems people are involved in projects that require computer technology, and a company that fails to educate its people in the ways and means of systems is asking for trouble. How can a user properly evaluate the best way to maximize the impact of information services if he or she doesn't understand the concepts involved? Information is the best asset a manager can have. Through a series of training programs for non-systems managers, we continue to make an investment in the future of our company and our people. If we don't, how can we expect to achieve any long-term growth? Mental nearsightedness is not profitable." ▪

Case: Sylvania Commercial Lighting Division of GTE

Sylvania supplies its sales personnel with portable computer terminals. The terminals can be used by the sales representatives to access a system that will analyze prospective customers' lighting requirements.

After requirements are analyzed, the system generates proposals that can be presented to customers on-site. Travel, elapsed time, and cost are all reduced, while customer service is increased. ■

Case: Holden Business Forms

Holden Business Forms is primarily a supplier of business forms for computer applications. At Holden, they recognize that a major difficulty in the business-forms business is differentiating oneself from the competition, that is, offering something better than what other form suppliers have to offer. George Holden, president of the company, puts it this way: "The primary way of conducting business in the business forms industry traditionally had been to compete for business on a price basis. Every time a company was looking for business-forms supplies they would go out on competitive bid and solicit bids from a variety of vendors. This is an expensive process both for the customer and for the supplier, as many vendors will bid but only one will be awarded the contract. We thought we had a strategy that could provide better service to our customers and reduce our cost to do business."

What Holden did was recognize that a major problem of business forms purchasers was that they tended to buy a larger volume than they needed in order to get unit cost down. This resulted in their having to deal with large bulk storage of forms until they were used. They also had to spend a great deal of time simply managing this inventory and making sure that they did not run out of stock.

To address this problem, Holden Business Forms developed information systems that could be used to manage inventory for customers. The proposal they gave to their customers was that in return for this service, the customers would give all of their business to Holden Business Forms. In return, Holden would reduce the customer's storage cost by approximately 75 percent and guarantee that forms would be available when needed and at a lower cost.

How could Holden afford to do this? Mr. Holden explains, "By not engaging our salespeople in the expensive process of doing competitive bids, and by using computer-based information systems to manage inventory for customers, we were able to focus on providing better service since we were not having to focus so much on preparing volumes of competitive bids, only a few of which would be awarded a contract."

In essence what Holden did was couple their information systems to their customer. This allowed them to manage their own inventory and their customers' inventory at the same time. Consequently, the customer did not have to store volumes of forms since Holden would provide these as needed. The result was a significant increase in business for Holden's. ■

Case: American Hospital Supply

American Hospital Supply did something similar to what Holden Business Forms did. They found that an expensive part of their doing business was to have salespeople go on calls to hospitals within their territories. Also, they found it difficult to handle rush orders to hospitals.

To handle these problems, American Hospital used a provocative approach. Rather than have salespeople take orders from hospitals, they installed on-line terminals that allowed customers to place orders directly to American Hospital Supply. The role of the salesperson became one of educating customers in the use of the terminals and handling any problems that the customer might have in using a terminal. Consequently, customers were able to be more self-sufficient and to more quickly place orders. Also, American Hospital Supply did not have the expense of sending salespeople out to take orders. Instead, salespeople could concentrate on new sales. ▪

Recognizing Opportunities

Clearly the days of "green-eye-shade" information systems, focusing only on accounting applications, are over. New opportunities to apply information systems to key company operations arise in organizations daily. As Steve Platt of Fingerhut says: "Invest in information systems. Anything you can cost justify even on a marginal basis today, will be even a better deal tomorrow, because the technology will cost less tomorrow. Too often, information systems are only used to reduce costs in clerical areas. But the real pay-offs come from using information systems for better decision making in key company operations. You have to look for opportunities."

Framework for Studying Problems and Opportunities

Two high-level, strategic frameworks for recognizing problems and opportunities—competitive strategy and customer resource life cycle (CRLC)—are covered in Chapter 13, "Strategic Planning for MIS." Both competitive strategy and CRLC are powerful tools for determining opportunities for information systems. However, they are easier to understand and appreciate in the context of the advanced techniques of Section III and are therefore deferred until later in the book. Nonetheless, a straightforward, practical framework is presented here that is easy to use and is useful for identifying problems and opportunities, which usually fall within one of the following six categories:

- Performance
- Information
- Economy
- Control
- Efficiency
- Services

Note that the first letters of the names of these categories can be combined to form the acronym PIECES. Indeed, each category can be viewed as a piece of a problem or opportunity. The categories occasionally overlap, but each is sufficiently different to warrant a separate discussion. The categories also interact. Consequently, a particular problem or opportunity may belong within one category or span several categories.

In the following discussion, problems and opportunities related to both object systems and information systems are examined.

Performance

Performance problems and opportunities primarily pertain to the throughput and response time of an organizational process. *Throughput* is the total volume of work performed over a given period of time, for example, the number of orders processed per day by the order-processing department. The *response time* of a system is the average time between submission of a request and return of the result. In an on-line system, response time is the time between the end of a block of user input and the display of the first character of the computer's response at the terminal.

Throughput and response-time dimensions should be both separately and collectively considered and then adjusted to achieve a desired performance level. For example, a bank may service a large number of customers (good throughput), but many customers may endure considerable delay before being serviced (poor response time). In any given situation, both throughput and response time can be poor, or one can be good while the other is poor. In either case, the result is a performance problem.

Even if existing performance is satisfactory, opportunities to improve performance may arise. Through such improvements, an organization may gain an important advantage over its competitors. For example, consider the first banks that installed electronic tellers for after-hours check cashing. In general, the management of these banks took an opportunistic-surveillance approach to improve their service beyond existing levels.

Information

Information is used in decision making; therefore, information problems and opportunities are generally related to improving the relevance, form, timeliness, or accessibility of information. First, to be useful in decision making, information must be *relevant* to the decision being made or it is unnecessary information and can be a deterent to decision making, since it has to be shifted through to locate the relevant information.

The more closely the *form* of relevant information matches the requirements of the decision maker, the more effective the information is. The language, format, accuracy, and presentation of the information are key variables in providing information that is comprehensible and usable to the decision maker.

To be used for decision making, information must be available prior to the decision-making process. Therefore, information should be available in advance of the *time* of need. Otherwise, the decision is delayed or made without the information. Besides being timely, information must be *accessible* to the decision maker. One of the significant contributions of on-line systems is the improvement of both the timeliness and the accessibility of information.

Economy

Economic problems and opportunities pertain to the costs of a process. The two key variables to be balanced when considering economy are service level and excess capacity. The more *excess capacity,* or slack resources, available for a process, the more quickly responses can be made to demands for goods or services. However, higher *service levels* are achieved with losses in economy because excess capacity is costly.

The proper balance between service level and excess capacity is contingent upon the organizational process. Generally, the more unpredictable the demands for service are, the greater the excess capacity required to ensure good service. For example, ambulance services and fire departments have considerable excess capacity so that they can quickly respond to demands for service. Conversely, an assembly line in a manufacturing plant attempts to have minimal excess capacity to keep down costs.

Though the appropriate level of excess capacity should be considered on an individual application basis, one guideline is reasonably universal: Excess capacity should be reduced to the point that it does not adversely affect the desired service level.

Control

Control is used to monitor or regulate processes. It is achieved by checking for deviations from planned performance. When deviations occur, some form of control mechanism is required to invoke corrective action. For example, a manager responsible for control of a process is primarily concerned with deviations outside allowable limits. Therefore, the manager's information system should provide exception reporting on problems. This reduces the information-processing requirements of the manager and allows the manager to concentrate on items requiring corrective action or further investigation.

Control can present problems or opportunities in one of two ways—lack of control or too much control. Lack of control can result in a system that is out of control. For example, a key factor in the design of an accounting system is the concept of auditing—the formal, periodic examination and checking of financial records to verify their correctness. Without auditing procedures, many opportunities exist for fraud and embezzlement. Fraud and embezzlement have been facilitated somewhat by computer technology due to some auditors' lack of computer expertise. This lack of expertise impedes the ability to audit properly that portion of the accounting system performed by the computer. Today, auditors have expanded their auditing skills to include an understanding of computer-based accounting records. By doing so, they have created a viable subfield in auditing known as EDP auditing and have put "teeth" back into the auditing process.

Too much control can result in diminishing returns. When an organization spends more on control than it can derive as benefits from the control, the control is no longer cost effective. Also, excessive control can become so-called red tape, negatively impacting productivity. Consider a university registration system that requires each student to get his or her class schedule approved by an advisor, department chairperson, and dean. Few deans or department heads have the time to review fully each student's schedule before signing it. Consequently, requiring additional signatures adds little if any control, though it adds inconvenience to the registration process.

When control is lacking, problems can be resolved or opportunities gained by increasing control. When control is excessive and unwarranted, reducing control can reduce costs and inconvenience.

Efficiency

Efficiency refers to producing a product or providing a service with minimum waste. It can be expressed as a ratio of effective work to energy expended in working. More simply, efficiency is output divided by input.

Understanding the significance of the efficiency ratio is a prerequisite to truly appreciating the meaning of efficiency. Efficiency is often confused with economy. *Economy* refers to the amount of resources committed to a process; *efficiency* refers to minimizing waste in the use of those resources.

For example, an organization might install slower, less expensive terminals to economize when setting up an on-line system. Given that the organization has slow-speed terminals, the organization should ensure that the terminals are used efficiently by locating them where they are most needed, ensuring proper training and productivity of terminal operators, and keeping the terminals in good working condition.

It is important in systems analysis to recognize the roles of economy and efficiency in the amount of output produced. For example, if management is concerned with increasing output, management must increase production resources (e.g., peoplepower and equipment) *and/or* increase the efficiency of the use of the resources. If the process is already efficient, additional output requires additional resources. If the process is inefficient, efforts to improve efficiency should be pursued before committing additional resources.

If management attempts to reduce costs by reducing resources, then either it must settle for less output, or efficiency must be increased to offset the loss of resources. If the process is already efficient, reduced output is inevitable when resources are reduced.

Services

A major area of potential problems and opportunities within an organization is that of services provided to its customers. Proper evaluation of service problems or opportunities requires that the focus be on the customer. The strategy is to evaluate the customer's needs and also the problems present for meeting those needs.

Once the difficulties that customers are encountering are understood, the next strategy is to evaluate ways that services could be expanded or improved to resolve those difficulties. There is no better way to endear a supplier of services to its customers than to help the customers solve a major problem they are encountering.

If you look at the industry examples that were provided earlier in this chapter, you will note that a major dimension of each opportunity pursued by those companies involves some form of improving or adding to services.

Defining Problems and Opportunities

It is not usually difficult to identify general problem areas or possible opportunities. Since perfection is rare, there is seldom an organizational process that is beyond improvement in some aspect of its performance, information, economy, control, efficiency, or services (PIECES).

It is generally more difficult to define with specificity the different dimensions or the various cause-and-effect relationships that are creating problems or preventing opportunities. For instance, late processing of customer orders may be identified as a problem. On the surface, this may appear to be a performance problem related to response time. However, there are many potential causes of the problem. These include

1. The salespersons fail to promptly turn in customer orders.
2. The order-processing department is managed incompetently.
3. The data-entry operators give too low a priority to the orders (e.g., they may give higher priority to payroll transactions).
4. The order-processing department has a morale problem.
5. There is an insufficient number of data-entry operators.
6. The computer programs that process orders have several "bugs" in them, resulting in some orders being processed incorrectly.
7. The computer capacity is insufficient to handle the work load.
8. The information on inventory fluctuations is not timely enough to prevent frequent stock-outs due to poor routing.
9. The inventory manager is not aware of late deliveries because this information is not reported.
10. The credit department does not review customer credit checks until just before orders are to be shipped. Therefore, an order has to be held if any questions on credit arise.

The preceding list indicates that to fully ascertain the cause of a problem or the prevention of an opportunity requires a thorough understanding of both the object system and the information system.

Causality

The actual cause of a problem can be quite elusive. The inexperienced systems analyst often confuses symptoms with problems. Just because something precedes something else does not mean that it caused it. For example, when a doctor gives you a physical, one of the things he or she will do is look at your fingernails to see if they curve downward. If they curve downward, this can be symptomatic of a respiratory problem. Note that the respiratory problem is not resolved by attempting to straighten the fingernails; rather, the respiratory problem has to be addressed directly.

Similarly, in addressing information systems symptoms for problems, great care has to be taken to identify what the real problem is. For example, when a new information system is implemented within an organization, a number of problems often occur. The organization may conclude that the information system has caused the problems. It may be, however, that the information system has merely revealed problems that the organization was previously unaware of.

A case study illustrates this phenomena. A wholesale distributor implemented a new on-line inventory management system. Prior to implementation of the system, various regional warehouses of the wholesale distributor had had manual information systems. One of the performance objectives that the distributors had had in managing their warehouses was never to make a shipment unless enough inventory existed in the shipment to make it profitable. Unfortunately, on occasion very important customers of these warehouses would insist on receiving a shipment even when inventory was not sufficient to make the entire shipment profitable. Not wanting to lose these important customers, the regional warehouse managers had gone ahead and made the shipments anyway. Then they "juggled" their reporting with other shipments so that all shipments, on the average, were profitable. This process worked fine until the new inventory system was implemented.

With the new system, it was not possible for the regional warehouse managers to juggle their reporting, and the computer system indicated shipments that were not profitable. The regional managers approached the information systems designer and explained that they needed the ability to "juggle" information as they had in the past. The systems designer felt that this was inappropriate and refused to provide such abilities. As the regional warehouse managers continued to have difficulty with management over their unprofitable shipments, they deliberately began to sabotage the new system by entering bad data. This ultimately led to erroneous reports that created credibility problems for the information system that the regional warehouse managers continued to complain about. After several months of this bickering, it was decided that the information system was simply no good; it was then removed, and the system designer was terminated.

Management had concluded that the system was causing problems. But what was the real problem? To properly identify a problem requires careful investigation of "what is causing what." Starting with the symptoms first, we can say that the system in this case was in fact putting out incorrect data. However, the cause of the bad data was the regional warehouse managers' entering of incorrect data. What was causing that? The regional warehouse managers were putting in bad data because the system was creating problems for them. But what was causing these problems? If one looks very closely at this situation, one would agree that for very important customers it is often in the best interest of the company to send out shipments even if they are not profitable. If that is the case, what is the real problem? In this case the real problem is top management's policy of not allowing regional warehouse managers to send out unprofitable shipments when necessary. Rather than ever really addressing this problem, however, everyone was focusing on the symptom—the information system—which was only *revealing* the problem, not causing it.

Therefore, when trying to identify problems, it is important to go through a careful assessment of what is causing what. This generally cannot be done casually. Determining causality requires an investigative mind on the part of the systems analyst.

Magnitude of Problem or Opportunity

Problems and opportunities vary in complexity. In many cases, a problem or an opportunity can be addressed with minimal difficulty or time involved. This usually occurs when the existing system is well understood, and new problems or opportunities are rather glaring. Consequently, the appropriate solution is apparent. For example, if late processing of customer orders is due to new salespersons' failing to promptly turn in orders, then this problem should be easy to resolve. The systems analyst may simply write a memo to the sales manager explaining to him or her the importance of promptly turning in orders and requesting that appropriate action be taken (see Figure 5.1).

For more complex problems and opportunities, a more comprehensive analysis is required. For example, if the information system is not providing needed information and/or the technology used in the information system is obsolete,

---

Figure 5.1   Use of Memo to Resolve Information System Problem

MEMORANDOM

October 15, 1989

To: Mr. Green
          Sales Manager

From: John Goode
          Systems Analyst

Subject: Delayed Processing of Customer Orders

During the past few months, Mr. Hickson (the company's president) has received numerous complaints about the tardiness of order processing. He requested that I review the work flow of the order processing system to isolate the cause of these delays.

The major bottleneck appears to be a delay in salespersons' turning in of customer orders. In some cases, orders are not being trurned in until a week after the orders are requested by customers. Most of these late orders appear to be caused by recently employed salespersons who are possibly not aware of the importance of turning in all orders promptly.

Continued delays in order processing can adversely affect customer relations and future sales. Your cooperation in resolving this problem will be greatly appreciated. If I can be of any assistance, please let me know.

cc: Mr. John Hickson, President
      Mr. Joe Bond, MIS Manager

---

problem resolution requires considerably more effort. In such cases, problem resolution or opportunity achievement requires a major overhaul of an existing system or the development of a new one.

Feasibility Study

When complex problems and opportunities are to be defined, it is generally desirable to conduct a preliminary investigation called a *feasibility study*. A feasibility study is conducted to obtain an overview of the problem and to roughly assess whether feasible solutions exist prior to committing substantial resources to a project. During a feasibility study, the systems analyst usually works with representatives from the department(s) expected to benefit from the solution.

The primary objective of a feasibility study is to assess three types of feasibility.

1. Technical feasibility. Can a solution be supported with existing technology?
2. Economic feasibility. Is existing technology cost effective (i.e., will the cost be offset by the benefit(s)?
3. Operational feasibility. Will the solution work in the organization if implemented?

By intent, the feasibility study is a very rough analysis of the viability of a project. It is, however, a highly desirable checkpoint that should be completed before committing more resources. The feasibility study answers a basic question: Is it *realistic* to address the problem or opportunity under consideration?

The final product of a successful feasibility study is a project proposal for management. The contents of this report may include, but are not restricted to, the following items:

1. Project Name
2. Problem or Opportunity Definition
3. Project Description
4. Expected Benefits
5. Consequences of Rejection
6. Resource Requirements
7. Alternatives
8. Other Considerations
9. Authorization

Figure 5.2 illustrates a complete project proposal.

Justifying Systems

With all of today's publicity surrounding the use of information technology for opportunistic or competitive-advantage applications, most information systems professionals are experiencing a combination of excitement and paranoia as they contemplate how to respond. Now that the opportunity is here—or the pressure is on—how exactly should they go about promoting competitive-advantage in-

Project Name

New Order Processing System

Problem or Opportunity Definition

Customers are complaining that they cannot get orders processed. They have to wait for sales representatives to come by to place an order which might take several days.

To send sales representatives by more often will increase our costs too much.

Project Description

Proposed is a new order-processing system that will allow customers to place orders directly using an on-line customer order-entry system. Customers can use their personal computers for entering orders.

Customers will be provided with information on inventory availability and alternatives, prices, and shipping schedules.

Besides placing orders, the system can also be used to let customers file complaints and we can use it to collect survey information from customers (e.g., product preferences, quality of services).

Benefits

Estimates are that we will be able to deliver to our customers an average of five days sooner than we or our competition currently can. This increase in customer service and reduction in sales costs (fewer sales calls) should be substantial.

Also, the ability to get complete, accurate, and timely information on customer complaints and surveys will allow us to better understand customer needs, preferences, and attitudes.

Consequences of Rejection

Initial estimates are that the new system will take three to six months to be completely operational. A project team of four to six peronnel will be required   half-time for the development effort. Budget could run from $100,000 to $300,000.

More accurate resource requirements can be determined once a working prototype of the system is complete. Prototyping could be complete within two to four weeks.

Alternatives

Besides doing nothing, the main alternative is to purchase a prewritten system or application package that could provide the same functionality as the one proposed. We have found two such systems, but they would require substanital revisions to adapt to our requirements. The expense of doing so exceeds the cost of customizing our own system.

Other Considerations

The credit department indicates that the proposed system could greatly enhance cash flow for accounts receivable.

Authorization

Manager Order Processing	_____
Sales Manager	_____
Shipping Manager	_____
Credit Manager	_____
Manager Information Systems	_____
Vice President of Operations	_____

formation systems (such as those discussed earlier in the chapter) to top management.

Perhaps the biggest obstacle to promoting them is that most systems analysts are shackled by early 1970s' thinking, whereby all computing applications had to be cost justified. Cost justification involves demonstrating that savings or revenue generated by the new system more than offset the cost of developing it. Common methods of cost justification are return on investment (ROI) and break-even analysis, depicted in Figure 5.3. ROI is the percentage return computed by ROI = net return ÷ investment (e.g., a $1,000 investment generating a $100 return after one year would net an annual ROI of 10 percent). Break-even analysis is concerned with the point in time at which revenue or savings from investment equals start-up cost. In other words, break-even analysis indicates when the original investment is recouped. Note in Figure 5.3 that when breakdown is reached, ROI is zero percent (i.e., there is no ROI until the original investment is recouped).

Unfortunately, as precise as the formulas for computing break-even analysis and ROI are, considerable judgment is usually required to come up with these numbers—especially the benefit numbers—which always makes the analysis somewhat suspect. For example, we all sense that getting an education is cost justified. But generating the actual costs (including the income foregone while going to school) and benefits (including intangibles such as greater meaning to life) are hard to quantify.

Cost-justification of information systems became popular in the late 1960s, when many companies were severely disappointed, and rightly so, with their investment in computer systems. Hard-nosed cost justification was employed to put the brakes on runaway information-technology budgets—and while this slowed some spending down, it also unfortunately created an infertile environment for future innovation.

It is clear now, however, that enhancements to the way a company operates, such as using competitive-advantage systems, are brought on by entrepreneurial, imaginative thinking that should not be dominated by premature preoccupation with exact cost and benefit justification.

American Hospital Supply's well-known customer order-entry system, for example, was initiated by marketing staff trying to solve a customer-service prob-

Figure 5.3 **Relationship between Cost and Benefits as Measured by Break-Even Analysis and Return on Investment (ROI).**

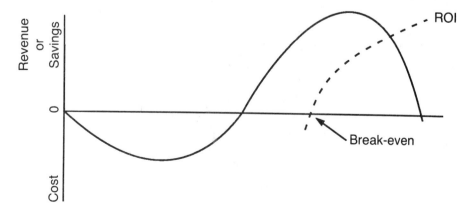

lem. Only later did information management realize the efficiency gains that emerged as a by-product—gains achieved by eliminating AHS's order-entry clerical function and allowing customers to conveniently place their orders directly.

The AHS example serves to support a fundamental concept: If the initial focus is on customer-oriented effectiveness issues, then efficiency gains are usually a fortunate by-product made possible by the power of information technology. Also, ideas such as improving customer service generally capture management's imagination. If, on the other hand, the initial focus is on cost-cutting efficiency issues (easier to cost justify), then customer service and effectiveness are usually compromised and the idea has less appeal. For example, it is much more exciting to talk about a system that allows a pharmacist to automatically check for improper drug combinations than it is to talk about reducing the accounts-receivable clerical staff by 20 percent. The excitement comes not from the appearance of a "cost-justification application"—but from the appearance of a "good idea." In the case of either AHS or the pharmacy in our example, before-the-fact estimates of cost and benefits were not precisely determinable. How many more customers will either gain because of these systems? How many will they keep? How do we know for sure?

With cost justification clearly lacking as a singular, effective way to promote competitive-advantage applications, it is necessary to develop a framework for successfully promoting such applications. One approach involves a three-phase program that can be likened to a political campaign:

1. *Selecting the candidate*. During this stage, the good idea must be developed and thoroughly refined. One approach that has proved successful at this stage is to conduct a *retreat*, where users and systems designers go off alone for a weekend or so to address the subject without distraction. It might help at such a retreat to include some customers and a *facilitator* who makes sure the focus remains on the competitive-advantage issues. The PIECES framework discussed earlier and the competitive strategy and CRLC techniques discussed in Chapter 13 are useful frameworks for this.

If one or more visionary concepts comes out of such a session, the battle is half over. If the concepts can be readily cost justified as well, so much the better; but remember that this aspect should not be the only criterion. The concepts selected for further development should be expressible in a crisp, succinct phrase, almost like a campaign slogan. For example, "We are going to be the first car rental company to sell our used cars through an on-line order-entry system instead of just advertising them in the newspaper."

2. *Establish support*. During this stage, management backing and grass-roots support is established. As in a political campaign, differences in opinion on strategic and logistical specifics should be worked out in small groups—"behind closed doors"—rather than exposing any weak link in public. Adjustments, even significant compromises, may be required to gain the constituency necessary for the concept's acceptance.

3. *The inauguration*. The final stage of promoting competitive applications involves the formal presentation—with all the appropriate fanfare—of the system. Ideally, an initial prototype of the system (as discussed in Chapter 7) should be available for demonstration purposes. By now, it should be said, the system has been "informally" wired for acceptance. Now the time is finally at hand for actual systems development and implementation to begin.

Project Team

If the project proposal is approved, management generally organizes a project team and assigns responsibility for the project to it. The specific personnel or type of personnel to be included on the project team should be defined under the "resource requirements" section of the project proposal.

The organization of a project team is a prerequisite to good systems analysis. Different types of expertise are required to thoroughly consider all system variables. For example, the system analyst provides primarily computer technology-oriented skills. Without adequate representation from the various departments affected by the information system, various dimensions of their needs and requirements are apt to be overlooked.

Project team participants should be given sufficient release time from their normal duties to allow them to contribute the necessary level of participation to the project. However, the release time should be provided only for the duration of the project. Upon project completion, the participants should return to their normal duties.

An interesting debate about project teams pertains to who should be the project leader. Some managers contend that a systems analyst should provide the leadership; others contend that this role is better filled by a management-level representative from the department(s) that will use the system. Projects tend to be completed faster when a systems analyst is in charge. However, users are generally more committed to, and more satisfied with, the results of projects when they are directly responsible. When a systems analyst is in charge, users have a tendency to give him or her too much latitude in designing the system (often because users do not understand much of the technology) and then decide whether or not they like the completed system. Unfortunately, the systems analyst may design a system that works right, but is not the right system (i.e., it does not adequately satisfy the user's information requirements). When users are directly responsible for a project, they are inherently more involved in the design of the system. Their involvement improves the probability that the right system is designed.

Project teams and project management are covered in depth in Chapter 11, "Project Management and Systems Development."

Summary

Problem and opportunity recognition is the starting point of analysis and design. Recognition may result from either reaction or opportunistic surveillance. Problems and opportunities may exist within the object system or the information system. The effective systems analyst should be both opportunistic and sensitive to the problems of object system, rather than simply to the problems with information systems.

Learning how to spot opportunities is an act of systems analysis. The systems approach and the PIECES framework are useful concepts for developing these abilities.

When defining problems and opportunities, care must be taken to assess causality, magnitude, and feasibility. A project team is usually organized to pursue the problem and/or opportunity.

Exercises

1. Differentiate between management actions that are reactionary, and those that are opportunistic, in orientation. Give examples of each.

2. Discuss the difference between an object system and an information system.

3. Refer to Table 3.7. Discuss how an information system can be used to improve the operation of the various mainstream processes. (A mainstream process is an object system.)

4. Identify and discuss how the different categories within PIECES (performance, information, economy, control, efficiency, and services) apply in the following situations. Remember that more than one category can apply. If additional information is required to more accurately pinpoint the problem, discuss this also.

   a. Employees are gaining unauthorized access to payroll information.

   b. Due to insufficient computer capacity, the response time for on-line credit checks is too slow.

   c. The cost of running the delivery department has increased considerably, with no increase in service.

   d. Salespersons do not receive timely enough information to determine if they have adequate inventory to satisfy customer orders.

   e. Poorly constructed products are getting past quality control and being shipped to customers.

   f. Important information on quality control activities (i.e., number of products checked, rejection rates, etc.) is not collected or processed.

   g. Though information on customer opinions of product quality is collected and processed, it is not made available to the production manager.

Selected References

Ackoff, Russell L. "Management Misinformation Systems." *Management Science*, December 1967, pp. 147–56.

Anderson, William S. "Bridging the Systems Expectation Gap." *Data Management*, November 1975, pp. 29–32.

Argyris, Chris. "Resistance to Rational Management Systems." *Innovation* no. 10: (1970) pp. 28–35.

Benbasat, Izak, and Schroeder, Roger G. "An Experimental Investigation of Some MIS Design Variables." *MIS Quarterly*, March 1977, pp. 37–47.

Benjamin, R. I. "When Companies Share, It's Virtually a New Game." *Information Systems News*, December 26, 1983, p. 24.

Benjamin, R. I., Rockart, J. F., Scott Morton, M.S., and Wyman, J. *Information Technology: A Strategic Opportunity*. Working Paper 108, MIT Center for Information Systems Research, Cambridge, Mass., December 1983.

Blumenthal, Sherman C. *Management Information Systems: A Framework for Planning and Development*. Englewood Cliffs, N.J.: Prentice-Hall, 1969.

Bott, Harold S., and Pasino, Jacque H. "How to Make a Strategic Move with Information Systems." *InformationWEEK*, May 1986, pp. 3–5.

Buday, Bob. "MIS Works to Bridge the Gap with Corporate Management." *InformationWEEK*, January 6, 1986, pp. 18–24.

———. "Other Firms Using Systems Strategically." *InformationWEEK*, May 1986, pp. 21–23.

———. "You Need a Strategy for Strategic Systems." *InformationWEEK*, August 4, 1986, p. 6.

Buday, Bob, and Ewing, Tom. "The Strategic Use of Information: Seizing the Competitive Edge." *InformationWEEK*, May 1986, pp. 1–2.

Business Week. "Business is Turning Data into a Potent Strategy Weapon." *Business Week*, August 22, 1983, pp. 92–98.

Business Week. "A Coupon Machine at the Supermarket." *Business Week*, March 5, 1984, p. 68.

Chervany, Norman L., and Dickson, Gary W. "An Experimental Evaluation of Information Overload in a Production Environment." *Management Sciences*, June 1974, pp. 1335–44.

Chester, Jeffrey A. "Technology Solution Sculptors." *InformationWEEK*, June 1987, pp. 46–47.

Davis, Gordon B. *Management Information Systems; Conceptual Foundations, Structure, and Development*. New York: McGraw-Hill, 1974.

Dickson, Gary W., and Simmons, John K. "The Behavioral Side of MIS." *Business Horizons*, August 1970, pp. 59–71.

Feltham, Gerald A. "The Value of Information." *The Accounting Review*, October 1968, pp. 684–94.

Hirsch, Rudolph E. "The Value of Information." *The Journal of Accountancy*, June 1968, pp. 41–45.

Johnson, Richard A.; Kast, Fremont E. and Rosenzweig, James E. *The Theory and Management of Systems*. New York: McGraw-Hill, 1973.

Layne, Richard. "MIS Can Cut Costs By Making Users Do the Work." *InformationWEEK*, April 27, 1987, p. 33.

Lucas, Henry C., Jr. "A Descriptive Model of Information Systems in the Context of the Organization," *Data Base*, Winter 1973, pp. 27–39.

McFarlan, F. W. "IS and Competitive Strategy." Note 0–184–055, Harvard Business School, Cambridge, Mass., 1983.

McFarlan, F. W., and McKenney, J. L. *Corporate Systems Management*. Homewood, Ill.: Richard D. Irwin, 1983.

McFarlan, F. W., McKenney, J. L., and Pyburn, P. "The Information Archipelago— Plotting the Course." *Harvard Business Review*, Volume 61, Number 1, January– February 1982, pp. 145–46.

Nolan, Richard L. "Systems Analysis for Computer-Based Information Systems Design." *Data Base*, Winter 1971, pp. 1–10.

Nugent, Christopher E., and Vollman, Thomas E. "A Framework for the Systems Design Process." *Decision Sciences*, 1972, pp. 84–109.

Schindler, Paul E. "How to Convert Information into a Competitive Weapon." *InformationWEEK*, July 1986, p. 26.

Simon, Herbert A. *The New Science of Management Decision*. New York: Harper & Brothers, 1960, pp. 54ff.

Wetherbe, James C. *Systems Analysis for Computer-based Information Systems*. St. Paul: West Publishing, 1979.

———. *Executive's Guide to Computer-Based Information Systems*. Englewood Cliffs, N.J.: Prentice-Hall, 1983.

_____. "Learning to Live in the Knowledge Society." *Corporate Report*, March 1981, 42.

_____. "Opportunity is a Computer." *Corporate Report*, April 1981, pp. 34 and 112.

Wetherbe, James C. and Dickson, Gary W. *MIS Management*. New York: McGraw-Hill, 1984.

Wetherbe, James C., and Whitehead, Carlton J. "A Contingency View of Managing the Data Processing Organization," *MIS Quarterly*, March 1977, pp. 19–25.

Systems and Information Analysis

Introduction

The identification and definition of a problem or an opportunity, and the organization of a project team, set the stage for a detailed analysis of the existing system. Concepts and techniques of analysis are presented in this chapter.

Don't Change It Until You Understand It

In the excitement of implementing new technology, it is often easy to overlook certain capabilities or functionalities of the old technology and to leave them out of the new system. Such actions result in disillusionment with the new system and nostalgia for the old system. Therefore, it is critical to conduct a thorough analysis of the old system before final design decisions are made on the new system.

Consider a simple, noncomputer example of inadequate analysis before implementation of a new technology—hot-air hand-dryers in public restrooms. Everyone seemed to like the idea immediately. It would eliminate the paper waste associated with paper towels, fewer trees would give their all to become paper towels, and it would be state-of-the-art technology. So it was done.

Unfortunately, it was not long before people discovered that certain functions served by those old paper towels were not served by the new system. Specifically, people found it difficult to:

- Dry their faces
- Clean up a spill
- Clean their glasses
- Blow their noses

So what do you normally find in a restroom that has hot-air dryers? Paper towels. Similarly, when new computer systems are implemented, remnants of the old system tend to linger on because the designers of the new system overlooked a function that was provided by the old system.

It may not have been the most professionally rewarding day of one's career, but if someone had just spent one day in a public restroom analyzing what people used paper towels for, he or she could have discovered what paper towels provide that hand dryers do not. Similarly, to design computer systems, we must take the time to do analysis of the old before bringing in the new.

Formal Systems Development Methodologies

Most organizations use a formal systems development methodology as a template or checklist to manage the systems development life cycle (SDLC). Several commercially available training programs and procedures manuals can help to carefully lead organizations through an SDLC. These methodologies include extensive checklists and checkpoints to facilitate the management of an SDLC. To illustrate the types of things covered in a systems development methodology, four commercial SDLC packages—PRIDE, SDM/70, CARA, and SPECTRUM—are listed in Table 6.1. Although systems development methodologies have similarities and differences, they are all based on the traditional systems development life cycle.

Computer-based support for managing the SDLC is available through CASE (computer-aided systems engineering) products such as Excelerator, PACBASE,

Table 6.1   Traditional Life Cycle and the Phases of PRIDE, SDM/70, CARA, and SPECTRUM

System Development Life Cycle	PRIDE Life Cycle Phases	SDM/70 Life Cycle Phases	CARA Life Cycle Phases	SPECTRUM Life Cycle Phases
Problem/ Opportunity Definition	Phase 1: System Study and Evaluation Activities	SRD: Systems Requirements Definition	Phase I: Feasibility Study	• Problem Definition/ Project Proposal
	Plus	SDA: System Design Alternatives	Phase II: Detail Systems Design and file for outputs	• User Requirements Definition
Analysis	Phase 2: Design of outputs and files only	SES: System External Specifications	Phase III: Programming of outputs	• System Definition • Advisability Study
Design	Phase 5: Programming of outputs only	Plus	Phase IV: Acceptance of outputs	• Preliminary Design • Program Design and testing for outputs
Technology/ Personnel Requirements	Phase 2: Complete System Design Activities for inputs and processes	PO: Programming of Outputs SIS: System Internal Specification	Phase II: Detail Systems Design of inputs and edits	• Preliminary Design • System/ Subsystem Design
Develop, Test, Validate	Phase 3: Subsystem Design Activities Phase 5: Program Design Activity for inputs and edits	PD: Program Development Phase for inputs and edits	Phase III: Programming and procedures for inputs and edits	• Program Design for inputs and edits • Programming/ Program Testing for inputs and edits
Implementation & Evaluation	Phase 4: Administrative and Computer Procedure Design Activities	TST, CNV, and Impl: System Testing,		• System Implementation Planning/User Manuals • System Test/ Start-up Preparation
	Phases 6 & 7: Computer Procedure and System Test Activities	Conversion, and Implementation Phases	Phase IV & V: System Acceptance and Implementation/ Support	• Operation Turnover • User Training/ Start-up • System Acceptance/ Project Wrap-up

CASE 2000 Designaid Workbench, Information Engineering Workbench, and Teamwork.

Note that in Table 6.1 the systems development life cycle is mapped against the formal systems development methodologies in the left-hand column of the table. Some of the formalized systems development methodologies do not focus on problem and opportunity recognition the way we have in this book. Accordingly, the methodologies are introduced in this chapter in association with the tasks for which they are most relevant. Research on the usefulness of using a formalized systems development methodology indicates that systems tend to be managed better, completed closer to schedule and budget, and better meet user requirements when a formal system development methodology is used.

Many organizations develop their own formal systems development methodology. Research indicates that these methodologies seem to work as well as the formalized development methodologies. The important thing is that a comprehensive template or checklist must be used to manage the process. One criticism of systems development methodologies is that they are often so comprehensive and require so much attention to details and procedures that they can get in the way of progress—particularly for short, simple projects. Accordingly, the steps and procedures within a systems development methodology should be considered only as a checklist, and not every step or procedure should necessarily be executed for each development project. In other words, some discretion should be used in evaluating which items on a checklist are appropriate for a particular project.

The objective of this book is not to provide detailed instruction in any specific systems development methodology. Rather, the objective is to provide a good conceptual and operational understanding of the types of tasks that have to be performed during systems analysis and the remainder of the systems development process. Therefore, this chapter and the remaining chapters focus on providing a good generic understanding of the types of concepts and techniques used during a systems development process, rather than going into detail about any specific systems development methodology. This will provide students who have completed the material with a comfortable transition into any organization that is using either a commercial or home-grown systems development methodology.

Fundamental Activities for Analysis

Analysis is the process of separating a whole into its parts to allow examination of the parts. This examination leads to an understanding of their nature, function, and interrelationships. Considerable effort is required to conduct an analysis of an information system. The overall system must be defined. Then it must be separated into its subsystems or processes for further analysis. For example, an order-processing system may be separated into subsystems of order collection, entry, editing, posting, shipping, and reporting.

The four basic activities used to gather information about an information system are listed in Table 6.2 and then discussed in more depth.

Documentation and Observation

A review of available documentation is a logical starting point when seeking insight into a system. A documentation review allows project team members involved in

Table 6.2   **Information-Gathering Activities**

Activity	Explanation
Review of Documentation	• This consists of reviewing recorded specifications that describe the objectives, procedures, reports produced, equipment used, etc., in an information system.
Observation	• This consists of watching the object system and/or the information system in process to note and record the facts and events about their operation.
Interviews	• This consists of meeting with individuals or groups to ask questions about their roles in and their use of an information system.
Questionnaires	• This consists of submitting questions in printed form to individuals to gather information on their roles in and use of an information system.

systems analysis to gain some knowledge of a system before they impose upon other people's time. Unfortunately, documentation seldom completely describes a system, and often it is not up to date. The current operation of the system may differ significantly from what is described. Therefore, after a review of available documentation, the next logical step is to observe the operation of the system (unless one does not yet exist). Observation provides a more tangible perspective of what is described in the documentation. It also brings to light aspects of the documentation that are incomplete and/or outdated.

Interviews and Questionnaires

Having gained as much knowledge as is reasonably possible without imposing on other personnel, those involved in systems analysis can intelligently use interviews and/or questionnaires to gather additional information. Techniques of interviewing and questionnaire design are topics for extensive study. A few key concepts, however, should be noted.

Interviews are generally preferable to questionnaires for information gathering because they allow thorough questioning and discussion. In an interview, not all questions have to be determined in advance (as they do in a questionnaire). However, when information has to be gathered from a large number of people and tabulated, questionnaires are usually preferable for efficiency reasons.

The two basic formats for questioning are categorized as open-ended and close-ended. The difference between the two types of questioning is similar to the difference between essay and multiple-choice exams.

Open-ended questions allow latitude to the persons responding. Consequently, information gathered from open-ended questions may be creative and rich in content. Examples of open-ended questions are the following:

1. Are there areas of dissatisfaction with the existing information system? If so, list these areas.

2. Do you have suggestions for improvements if a new system is designed? If so, list your suggestions.

Close-ended questions are more restrictive in the latitude allowed to a respondent. The respondent must select an answer from available choices. The major advantage of close-ended questions is that responses are easily quantified for tabulation. This greatly expedites analysis. The most common application of close-ended questioning involves scaling techniques. Responses are categorized on a scale between two extremes. Examples of such questions are:

1. Do you receive your inventory reports on schedule?

☐ ☐ ☐ ☐ ☐

Never   Almost   Usually   Almost   Always
*   Never   Always*

2. Is the information in your inventory reports accurate?

☐ ☐ ☐ ☐ ☐ .

Never   Almost   Usually   Almost   Always
*   Never   Always*

A final point about close-ended questioning is that only one question should be asked at a time. For example, asking if an inventory report is accurate and on schedule in one question places the respondent in an awkward position if the report is on time but is not accurate.

Application of Analysis Activities

The techniques of documentation, observation, interviews, and questionnaires are used in varying degrees during the following steps of analyzing the existing system:

1. Review object system.
2. Define decision making associated with object system.
3. Determine information requirements.
4. Document and analyze existing information systems.
5. Isolate deficiencies in the information system.

These steps of systems analysis are discussed in the remaining sections of this chapter.

Review Object System

During this step, project team members have the opportunity to develop a working knowledge of the physical processes associated with the object system under consideration. A clear understanding of the purpose of the object system and of the means used to achieve them should be developed.

The systems analyst should make a special effort to become familiar with the vocabulary associated with the object system. This minimizes communication problems as the systems analysis progresses.

Some type of procedures manual that describes the operation of the object system is usually available. The copy of the manual should be requested from the manager(s) responsible for the object system(s). After the manual has been reviewed, a tour that allows initial observation of the object system in operation should be arranged. Interviews and, if necessary, questionnaires can be used to complete the review.

Define Decision Making Associated with Object System

After an operating knowledge of the object system has been acquired, the next logical step is to define the decision making associated with managing the object system. A definition of the decision-making system (i.e., decision system) provides the framework for determining what information is required. This is one of the most neglected aspects of systems analysis. Since the utility of information is its ability to improve decision making, such negligence is somewhat surprising.

Managers are frequently asked what information they would *like* to have, or they are offered copies of reports that are currently being produced or will be produced for other managers. This approach tends to encourage managers to ask for *more* information than they *need*. Research in the area of decision making and the use of information indicates the following:

1. Decision makers tend to ask for and feel more comfortable with more detailed information than they need. However, they appear to make better decisions with summarized information and exception reporting.[1]

2. The less knowledgeable decision makers are about the decisions required to properly manage a process, the more information they tend to request (presumably, hoping to find something of value). However, much of the information they request is irrelevant to their decision making.[2]

By basing information requests on the decisions they have to make, managers can be more discriminating in their information requests. This reduces the tendency of managers to create information overloads for themselves. Such overloads are both dysfunctional and expensive.

Difficulty in Defining Decision System

Defining the decision system is a desirable step of systems analysis, and it is not a trivial one. Considerable discipline and effort on the part of managers may be required to define such a system. Managers often make decisions so routinely that it seldom occurs to them how often they make decisions and what types of decisions they make. Consequently, they are understandably tempted to take a "shotgun" approach to defining their information requirements, rather than a more time-consuming, but more specific, "rifle" approach.

Decision-Making Process

In order to define decision-making processes, it is necessary to understand how decisions are made. The most useful and well-known model of decision making

1. Norman L. Chervany and Gary W. Dickson, "An Experimental Evaluation of Information Overload in a Production Environment," *Management Science*, June 1974, pp. 1335–44.

2. Izak Benbasat and Roger G. Schroeder, "An Experimental Investigation of Some MIS Design Variables," *MIS Quarterly*, March 1977, pp. 37–47.

Table 6.3   Phases of Decision Making

Phase of Decision Making	Explanation
1. Intelligence	Recognizing a problem or an opportunity that calls for a decision Gathering and structuring information for enlightenment about the problem or opportunity
2. Design	Developing and analyzing alternative solutions
3. Choice	Selecting a particular solution from those available for implementation

is that proposed by Herbert Simon.[3] His description of a three-phase process is summarized in Table 6.3.

The transaction-processing, information-providing, decision-support, and programmed decision-making functions of an information system identified in Chapter 3 are used to support each phase of the decision-making process. During the intelligence phase, the information system may call problems or opportunities to the decision maker's attention. Once identified, information relevant to the problem or opportunity may be gathered from the information system in both predetermined and ad hoc ways. That is, normally produced reports or special studies generating new reports may be used. These studies may use existing transaction data, or special data (e.g., results of customer surveys) may be required.

The design phase may be supported by the decision-support capabilities of the information system. Decision-support systems allow the decision maker to play "what if" with respect to the decision under consideration. This is accomplished by using a model of the object system. The decision maker can test different decisions using different assumptions. The model then generates the likely outcomes. The design and use of such models are covered in Chapter 13.

Once a decision has been made, the information system provides information based on transaction processing or special studies as feedback to management about the effectiveness of the decision. In some cases, decisions may have to be made on an ongoing and routine basis. If the decision is well structured and defined, it can become a programmed decision, automatically handled by the information system. For instance, inventory can be reordered automatically when stock is reduced to a certain level.

The organizational concepts discussed in Chapter 3 are useful in explaining the difficulty of defining decision-making processes. The more closed/stable/mechanistic the operation of an organizational process, the easier it is to define the decision-making process and also the greater the potential for programmed decision making. The more open/adaptive/organic the organizational process, the more difficult it is to define the decisions that have to be made. Consequently, it is more difficult to predetermine information requirements. An information system in an open/adaptive/organic environment should have great flexibility in providing information and should place great emphasis on decision support capabilities.

---

3. Herbert A. Simon, *The New Science of Management Decision* (New York: Harper & Brothers, 1960), p. 54.

Decision Centers

The decisions made in an organization tend to be clustered into decision centers. A decision center generally consists of a decision maker(s), decision procedures, and an activity for which decisions must be made.[4] Accordingly, the decisions made in a decision center tend to pertain to the management of a particular organizational process.

Viewing the organization in terms of decision centers is particularly useful in systems analysis. Decision centers are potential areas where organizational processes may be improved by the provision of more relevant, timely, accurate information. In organizational terms, a decision center combined with an activity center constitutes a functional unit or department.

Table 6.4 assumes that an existing order-processing system is being analyzed. It shows the decision centers and major decisions likely to exist in the organization.

Another advantage of defining decision centers and major decisions is that doing so provides an overall profile of decision making and where decisions are made. This profile often reveals discrepancies in decision making. For example, if salespersons are under the impression that they can commit inventory, but they really cannot, this discrepancy should be cleared up. Otherwise, inventory may be committed to the wrong customers.

If the same decision is being made by two or more decision centers, this too should be resolved. Consider, for instance, a situation where all orders must be approved for credit by the credit department before they are sent to the order-processing department. However, the order-processing department is unaware of

Table 6.4	**Decision Centers Involved in Order Processing**	
Decision Center	*Activity*	*Examples of Major Decisions*
Salespersons	Selling Merchandise	• Which customers to call on • What to sell customers • What is available to sell
Credit Department	Accounts Receivable Management	• Which customers to allow credit to • How much credit to allow • Which customers need past-due notices • Which customers' credit should be discontinued
Ordering Department	Inventory Management	• What inventory to stock • How much inventory to stock • When to reorder stock • When to unload slow-moving inventory • Which customers to allocate available inventory to
Shipping Department	Packing and Shipping Orders	• What merchandise to send to what customers • What orders can be shipped together to save delivery cost

---

4. Sherman C. Blumenthal, *Management Information Systems: A Framework for Planning and Development* (Englewood Cliff, N.J.: Prentice-Hall, 1969), n.p.

this procedure so a second credit check is made, using an exact copy of the credit-rating report used by the credit department. Such duplication of effort is not uncommon in organizations. It may go on, unnoticed, until identified by someone aware of what both departments are doing. Requiring that information received by a department be justified for decision making should expose discrepancies and duplication of decision-making processes.

Determine Information Requirements

The determination of information requirements is a critical stage in the systems development process. Whether or not the "right" system is to be developed is settled at this stage. Subsequent stages of systems development are concerned with making the system "work right." If the "wrong" system is defined during requirements determination, subsequent effort is in vain. The system will not be fully used or will require expensive, time-consuming modification.

Estimates are that it costs fifty to one hundred times more to change an information system after it is completed than it costs to change it during the analysis and design stages. Those numbers sound like an exaggeration, but consider the analogy of building a house. Adding an additional bathroom to a floor plan during the architectural (design stage) costs a fraction of what it would cost to add it to a home after it is completed.

The Five Most Common Mistakes

There are five common mistakes made in the determination of information requirements:

1. The system analysts assume managers know what information they need. This fallacy was discussed in the section on misconceptions about information in Chapter 3.

2. The systems analysts do not determine requirements for the complete set of decision makers. Table 6.4 reveals that more people than those in the ordering department make decisions about order processing, and, therefore, managers outside the ordering department will need information from a new order-processing system. There is a tendency for analysts to only focus on the business function requesting a new information system. But, as pointed out in Chapter 3, business processes and information systems transcend departmental boundaries.

3. Systems analysts tend to determine requirements for managers one at a time whereas they should do it as a group process. Consider this scenario: You are in a classroom; the instructor looks at you and says, "Please tell the class ten good jokes." Even though you probably know ten jokes, would you have difficulty recalling them? Most people would. Let us change the scenario. What if the instructor invited the entire class to spend some time telling jokes (a preferable activity to normal class activities), and the instructor even told a couple to get the process started. Likely these two jokes would stimulate the memory of jokes known by others and within an hour close to one hundred jokes might be generated. Most people in the room would have heard 80 to 90 percent of those jokes before. In other words, most people are familiar with many jokes but cannot recall them easily when individually asked to. The reason for this difficulty is the limits of short-term memory (discussed in Chapter 3). Similarly, managers cannot

think of everything they need when individually asked to, but a group process can alleviate the problem.

4. Systems analysts ask the wrong questions to determine information requirements. Most systems analysts ask the obvious question, "What information do you need from the new system?" This obvious question is generally not at all helpful to managers. As an analogy, in marketing a distinction is made between an "order taker" and a "problem solver." To illustrate the difference in determining requirements for the following items, consider the difficulty in answering the following order takers' questions versus the problem solvers'.

Requirements Determined for	Order Taker	Problem Solver
Lawn Mower	How much horsepower do you want?	How large is your yard? How thick is your grass? How steep is your yard?
Video Recorder	Do you want Betamax or VHS?	Do or will you exchange tapes with friends, relatives, etc.? Do they have Betamax or VHS?
Camera	Do you want shutter, aperture priority features, or both?	Will you be taking action shots, still pictures, or both?
Sailboat	Do you want a fixed or retractable keel?	Do you have a dock available for your boat, or will you be using a trailer each time you go sailing?

In the preceding examples, the problem solver has creatively determined how to obtain answers to requirements determination questions through less obvious, indirect questions. The systems analyst must do the same thing for information requirements determination.

5. The systems analyst expects managers to analytically determine their exact detail requirements and get it right the first time. In the discussion of cognitive style in Chapter 3, the trial-and-error nature of most problem solvers was discussed. For the same reason people move furniture around in a room in a trial-and-error process to satisfy requirements, trial and error must be provided for the process of determining information requirements.

Information Requirements: Getting it Right

In order to determine information requirements as precisely as possible, systems analysts must not assume managers know their requirements. Systems analysts are not order takers; they must creatively and cooperatively work with managers to determine information requirements. Specific actions toward this end include the following:

1. During the requirements-determination exercise, involve all managers who have a decision-making relationship to the business process for which the new system is being developed. For example, using Table 6.4, decision makers from

sales, credit, ordering, and shipping should be included in requirements determination.

2. Make requirements determination a group or joint effort. This is referred to as *joint-application design* (JAD). The managers should collectively meet with members of the project team to conduct their information-requirements exercises.

3. A well-designed, structured interview should be used to facilitate group discussion and resolve information requirements. The next topic of this chapter—structured interview—provides such an interview.

4. The final detailed reports and screen displays should be generated through a trial-and-error process using heuristic designs and prototyping tools as discussed in Chapter 7.

Structured Interview

The results of a structured interview provide a conceptual definition of the types of information (reports and screen displays) needed from a system. The interview does not provide the exact contents and format of the information; those aspects are addressed using prototyping and heuristic design techniques (Chapter 7). As discussed earlier, the tendency of the systems analyst is to ask direct questions, such as, "What are your information needs?" and "What are your information system requirements?" This direct approach is very natural to the analyst conducting the interview; it tends to be unnatural, however, to the manager being interviewed. The question requires the manager to mentally examine business problems and convert them into information requirements. A good mental model of this process may not exist with the manager, and the requirements from an ad hoc attempt are usually not complete or accurate.

The first principle of the information requirements interview is, therefore, not to force the manager to answer the direct questions about information requirements. Instead, ask questions that correspond to the mental maps or mental structures of the manager and then derive the information requirements from these answers. This simple change in interviewing approach results in a powerful change in the ability of managers to respond.

The principle of using indirect questions to arrive at the direct information being sought requires the interviewer to formulate indirect questions that correspond to the mental structures or mental processes of the executive. The *comprehensive interview method* presented in this chapter provides you with these questions. This method of indirectly arriving at information requirements is based on three observations:

1. Not all managers have the same cognitive or mental structures, so it is difficult to know in advance the one that fits a given executive.

2. Specifying requirements, even indirectly, is difficult mental work. The level of difficulty changes depending on the type of requirements question that is asked.

3. No single way of asking a requirements question is complete. Each additional way that requirements are elicited will yield yet additional requirements.

A number of indirect eliciting approaches have been developed and tested. Each has proven helpful because it fits the cognitive structure of a number of people who have used it, and others find they can adapt to it even though it may

Table 6.5 **Comprehensive Interview Approaches, Implementations, and Developers**

Comprehensive Approach	Information System Implementation	Developers
Specify problems and decisions	The executive interview portion of Business Systems Planning (BSP)	IBM
Specify critical factors	Critical Success Factors (CSF)	Rockart (also Zani)
Specify effectiveness criteria for outputs and efficiency criteria for processes used to generate outputs	Ends/Means Analysis (E/M analysis)	Wetherbe and Davis

not correspond to their own preferred mental processes. Three different approaches are combined to form a comprehensive interview approach. The method is thus a synthesis of proven techniques. The comprehensive approaches, implementations, and developers are shown in Table 6.5.

Each of these approaches provides a mental structure for the executive being interviewed. A model of the structure they provide is illustrated in Figure 6.1. Note that the paths for each of the approaches lead to information requirements.

Managers differ in their preferences for the three approaches, but when more than one approach is used, the first will provide the majority of the requirements. Each additional approach used results in additional requirements being identified.

The comprehensive interview method consists of asking three sets of indirect questions based on concepts drawn from BSP, CSF, and E/M analysis. The responses are then used (with the manager's help) to define information requirements. The specific questions asked are as follows:

Figure 6.1 **Framework for Information Requirements Interview**

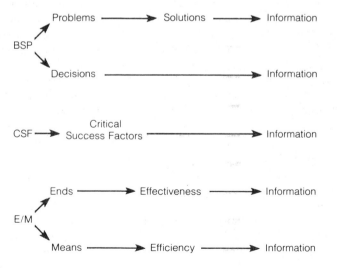

Business Systems Planning (BSP) (Problems and Decisions)

1. What are the major problems encountered in accomplishing the purposes of the organizational unit you manage?
   a. What are good solutions to those problems?
   b. How can information play a role in any of those solutions?
2. What are the major decisions associated with your management responsibilities?
3. What improvements in information could result in better decisions?

Critical Success Factors (CSF)

1. What are the critical success factors of the organizational unit you manage? (Note that most managers have four to eight of these.)
2. What information is needed to ensure that critical success factors are under control?

Ends/Means (E/M) Analysis

1. What makes goods or services provided by the organizational unit you manage effective to users?
2. What information is needed to evaluate that effectiveness?
3. How do you define efficiency of the organizational unit in providing goods or services?
4. What information is needed to evaluate that efficiency?

The method of using three sets of indirect questions as the basis for obtaining a reasonably correct and complete set of information requirements is both simple and powerful. It is simple because it consists of simple components that can be learned by an analyst and a manager in a relatively short time. It is powerful because it is based on fundamental theories of human information processing and human strengths and limitations. It provides a comprehensive set of approaches to eliciting information that are additive in their result. Figure 6.2 provides an illustration of the results generated from conducting an information requirements interview. The interview was conducted for an order processing system.

Research in the use of this interview technique indicated the following:[5]

- No single approach to eliciting information was dominant. Different managers liked and were able to respond more easily to different approaches.
- The eliciting approaches were additive. Each provided additional insight and requirements.
- Managers agreed it was best to ask questions using all three approaches because they found it helpful to have the three different lines of questioning.

The practical application of the comprehensive interview method thus supports the theoretical arguments for it. The net result is a simple yet powerful addition to the tools and techniques for information systems analysis and design.

---

5. James C. Wetherbe and Gordon B. Davis, "How to Ask Executives for Information Requirements and Get Good Answers," Working paper, *Management Information System Research Center*, University of Minnesota, Minneapolis, Minn., 1983.

Figure 6.2(a)   **Requirements Interview for Order-Processing System**

Problems	Solution	Information
Out of stock too often	Better inventory management	Out-of-stock, below-minimum report; automatic reordering of inventory
Ordering department often allocates limited inventory to the least important customers and/ or customers who have credit problems.	Let ordering department know relative importance and credit status of different customers	Customer-importance rating and credit rating
Shipping department often sends off a truck, unaware that another order going to the same destination will be coming to the dock within an hour.	Let shipping department know the destination of orders that are being processed through credit and warehouse	Shipping destination of orders provided when orders received from customers

Figure 6.2(b)   **Requirements Interview for Order-Processing System**

Decision	Information
Which customers to call on and what to sell them?	Customer-order history; inventory available
Credit for whom? How much? When to discontinue?	Credit rating; current status of account
What and how much inventory to stock?	Inventory on hand; sales trends on inventory items
When to reorder?	Lead times to replenish inventory items; safety stock
How to allocate limited inventory?	Priority of order; importance of customer; credit status of customer; shipping schedule
When to unload slow-moving inventory?	Sales trends of inventory items
Destination of ordered inventory?	Customers' addresses
What orders can be shipped together to save delivery costs?	Shipping schedule and customers' destinations for orders awaiting shipment

Document and Analyze Existing Information Systems

Once information required from the system is defined, the next step is to thoroughly document and analyze the existing system. This activity consists of analyzing the existing information system and its support of the decision system, and operations, including the collection, processing, storing, and reporting of information.

Some form of documentation of the existing information system is generally available and is the logical starting point for analysis. Narrative documentation

Figure 6.2(c)

Requirements Interview for Order-Processing System

Critical Success Factor	Information
Adequate inventory to fill customer orders	Percentage of orders filled on time—overall and also categorized by customer
Prompt shipment of orders	Delivery time—overall and also categorized by customer
High percentage of customer payments made	Delinquency report on nonpaying customers
Vendors (suppliers) promptly fill reorders	Exception report of vendor reorders not filled on time

Figure 6.2(d)

Requirements Interview for Order-Processing System

Ends	Effective	Information
Fill customer orders	Customer orders delivered as ordered, when expected, and as soon or sooner than competition	Summary and exception reports on customer deliveries; number of order corrections made; comparative statistics on delivery service vs. competition's
Good customer service	Promptly provide credit to qualified customers	Customer credit status and payment history
	Quick response to and reduction of customer complaints	Report of number and type of complaints by customers and average time to resolve complaint
	Customers are satisfied	Customer attitudes toward services perhaps determined by customer surveys

Means	Efficiency	Information
Process orders	Low transaction cost	Cost per transaction with historical trends
Process credit request	Low transaction cost	Cost per transaction with historical trends
Make shipments	Minimize shipment costs	Shipment cost categorized by order, customer, region, and revenue generated

is usually too vague and imprecise to ensure accurate understanding of the system. Therefore, graphical documentation is preferred. To illustrate, when receiving complex directions to a geographic location, do you generally prefer verbal directions or a map? The most useful forms of documentation for reviewing an information system are *systems flowcharts*, *data-flow diagrams*, and presentation graphics using icons—all of which are structured techniques.

A flowchart is a graphical representation of the system. Flowcharting symbols are used to represent operations, data-flow, files, equipment, and so on. Figure 6.3 illustrates the various symbols used on systems flowcharts. (Program flow-

Figure 6.3   **Systems Flowcharting Symbols**

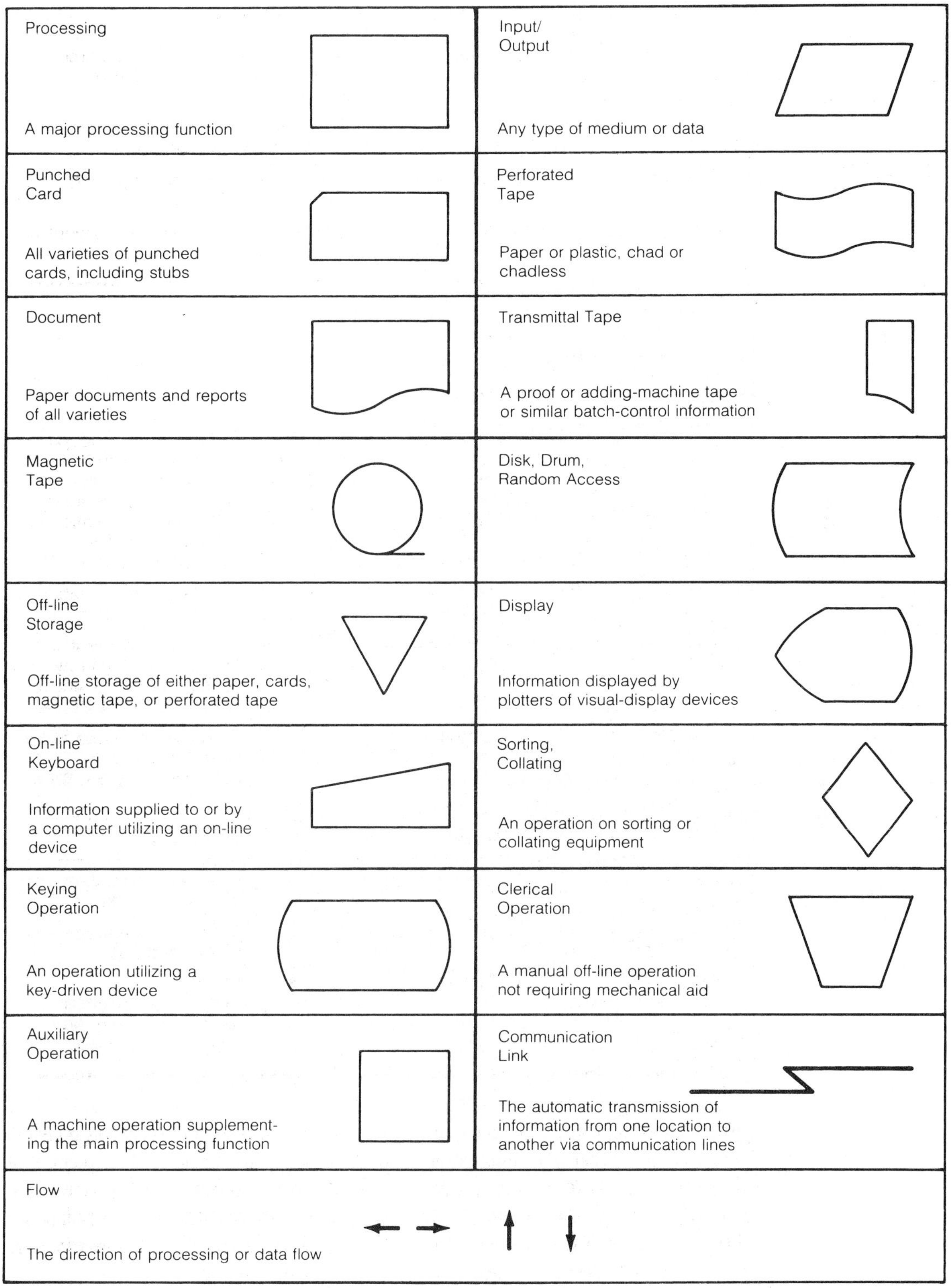

Processing A major processing function	Input/ Output Any type of medium or data
Punched Card All varieties of punched cards, including stubs	Perforated Tape Paper or plastic, chad or chadless
Document Paper documents and reports of all varieties	Transmittal Tape A proof or adding-machine tape or similar batch-control information
Magnetic Tape	Disk, Drum, Random Access
Off-line Storage Off-line storage of either paper, cards, magnetic tape, or perforated tape	Display Information displayed by plotters of visual-display devices
On-line Keyboard Information supplied to or by a computer utilizing an on-line device	Sorting, Collating An operation on sorting or collating equipment
Keying Operation An operation utilizing a key-driven device	Clerical Operation A manual off-line operation not requiring mechanical aid
Auxiliary Operation A machine operation supplement- ing the main processing function	Communication Link The automatic transmission of information from one location to another via communication lines
Flow The direction of processing or data flow	

charts are covered in Chapter 9.) Flowchart templates with cutout forms of these symbols can be purchased and used in constructing such flowcharts.

Flowcharts can be used to convey varying degrees of detail. In Chapter 1, a basic overview flowchart of an order-processing system was presented (see Figure 1.2). A more detailed flowchart of the same system is provided in Figure 6.4. It is important to note that Figure 6.4 represents only the order-processing system (subsystem) of an information system. The order-processing system uses the inventory master file in the processing of orders. However, the order-processing flowchart does not include all the processing necessary to control inventory. Another subsystem is required to maintain (add, delete, or modify) the inventory master file. This subsystem is illustrated in Figure 6.5. The inventory master file updated in the inventory subsystem is the file that is used in order processing.

An alternative that is replacing flowcharting in many organizations is called a *data-flow diagram* (DFD). DFDs use the symbols defined in Figure 6–6 and illustrated by combining the processing from Figures 6–4 and 6–5 in Figure 6–7. DFDs offer a more efficient and, in some ways, more profound way of showing data flow in relationships than can be made with a flowchart. They are easy to work with, since they are less formal than systems flowcharts; and, like flowcharts, DFDs can be used as analysis tools if existing documentation does not provide them, or they can be used as a design tool. In either case it is a very straightforward process to communicate with DFDs.

Presentation, or custom, graphics are a more elaborate form of documentation that involves the use of pictures that are representative of the objects being portrayed. Figure 6.8 provides examples of icons used in presentation graphics. Presentation graphics are popular for management presentation and documentation.

Preparing flowcharts, data-flow diagrams, and presentation graphics originally was a manual task. Consequently, it was tedious, time-consuming, and particularly frustrating when corrections were required. The illustrations provided in Figures 6.5, 6.6, and 6.7 went through several messy versions before arriving at the publishable forms. Exercises provided in the case book for this text provide experience in "shaping" narrative descriptions into presentable graphics.

One of the tools provided by CASE technology is computer-based graphics to ease the clerical tasks associated with preparing these graphics. Products such as PACBASE, CASE 2000 Designaid Workbench, Information Engineering Workbench, and Teamwork can be run on personal computers. Most of the products include syntax verifiers to check the proper use of diagramming symbols.

Flowcharts or DFDs may not exist for poorly documented systems or for systems that have not been implemented on a computer system. In such cases, flowcharts or DFDs must be developed from scratch, that is, from information gathered by observing the operation of the system and by interviewing various persons involved in the operation of the system.

Systems flowcharts or DFDs indicate the various reports, files, and transaction documents used in an existing system. (If the system is manual, the files may be in filing cabinets.) Collecting and analyzing copies of the various reports, file contents, and transaction documents in the context of the systems flowchart leads to a thorough understanding of the actual processing involved in transforming data into usable information.

Constructing Data-Flow Diagrams

Systems flowcharts are straightforward to construct, but a well-constructed DFD (which is more informative than a flowchart) requires some understanding of

Figure 6.4   Systems Flowchart of an Old Order-Processing System

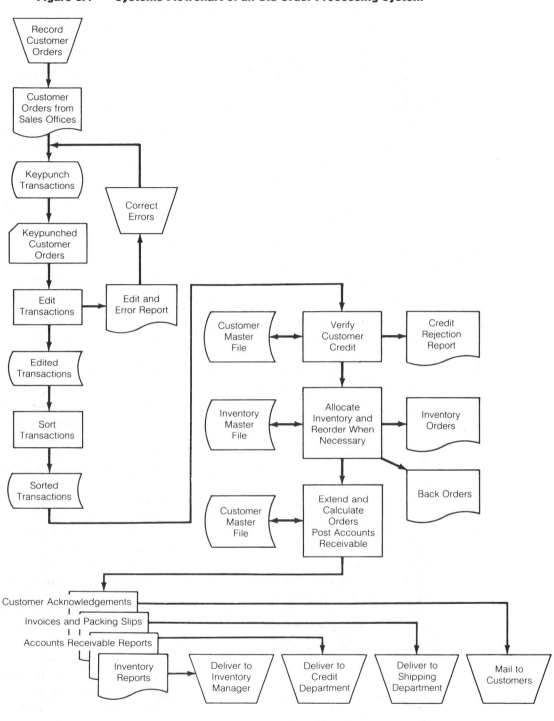

Figure 6.5   **Systems Flowchart of an Old Inventory Subsystem**

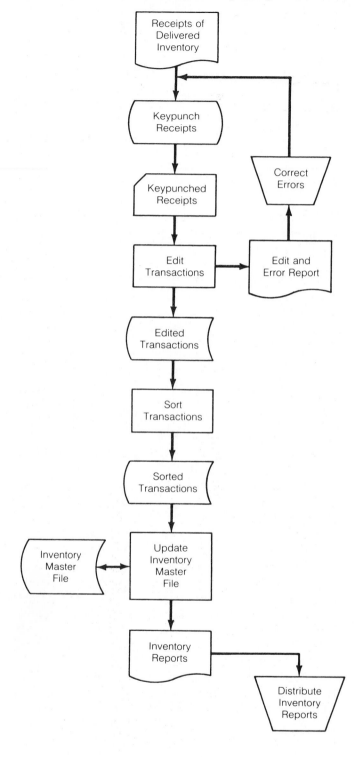

Figure 6.6 **Symbols Used in Data-Flow Diagrams**

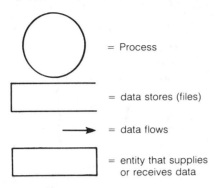

= Process

= data stores (files)

= data flows

= entity that supplies
or receives data

DFD techniques. Since DFDs are emerging as the "tool of choice" for analysis, techniques for constructing them are presented here. Although a number of symbols and techniques are used to construct DFDs, an understanding of those used in this book will allow easy adaption to others.

DFDs emphasize data flow and processes. They do not convey timing or control. They can be used to represent entire systems or portions of systems. If constructed properly, they are for the most part self-explanatory. For example, Figure 6.7 is easy to interpret.

As discussed in general systems theory (Chapter 3), we cope with complexity by factoring large systems into subsystems. Similarly, a final DFD or set of DFDs can be quite complex and detailed. We don't start with the detail; rather, we start with a conceptual model and factor down as far as is necessary (a judgment call) to properly portray the system. The conceptual model is depicted by surrounding a large circle (labeled by the system name or process) and surrounding it with entities. The major transactions or information flows between the entities and the circle are determined through analysis (i.e., interviews, observation, documentation) and are labeled, as illustrated in Figure 6.9.

Once the conceptual model is created, the next step is to identify the major data stores. They can be determined by examining the current information system and the files or data stores currently in use. For example, in an order-processing system, information would likely be stored about customers, orders, inventory, vendors, and vendor orders.

Once the data stores are identified, the major processes can be identified by clustering events or transactions by "what they make happen" or "what makes them happen." Figure 6.10 illustrates how the transactions from Figure 6.9 can be clustered into fragmented processes. Note that the processes always use an imperative verb to begin the label. Also note that each arrow in Figure 6.9 is in some form individually represented in Figure 6.10. Bubbles in the figure are numbered to facilitate documentation.

The next step is to assemble the fragments from Figure 6.10 into a single DFD. Completing a DFD that is straightforward and understandable usually takes several revisions. Fortunately, CASE graphics support makes revisions easier.

When constructing a DFD, the following guidelines can be used to check for correctness:

1. Be sure data stores, data flows, and processes have descriptive titles. Processes should use imperative verbs to project action.

Figure 6.7   **Data-Flow Diagram of Order-Entry and Inventory Management as Alternative to Flowcharts (in Figures 6.5 and 6.6)**

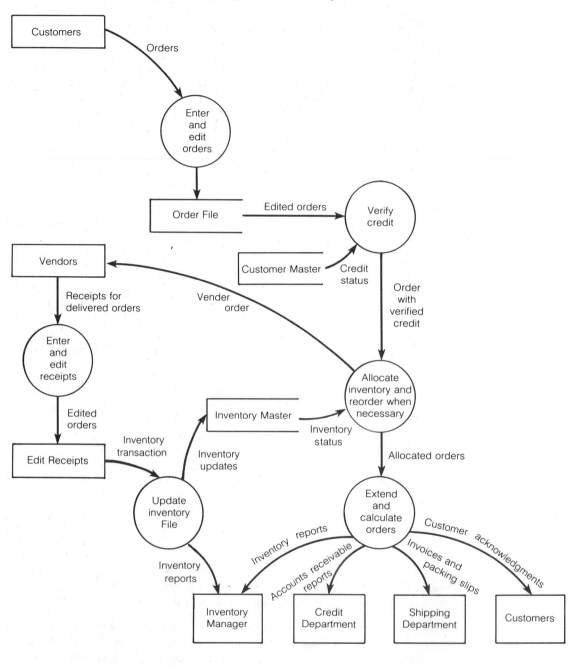

Figure 6.8   **Examples of Presentation Graphics**

TERMINAL   PERSON   FLOPPY   ACTIVITY   HEAP

Figure 6.9 **Conceptual Model of Order-Processing System**

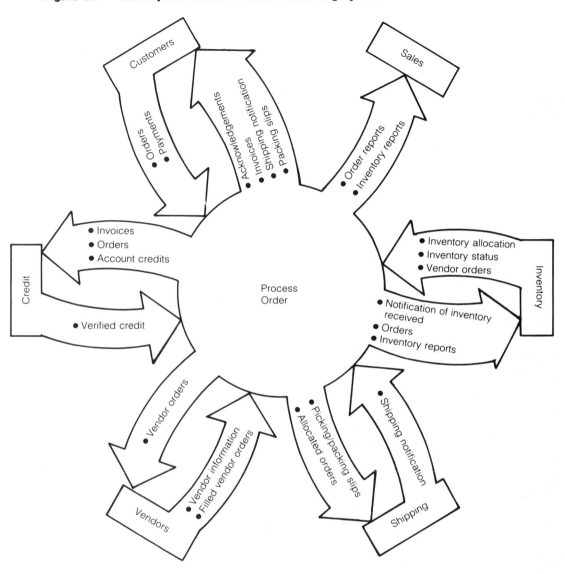

2. Be sure all processes receive and generate at least one data flow.
3. Begin and/or end data flows with a process bubble.

CASE diagramming products usually include syntax checkers to enforce steps 2 and 3. Figure 6.11 illustrates how the process fragments of Figure 6.10 can be assembled into a single DFD. The numbering scheme provides a check to ensure that all bubbles are represented in the diagram. Compare the order-processing DFD of Figure 6.11 with that of Figure 6.8. Note the additional detail and subtleties in Figure 6.11 that are made available from using a structured approach to DFD construction. The more in-depth documentation and deeper levels of detail that can be generated from DFDs explain why they are increasingly re-placing flowcharting.

Also, further factoring of the DFD can be done. Any one process can be decomposed into sub-bubbles, as was done with Figure 6.9 (which was decom-

Figure 6.10   Data-Flow Fragments Based upon Events or Transactions Identified in Figure 6.9

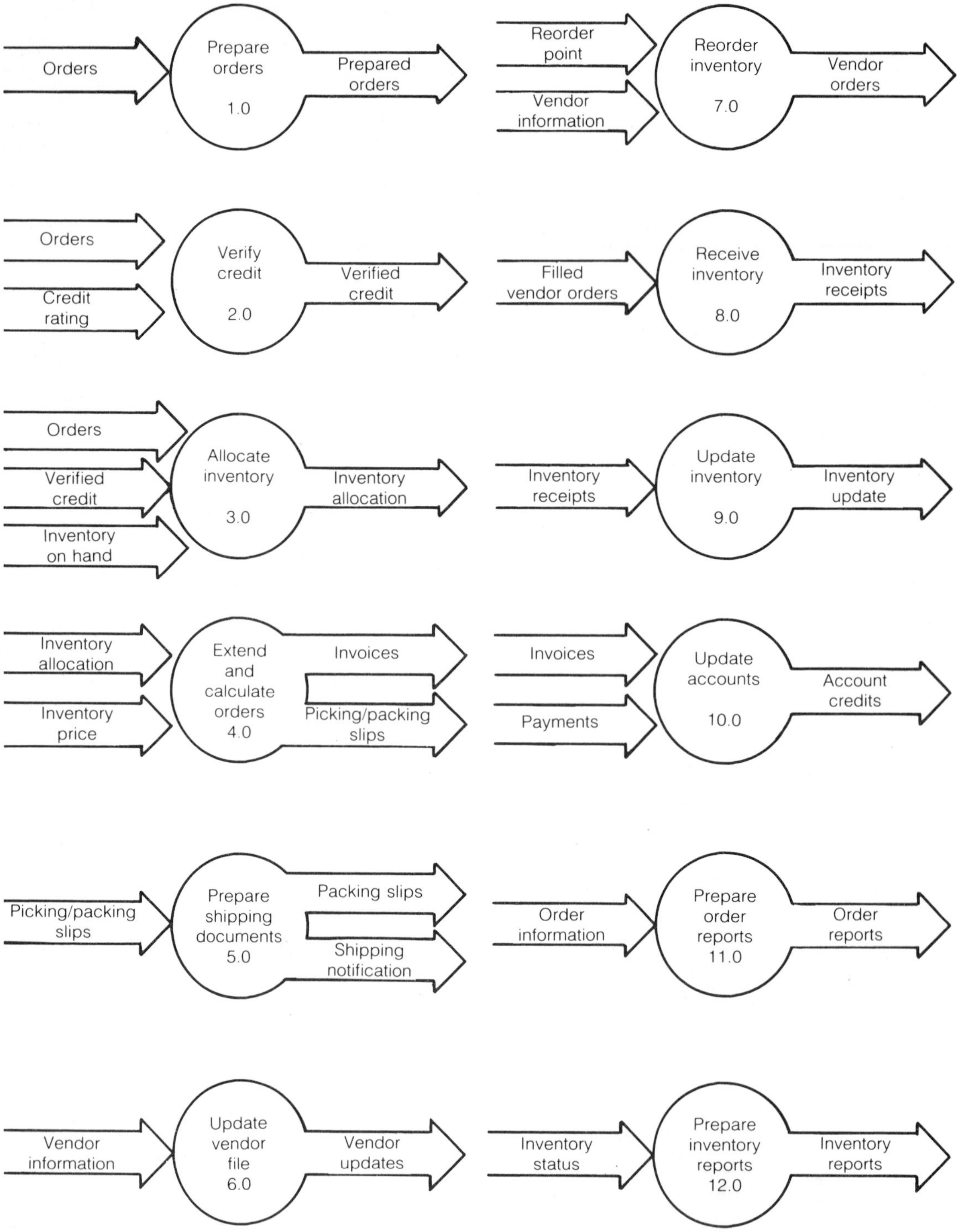

Figure 6.11   Data-Flow Diagram for Order-Processing

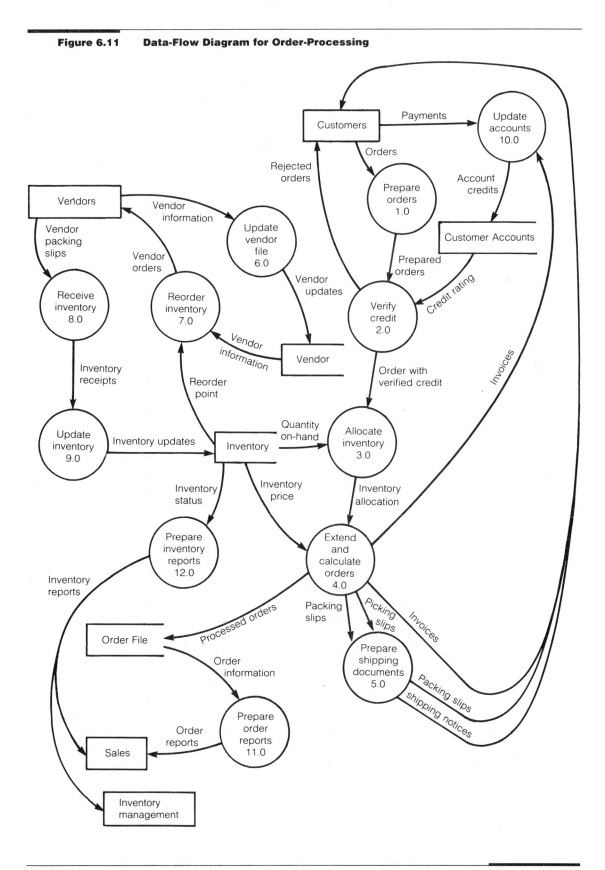

posed into Figure 6.11). The number of levels of detail needed to analyze the system is a judgment call. For example, the "allocate inventory" process of Figure 6.11 can be expanded, as done in Figure 6.12. Note the sub-numbering scheme (i.e. 3.1, 3.2, etc.) used for documentation purposes. Too much detail is generally not needed for good analysis, but as a design tool for software development it can be useful. When a lot of detail is desired, several pages of DFDs are created, beginning with the conceptual model and factoring down to lower-level DFDs with more detail, as portrayed in Figure 6.13.

Figure 6.12   Expansion of Inventory Allocation Process

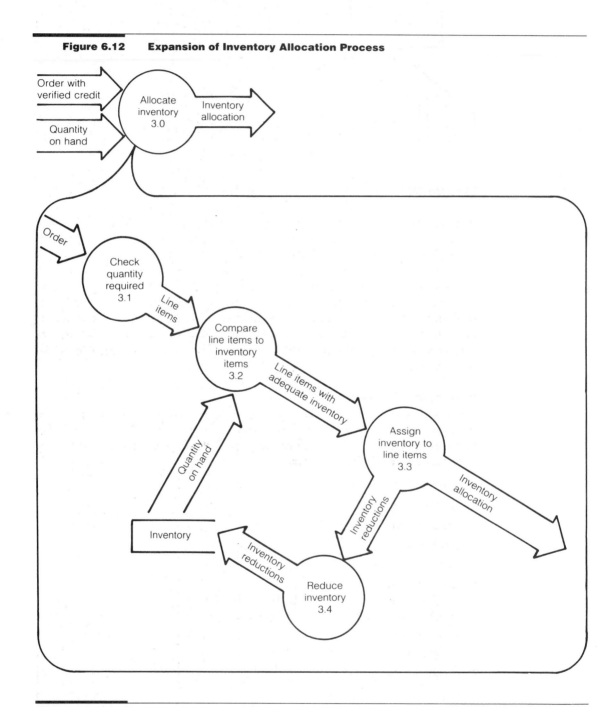

Figure 6.13   Decomposing of Data-Flow Diagrams

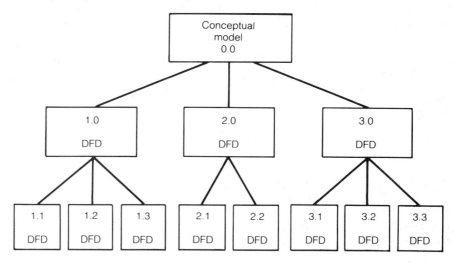

Isolate Deficiencies in the Information System

The final step of analysis involves contrasting the decision system with the information generated from the information system. The objective is to identify discrepancies between the information required and the information available. Deficiencies in the information system can then be identified.

In Chapter 1, deficiencies in information systems were categorized as inclusion and/or structure deficiencies. Inclusion deficiencies pertain to *what* information, technology, and personnel are included or lacking in the system. Structure deficiencies pertain to *how* the information, technology, and personnel are organized and interrelated throughout the system. (To review this framework, refer to Table 1.1.)

Information Deficiencies
The decision makers, of course, must decide what information is relevant to the decision associated with their management responsibilities. For example, the manager responsible for inventory may say that the following information is necessary for reordering decisions:

1. Date of last order

2. Quantity on hand

3. Quantity on order

4. Back order

5. Date of last purchase order

6. Sales of item
   a. This month
   b. One month ago
   c. Two months ago
   d. Three months ago
   e. Year-to-date

7. Suppliers

8. Suggested order size

Given the preceding information requirements, consider the following inventory information system deficiencies:

1. Information on back orders and date of last purchase order are not currently processed.

2. Information on sales of item last year is provided, but it is not used in decision making.

3. Necessary information is provided, but it is spread through several reports and is, therefore, time-consuming to retrieve.

4. Year-to-date sales information is processed, but it is not distributed to the manager responsible for inventory.

5. The timeliness of inventory information is inadequate. Because the existing computer system is too small to handle the daily work load, inventory reports have to be processed over the weekend.

6. Inventory information is stored on magnetic tape. Consequently, inventory transactions that arrive after the inventory file is updated, but prior to printing of reports, are not processed before the printing. (The entire inventory file would have to be processed to update a few records.) Therefore, inventory reporting on these records is not current until the next processing-and-reporting cycle.

7. Inventory transactions are keypunched but, due to lack of control, verification is often bypassed. This results in frequent errors in inventory information.

8. The inventory program contains a "bug" that results in the erroneous computation of the quantities on hand.

9. The decision maker is unqualified to use information produced by the system.

10. The programmer who works on the system is incompetent.

11. The least-experienced keypunch operators are verifying inventory transactions, and the error rate is high.

12. The manager of the order-processing department is not defining, encouraging, or rewarding good performance by order-processing clerks. Both the productivity and the morale of the department are low. This leads to a high error rate and delayed processing.

These information system deficiencies can be categorized individually as inclusion deficiencies and structure deficiencies using the framework provided in Table 6.6. (See Exercise 3 in this chapter.)

PIECES

In the discussions of problem and opportunity identification and definition in chapter 5, it is indicated that management's expression of problems or opportunities can generally be categorized into the PIECES framework. For thorough analysis, PIECES problems or opportunities must be translated into inclusion and/or structure deficiencies. For example, the performance of an object system may be poor for one or more of the following reasons:

1. Information that could improve performance
   a. Is not processed by the information system (information inclusion)
   b. Is in the wrong form (information structure)
   c. Cannot be processed fast enough (technology inclusion)
   d. Is computed incorrectly (technology structure)

Table 6.6   Categorization of Inventory Information System Deficiencies

	Inclusion Deficiency	Structure Deficiency
Information	Appropriate information is lacking, and/or inappropriate information is included in the system.	The manner in which information is collected, stored, and/or reported in the system is difficult or inconvenient to utilize.
Technology	Appropriate technology is lacking, and/or inappropriate technology is included in the system.	The hardware and software included in the system are not organized to most effectively and efficiently process the available information.
Personnel	Appropriate personnel are lacking, and/or inappropriate personnel are included in the system.	The personnel associated with the system are poorly organized and/or managed.

2. Decision maker is unqualified (personnel inclusion)

3. Personnel are unaware of how to properly process information (personnel structure)

Isolation of the specific cause(s) of a system deficiency requires acute perceptive skills on the part of those involved in systems analysis. They must avoid improperly analyzing the cause of the problem. A problem or an opportunity must be continually refined until it can be uniquely identified and categorized. The completed analysis then sets the stage for systems design, which is covered in the next chapter.

Summary

A common mistake in analysis is to be less thorough than necessary. Often systems analysts, in their eagerness to implement new technology, fail to understand the old system before they change it. This usually results in key capabilities of the old system being left out of the new system.

The fundamental activities for analyzing a system consist of reviewing documentation, observation, interviews, and questionnaires. The five steps of the analysis are the following:

1. Review object system

2. Define decision making associated with object system

3. Conduct information-requirements interviews

4. Analyze existing information system support of decision making

5. Isolate deficiencies in the information system

Structured interviews based upon BSP, CSF, and E/M analysis can be used to thoroughly define conceptual information required by decision centers. Flowcharts, DFDs, and pictorial graphics can be used to document the existing system.

The result of the analysis defines inclusion and structure deficiencies as they pertain to the information, technology, and personnel aspects of the information system.

Exercises

1. Discuss the fundamental activities of analysis (reviewing documentation, observation, interviews, and questionnaires) as they may be used for
   a. Reviewing the object system
   b. Defining decision making associated with an object system
   c. Analyzing existing information system support of decision making
   d. Isolating deficiencies in the information system

2. Develop a systems flowchart and a data-flow diagram for a payroll system described as follows:
   a. Payroll transactions (employee data, pay rate, hours worked, etc.) are prepared by deparatments and sent to the payroll department.
   b. The payroll department reviews payroll transactions and sends them to the data-processing department for keypunching.
   c. Transactions are keypunched.
   d. Keypunched transactions are validated (checked for errors) by a computer program that creates an error report and also a validated transaction file on magnetic disk.
   e. The error report is sent back to the payroll deparatment for corrections. Corrected transactions are then sent back to the data-processing department, and steps *c* and *d* are repeated.
   f. When all errors have been corrected, the validated transaction file is sorted into social-security-number order by a computer sort program. It is then inputted along with the payroll master file (also a disk file) to a computer program that updates the payroll master with new employee data and then performs the payroll computations (e.g., multiplies hours worked by pay rate). The same program prints the following payroll output:
      Paychecks
      Pay register
      Check reconciliation journal
   g. The paychecks are sent to employees. The other documents are sent to the payroll department.

3. Using a system deficiency table, categorize the twelve problems and opportunities listed on page 140 into Table 6.6. Remember that personnel deficiencies caused by negligence, lack of training, poor morale, and the like are structure deficiencies. Personnel deficiencies caused by personnel being incapable of doing a job are inclusion deficiencies.

4. Conduct a structured interview for a manager using the BSP, CSF, and E/M analysis questions described in this chapter. Note: Be sure to be patient—a good interview may take one to four hours.

Selected References

Ackoff, Russell L. "Management Misinformation Systems." *Management Science*, December 1967, pp. 147–56.

———. "Management Misinformation Systems." *Management Science*, December 1967, pp. 147–56.

Alavi, M. "The Evolution of Information Systems Development Approach: Some Field Observations." *Data Base*, 15(3), Spring 1984a, pp. 19–24.

———. "An Assessment of the Prototyping Approach to Information Systems Development," *Communications of the ACM*, 27(6), June 1984b, pp. 556–63.

Anderson, William S. "Bridging the Systems Expectation Gap." *Data Management*, November 1975, pp. 29–32.

Andrews, W. "Prototyping Information Systems." *Journal of Systems Management*, September 1983, pp. 16–18.

Argyris, Chris. "Resistance to Rational Management Systems." *Innovation*, no. 10 (1970) pp. 28–35.

Athey, T. H. "Information Gathering Techniques." *Journal of Systems Management*, January 1980, pp. 11–14.

Bariff, M. L. "Information Requirements Analysis: A Methodological View." Working paper #76-08-02, The Wharton School, January 1977.

Bariff, M. L. and Lusk, E. J. "Cognitive and Personality Tests for the Design of Management Information Systems." *Management Science*, April 1977, pp. 820–29.

Benbasat, Izak, and Dexter, A. S. "Value and Events Approaches to Accounting: An Experimental Evaluation." *Accounting Review*, October 1977, pp. 735–49.

Benbasat, Izak, and Schroeder, Roger G. "An Experimental Investigation of Some MIS Design Variables." *MIS Quarterly*, March 1977, pp. 37–47.

Berrisford, T. and Wetherbe, J. "Heuristic Development: A Redesign of Systems Design." *MIS Quarterly*, March 1979, pp. 11–19.

Blaylock, B. K. and Rees, L. P. "Cognitive Style and the Usefulness of Information." *Decision Sciences*, 15, 1984, pp. 74–91.

Blumenthal, Sherman C. *Management Information Systems: A Framework for Planning and Development.* Englewood Cliffs, N.J.: Prentice-Hall, 1969.

Bostrom, R. P. "Development of Computer-Based Information Systems: A Communications Perspective." *Computer Personnel*, 9(4), August 1984, pp. 17–25.

Burns, R. N. and Dennis, A. R. "Selecting the Appropriate Application Development Methodology." *Data Base*, Fall 1985, 17(1), 19–23.

Carey, T. T. and Mason, R. E. A. "Information Systems Prototyping: Techniques, Tool and Methodologies." *INFOR*, 21(3), August 1983, pp. 177–91.

Cerveny, R. P., Garrity, E. J. and Sanders, G. L. "The Application of Prototyping to Systems Development: A Rationale and Model." Working paper 649, School of Management, State University of New York at Buffalo, October 1985.

Chandler, E. W. and Nador, P. "A Technique for Identification of User's Information Requirements: The First Step in Information Systems Design." *Information Processing 17*, North-Holland Publishing Co., 1972.

Chervany, Norman L., and Dickson, Gary W. "An Experimental Evaluation of Information Overload in a Production Environment." *Management Science*, June 1974, pp. 1335–44.

Davis, Gordon B. *Management Information Systems: Conceptual Foundations, Structure, and Development.* New York: McGraw-Hill, 1974.

Davis, Gordon B. "Strategies for Information Requirements Determination." *IBM Systems Journal*, 21(1), 1982, pp. 4–30.

Dickson, Gary W., and Simmons, John K. "The Behavioral Side of MIS." *Business Horizons*, August 1970, pp. 59–71.

Earl, M. J. "Prototype Systems for Accounting, Information and Control." *Accounting, Organizations, and Society*, 3(2), 1978, pp. 161–70.

Feltham, Gerald A. "The Value of Information," *The Accounting Review*, October 1968, pp. 684–94.

Grudnitski, G. "A Methodology for Eliciting Information Relevant to Decision Makers." Proceedings of the First International Conference on Information Systems, 1980.

Henderson, J. C. and Nutt, P. C. "On the Design of Planning Information Systems." *Academy of Management Review*, 3(4), October 1978, pp. 774–85.

Hirsch, Rudolph E. "The Value of Information." *The Journal of Accountancy,* June 1968, pp. 41–45.

Holland Systems Corporation. *Strategic Systems Planning*, Ann Arbor, Michigan, document # MO154-048611986.

Howard, R. A. "An Assessment of Decision Analysis." *Operations Research*, 28(1), January–February, 1980, pp. 4–27.

Huber, G. P. "Cognitive Style as a Basis for MIS and DSS Design: Much Ado about Nothing," *Management Science*, May 1983, pp. 567–79.

Hudson, M. H. "Determining Organizational Information Requirements." *Journal of Systems Management*, pp. 6–10.

IBM Corporation. *Business Systems Planning—Information Systems Planning Guide*, Publication # GE20-0527-4, 1975.

Janson, M. A. and Smith, L. D. "Prototyping for Systems Development: A Critical Appraisal." *MIS Quarterly*, December 1985, pp. 305–16.

Jarvenpaa, S. L., Dickson, G. W. and DeSanctis, G. "Methodological Issues in Experimental IS Research: Experiences and Recommendations." *MIS Quarterly*, June 1985, pp. 141–56.

Jenkins, M. A. "Prototyping: A Methodology for the Design and Development of Application Systems," Discussion paper #227, Indiana University, April 1977.

Johnson, Richard A.; Kast, Fremont E.; and Rosenzweig, James E. *The Theory and Management of Systems*. New York: McGraw-Hill, 1973.

Kennedy, M. E. and Mahapatra, S. "Information Analysis for Effective Planning and Control." *Sloan Management Review*, 16(2), Winter 1975, pp. 71–83.

Kent, William. "A Simple Guide to Five Normal Forms in Relational Database Theory." *Communications of the ACM*, February 1983, pp. 120–25.

King, W. R. and Cleland, D. I. "The Design of Management Information Systems: An Information Analysis Approach." *Management Science*, 22(3), November 1975, 286–97.

Kraushaar, J. M. and Shirland, L. E. "A Prototyping Method for Applications Development by End Users and Information Systems Specialists." *MIS Quarterly*, 9(9), September 1985, pp. 189–97.

Langle, G. B., Leitheiser, R. L., and Naumann, J. D. "A Survey of the Application Systems Prototyping in Industry." *Information and Management*, 7, 1984, pp. 273–84.

Lederer, A. L. "Information Requirements Analysis." *Journal of Systems Management*, December 1981, pp. 15–19.

Lederer, A. L. and Mendelow, A. L. "Issues in Information Systems Planning." *Information and Management*, May 1986, pp. 245–54.

Lederer, A. L. and Sethi, V. "The Implementation of Strategic Information Systems Planning Methodologies." Unpublished paper, University of Pittsburgh, 1987.

Lucas, Henry C., Jr. "A Descriptive Model of Information Systems in the Context of the Organization." *Data Base*, Winter 1973, pp. 27–39.

Martin, J. *Strategic Data-Planning Methodologies*. Englewood Cliffs, N.J.: Prentice-Hall, 1982.

McKeen, J. D., Naumann, D. J., and Davis, G. B. "Development of a Selection Model for Information Requirements Determination." Working paper MISRC-WP-79-06, Graduate School of Business, University of Minnesota, June 1979.

Montazemi, A. R. and Conrath, D. W. "The Use of Cognitive Mapping for Information Requirements Analysis." *MIS Quarterly*, 10(1), 1986.

Munro, M. C. "Determining the Manager's Information Needs." *Journal of Systems Management*, June 1978, pp. 34–39.

Munro, M. C. and Davis, G. B. "Determining Management Information Needs: A Comparison of Methods." *MIS Quarterly*, June 1977, pp. 55–67.

Munro, M. C. and Wand, Y. "A Systems Theory Approach to Understanding Information Requirements Analysis." Working paper WP-05-81, Faculty of Management, University of Calgary, September 1981.

Naumann, J. D. and Jenkins, M. A. "Prototyping: The New Paradigm for Systems Development." *MIS Quarterly*, September 1983, pp. 29–44.

Newell, A. and Simon, H. A. *Human Problem Solving*, Englewood Cliffs, N.J.: Prentice-Hall, 1972.

Nolan, Richard L. "Systems Analysis for Computer-Based Information Systems Design." *Data Base*, Winter 1971, pp. 1–10.

Nugent, Christopher E., and Vollman, Thomas E. "A Framework for the Systems Design Process." *Decision Sciences*, 1972, pp. 84–109.

Nutt, P. C. "Evaluating MIS Design Principles." *MIS Quarterly*, June 1986, pp. 138–56.

Pollack, S. L., Hicks, H. and Harlon, W. "A Decision Table Approach to Systems Analysis." *Data Base*, Spring 1970.

Rockart, J. F. "Chief Executives Define Their Own Data Needs." *Harvard Business Review*, March–April 1979, pp. 215–29.

Segall, M. J. "The Use of Prototyping to Aid Implementation of an On-Line System." *Systems, Objectives, and Solutions*, 4, 1984, pp. 141–56.

Sethi, V. and Teng, J. M. C. "Choice of an Information Requirements Analysis Method: An Integrated Approach." *INFOR* (forthcoming), 1987.

Simon, Herbert A. *The New Science of Management Decision*. New York: Harper & Brothers, 1960, pp. 54ff.

Slovic, P., Fleissner, D. and Bauman, S. "Analyzing the Use of Information in Investment Decisionmaking: A Methodological Proposal." *Journal of Business*, April 1972.

Specht, P. H. "Job Characteristics as Indicants of CBIS Data Requirements." *MIS Quarterly*, September 1986, pp. 270–87.

Taggart, W. M., Jr., and Tharp, M. O. "Dimensions of Information Requirements Analysis." *Data Base*, 7(1), Summer 1975, pp. 5–13.

————. "A Survey of Information Requirements Analysis Techniques." *Computing Surveys*, 9(4), December 1977, pp. 273–90.

Teng, J. T. C. and Galletta, D. F. "MIS Researchers' Perspective on Methodological and Content Issues of MIS Research." Unpublished paper, University of Pittsburgh, 1987.

Valusek, J. R. and Fryback, D. G. "Information Requirements Determination: Obstacles within, among, and between Participants." Proceedings of the SIGCPR ACM Conference on End User Computing, May 2–3, Minneapolis, 1985.

Wetherbe, James C. *Systems Analysis for Computer-Based Information Systems*. St. Paul, Minn.: West Publishing, 1979.

————. *Executive's Guide to Computer-Based Information Systems*. Englewood Cliffs, N.J.: Prentice-Hall, 1983.

————. "Traditional Approaches to Systems Development." *Auerbach Information Management Series*, December 1982.

————. "Advanced Approaches to Systems Development." *Auerbach Information Management Series*, December 1982.

————. "System Development: Heuristic or Prototyping?" *Computerworld*, April 1982.

————. "Taking the Systems for a Test Drive." *Corporate Report*, April 1982, p. 38.

————. "Evolution in Systems Analysis and Design." *Proceedings of the 1982 Annual Conference of the Association for Systems Management*, Kansas City, May 1982.

Wetherbe, James C. and Berrisford, Thomas R. "Heuristic Development: A Redesign of Systems Design." *MIS Quarterly*, March 1979, pp. 11–19.

Wetherbe, James C. and Davis, Gordon B. "Information Requirements Determination Tools: Borrowing from the Best." *AIDS 1982 Proceedings*, San Francisco, November 1982.

Wetherbe, James C. and Dickson, Gary W. *Management of Information Systems*. New York: McGraw-Hill, 1984.

Wetherbe, James C., and Whitehead, Carlton J. "A Contingency View of Managing the Data Processing Organization." *MIS Quarterly*, March 1977, pp. 19–25.

Wetherbe, James C.; Bowman, Brent; and Davis, Gordon B., "Three Stage Model of MIS Planning." *Information and Management*, vol. 6, March 1983, pp. 11–25.

Yadav, S. B. "Determining an Organization's Information Requirements: A State of the Art Survey," *Data Base*, Spring 1983, pp. 3–20.

Zani, W. M. "Blueprint for MIS" *Harvard Business Review*, November–December 1970, pp. 95–100.

Information Systems Design

Introduction

This chapter presents the concepts and techniques that are the core, and perhaps the glamorous aspect, of systems analysis—the design of information systems. During this stage, the systems analyst begins designing the means of solving the problems or capitalizing on the opportunities discovered during the analysis stage. The systems analysis stage primarily defines the way things *are;* the systems design stage defines the way things *should be.*

Framework for Systems Design

Systems design is a process wherein the inclusion and structure aspects of the information system are redesigned in an effort to more closely align the information system with the information requirements of the organization. Though the systems analyst plays a key role in performing this task, he or she does not single-handedly accomplish it. Figure 7.1 illustrates the roles and relationships of those involved in systems design with respect to the inclusion and structuring of new or revised information systems.

Information

Managers must first decide what information has to be provided by an information system. Systems analysts assist in this activity by reviewing and further refining information requirements, and by determining how that information can best be made available. Once the information to be provided by the system has been determined, systems analysts play the lead role in structuring the information in terms of its collection, processing, storage, and reporting.

Technology

The selection of the technology is generally a joint effort between managers and systems analysts. Computer vendors are called upon to provide information on

Figure 7.1 **Roles of Systems Design Participants**

	Inclusion	Structure
Information	1. Manager(s)	1. Systems Analyst(s)
Technology	1. Manager(s) and Systems Analyst(s) 2. Computer Vendor(s)	1. Programmer(s) 2. Systems Analyst(s)
Personnel	1. Manager(s) 2. System Analyst(s)	1. Manager(s) 2. Systems Analyst(s)

Legend: 1. = Lead role.
2. = Support role.

the current hardware and software available. The structuring of the technology is primarily a computer programming activity in which systems analysts assist by providing detailed specifications as required.

Personnel

Managers generally make the final decisions pertinent to the selection and structuring or organization of personnel. Systems analysts, however, are often called upon to make recommendations as to personnel and their training.

Systems Design and Specification

The structuring of information in the information system is the primary domain of the systems analyst and the major thrust of this chapter. The inclusion and structuring of technology and personnel that represent the means to support the information structure are discussed in Chapters 8 and 9.

When structuring information for an information system, a systems analyst may not know what specific hardware and software will be best suited to support the system. Therefore, systems design techniques and procedures should be as hardware- and software-independent as possible. This allows the systems specifications to be implemented using whatever hardware and software turns out to be most cost effective in supporting the information structure. The relationship between information structure design and hardware and software is similar to the relationship between architectural design and building materials. In both cases, the design specifications should be precise, without unduly restricting the selection of the best available means for constructing the final product.

In this chapter a number of sophisticated techniques and technologies are covered. Before tackling them it is helpful to have a firm understanding of the outcomes of deliverables to be generated from the systems-design phase of the SDLC. The development of each of these deliverables is greatly facilitated by the use of CASE technologies' products, such as Excelerator, PACBASE, CASE 2000 Designaid, Information Engineering Workbench and, Teamwork. The role played by CASE technology in the creation of design specification is discussed where appropriate throughout the chapter.

Systems design should generate the following deliverables:

1. *Output definitions* describe the printed reports or terminal displays to be provided by the system. Outputs can be categorized as either formal, predefined reporting or informal, ad hoc reporting. The latter requires good anticipation of requirements through the data-modeling techniques discussed in this chapter.

2. *Logical data structure* uses data modelling to define the various entities, attributes, and relationships within the database designed to support the information system.

3. *Data dictionary* defines all data elements that are inputted, computed, stored, and reported in the information system. Definition includes the source of each data element, validation logic, processing or computation logic, where each data element is used and where it is stored.

4. *Logic definitions* graphically define complex processing rules necessary for input, computation, processing, and storing of data. Techniques include decision tables and decision trees.

5. *Input definitions* describe input documents and input screens.

6. *Presentation graphics of system* portray in a macro sense how the overall system is put together. These graphics illustrate information flow, reporting, files, and so forth.

The various design specifications (with the exception of presentation graphics) are linked together as illustrated in Figure 7.2 such that the input, computation, logical processing, storage, and reporting of each item of information can be traced forward or backward from its point of origin to its final use(s). As depicted in Figure 7.2 and as will be illustrated later in the chapter, the data dictionary is the core of design specifications. Upon completion, the specifications should provide sufficient documentation to guide the generation of software required to operate the system.

Though organizations may have systems design procedures and CASE technologies that differ from the procedures presented in this chapter, the differences will generally be cosmetic rather than conceptual in nature. Readers who master the techniques presented in this chapter should experience a comfortable transition to other systems design procedures and techniques.

Output Requirements Drive Design Process

As discussed in Chapter 3, the outputs of any system determine the inputs and processes necessary to support the system. Once the output requirements are

Figure 7.2   Relationship of Systems Design Specifications

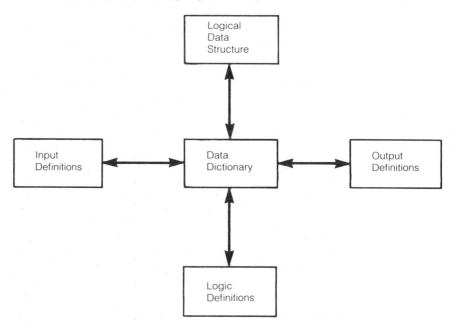

determined, the system designer can determine what to *include* in the system and how to *structure* it so that outputs can be provided.

The process of translating output requirements to determine information inclusion and structure is a two-phase process consisting of a conceptual phase and a detail phase. Figure 7.3 portrays the conceptual and detail stages of the translation process and indicates the techniques and technologies to support each stage.

Conceptual/Logical Phase

The conceptual stage of information requirements was covered in Chapter 6. The use of JAD (joint application design) or group interview in conjunction with the structured interview (BSP, CSF, and E/M analysis) provides a list of the conceptual types of information needed from a system (see Figures 6.2*a* through 6.2*d*). Combining the results of the interviews with results of DFDs or flowcharts is sufficient for constructing a conceptual or logical data structure.

Conceptual Data Structure

A conceptual or logical data structure defines the subjects or entities about which data is to be stored and the relationships among those entities. The first step for constructing a conceptual data model is to identify the entities to be included in the data structure. Entities are defined as a person, place, or thing about which data is to be stored. Entities must always have at least one attribute or describing characteristic. For example "price" does not have any attributes and therefore is an attribute of some entity rather than an entity itself.

Figure 7.3   Conceptual and Detail Phases of Translating Information Requirements into Data Structures

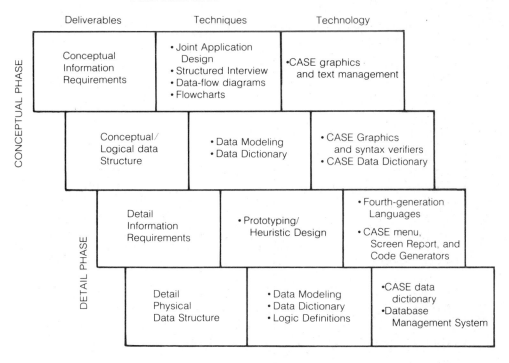

The order-processing application discussed in Chapter 6 will be used to illustrate how this is done. Using the structured interviews illustrated in Figures 6.2a and 6.2d, we can determine entities by searching the information column of each figure. We can also determine attributes or characteristics of those entities.

For example, in Figure 6.2a, the following entities and attributes can be derived:

Inventory Item
Item #
Reorder Point

Customer
Customer #
Importance Rating
Credit Rating

Order
Order #
Shipping Destination

Entity identifiers (i.e., item #) are automatically added to each entity. Some judgment and interpretation is required to establish entities and attributes. For example in Figure 6.2a, the terms *stock* and *inventory* are used as information items to solve the first problem, "Out of stock too often." *Stock* and *inventory* are synonymous, so *inventory item* was selected to represent both concepts. The same was done for *below minimum* and *reorder point*, which mean the same thing. The correctness of such interpretations should be confirmed by user management.

Continuing with this procedure for Figures 6.2b through 6.2d results in entities and attributes as illustrated in Figure 7.4. As a way of double-checking entity identification, the DFD or flowchart can be reviewed to see if any entities have been overlooked. Reviewing those charts (see Figure 6.4, 6.5, and 6.7) for the order-processing system provided in Chapter 6 indicates entities of customers, orders, inventory, and vendors. Entities are identified by reviewing files. A file always represents at least one entity. Since entities in Figures 6.4, 6.5, and 6.7 were identified from the structured interview, no additions are necessary for Figure 7.4. Another way to double-check entity identification is to review whatever reporting is provided by the existing system. The content of reports can be used to determine entities.

The interpretation of interviews, DFDs, flowcharts, and old reports into entities and attributes need not be perfect. In subsequent steps, deficiencies in the interpretation will be revealed and can be corrected at that time.

The exact nature or details of the attributes is often not available from the interviews. For example, from Figure 6.2d the information about customer attitudes is not very specific. Accordingly, it is simply listed as "attitudinal data" under the "customer surveys" entity in Figure 7.4. The specifics of "attitudinal data" are determined later.

When the conceptual information will likely be *computed* from other data or attributes, it is not listed as an attribute. This generally happens when the words *report* or *statistics* are used to describe the information. For example, in Figure 6.2c, information required includes "percentage of orders filled on time." To provide that information involves computation based upon "time promised" and

Figure 7.4

Conceptual Entities and Attributes as Derived from Figure 6.2

Inventory Item	Customer	Order
Item # Reorder point Quantity on hand Lead time	Customer # Credit rating Address	Order # Shipping destination Priority Time promised Time delivered Correction required Shipping cost
Account	**Shipment**	**Vendor**
Customer # Account status Deliquency date Payment history	Shipment # Schedule Shipping cost	Vendor-item #
Vender Order	**Competitor**	**Competitor Item**
Vender order # Time promised Time delivered	Competitor #	Item # Delivery time
Complaints	**Customer Surveys**	
Customer # Time to resolve complaint	Survey # Attitudinal data	

"time delivered," which are carried as an attribute under the term *order* (see Figure 7.4).

As you become more experienced, interpretations such as those illustrated above will become apparent. But remember, mistakes will be caught by subsequent analysis techniques, so do not fret over an inability to achieve perfection during conceptual interpretation.

Data Modeling

Now that the logical entities are identified, data modeling can be used to construct a conceptual data structure. Figure 7.5 provides symbols to use for data modeling.

Figure 7.5   **Symbols for Data Modeling**

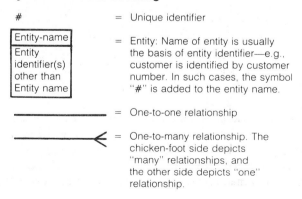

#	= Unique identifier
Entity-name / Entity identifier(s) other than Entity name	= Entity: Name of entity is usually the basis of entity identifier—e.g., customer is identified by customer number. In such cases, the symbol "#" is added to the entity name.
———————	= One-to-one relationship
———————<	= One-to-many relationship. The chicken-foot side depicts "many" relationships, and the other side depicts "one" relationship.

The rules for data models are as follows:

1. Every entity must be represented in the model.
2. Every entity must have a unique identifier.
3. Every entity must have at least one relationship (i.e., a one-to-one or one-to-many relationship).
4. Many-to-many relationships are not allowed.

The last rule is often the most difficult to understand and adhere to. Initially there are many cases in which many-to-many relationships appear to exist. For example, CUSTOMERS buy many ITEMS and many ITEMS are bought by many CUSTOMERS. Therefore, it seems logical that

However, when one of these many-to-many relationships appears to exist, it is because another entity (called an *intersecting entity*) has been overlooked.

This intersecting entity can generally be determined by asking the question, what event or transaction causes the many-to-many relationship to occur? In the case of CUSTOMERS and inventory ITEMS, the event or transaction is an ORDER. Therefore, the relationship between CUSTOMERS, ORDERS, and inventory ITEMS should be modeled as follows:

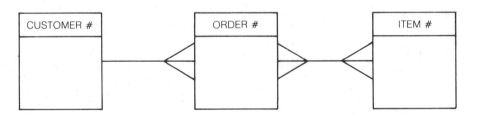

The preceding illustration says CUSTOMERS have many ORDERS, but an ORDER is for only one CUSTOMER. But we still have a problem. Since an ORDER can require many ITEMS, and ITEMS can fulfill many ORDERS, we still have a many-to-many relationship. What is the event or transaction intersecting between an ORDER and an ITEM? The answer is a LINE-ITEM of an ORDER. By adding LINE-ITEMS to the model we have

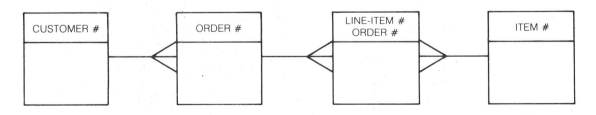

We have now completely corrected the problem. An ORDER can have many LINE-ITEMS, but a LINE-ITEM only has one order. An inventory ITEM can be used in many LINE-ITEMS, but a LINE-ITEM only refers to one inventory ITEM. Note that it takes both ORDER # and LINE-ITEM # to create a unique identifier for a LINE-ITEM.

Note that LINE-ITEM is an entity that was not created during the interpretation of the structured interviews and therefore was not originally included in Figure 7.4. Determining the need for LINE-ITEM through data modeling is an example of how conceptual errors or omission get rectified as we proceed through subsequent steps of this process.

Figure 7.6 provides a conceptual model based upon the entities listed in Figure 7.4. Note that to resolve many-to-many relationships, the additional entity of VENDOR-ORDER-LINE-ITEM had to be added.

The determination of the one-to-one and many-to-one relationships is a combination of common sense and intuition and should be validated by user/management review. For the order-processing system, sales, credit, ordering, and shipping management should review the data model. For example, it might turn out during the review that an order can be for more than one customer, in which case the model would be changed as follows:

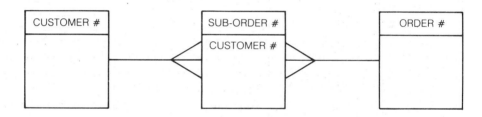

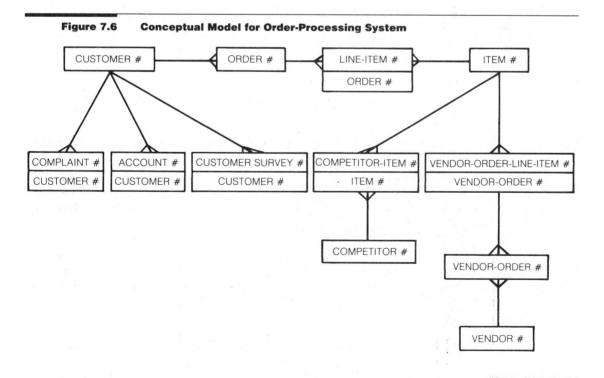

Figure 7.6   Conceptual Model for Order-Processing System

The reason that many-to-many relationships are not allowed is that without an intersecting entity, such as SUB-ORDER, it is not possible to retrieve information about what part of the order is for which customers. Consequently, it would not be possible to use the computer system to accurately respond to customer inquiries when there is more than one customer per order.

Data Dictionary

With a conceptual data structure established and some data elements identified, it is possible to begin the data dictionary. The data dictionary is the core documentation for systems design, development, and subsequent maintenance or modifications.

Data dictionaries are supported by CASE products, which greatly ease the task of creating and maintaining data dictionaries. Also, as will be illustrated later, data dictionaries are the cornerstone of automatic program-code generation.

The exact content and format of data dictionary items vary. Most products have a fixed set of entries, such as data definition and format. They also allow for *free-text* documentation for other entries desired by the designer. Although data-dictionary entries, as depicted in Figure 7.7, are for the most part self-explanatory, a detailed explanation of entries follows:

Name
The data-element name is the unique name assigned to a data element. This name can be used to track an element referred to in an output, input, or decision table back to the data-element dictionary for a complete description.

Because the data elements will be referred to in computer programs, it is advisable to choose data-element names that can be used as names in computer programs. This allows data names in computer programs to be directly cross-referenced to the data-element dictionary.

Through the use of hyphens and abbreviations, data names can be created and used in computer programs. For example, a social security number may have one of the following data names:

SOC-SEC-NUM
SS-NUM
SSN

Such abbreviations save time and space in coding specifications and programs. However, an abbreviation should not be carried to the point where it becomes difficult to determine the original full name of the data element. If the programming language to be used is not known when the data names are created, the names can be modified, if necessary, after the language has been determined.

Keywords
Often data elements will be familiar to users by names other than the unique names assigned for data-dictionary use. This can complicate matters when a programmer, user, or other analyst is looking up a definition and does not know the unique name assigned. To alleviate this problem, data dictionaries provide a *keyword* feature that allows the analyst to list other names by which a data element might be referred to. The data-dictionary function of CASE products can then perform a keyword search and try to locate the data element.

For example, in Figure 7.7, if the name ITEM-NUMBER is not known by the searcher, the data-dictionary facility could ask for keywords. If the searcher

Figure 7.7   **Illustration of a Data-Element Dictionary Screen Display**

```
NAME: ITEM-NUMBER
KEYWORDS: STOCK, STOCK-NUMBER, INVENTORY-NUMBER
DEFINITION: UNIQUE IDENTIFIER ASSIGNED
            TO EACH INVENTORY ITEM

FORMAT: INPUT         9(8)
        INTERNAL      9(8)
        OUTPUT        9(Z)

SOURCE: INPUT FROM INVENTORY UPDATE SCREEN

VALIDATION RULES: MUST BE NUMERIC AND UNLESS A NEW ITEM, ITEM NUMBER-MUST BE
                  EQUAL TO AN EXISTING ITEM NUMBER

                          OUTPUTS
WHERE USED: OUT OF STOCK/BELOW MINIMUM REPORT
            INVOICE
            PICKING LIST
            PACKING SLIP

                     COMPUTATIONS
                         NONE

                       PROGRAMS
                         OP008
                         OP010
                         OP012

ENTITY: INVENTORY-ITEM
        LINE-ITEM*
STORED: INVENTORY MASTER
        ORDER FILE*
MAINTENANCE: ASSIGNED BY INVENTORY DEPARTMENT
             SELDOM IF EVER CHANGED
```

*LINE-ITEM is a logical entity that is stored in the physical file ORDER-FILE.

can come up with a response approximating the keywords (e.g., STOCK, IN-VENTORY) the system can match it up with ITEM-NUMBER.

Format

The format of the field is defined by indicating both the field length and its characteristics (e.g., numeric or alphabetic). The input and output formats define how the data element is formatted for input and output, respectively. Editing symbols, their meanings, and examples are provided in Figure 7.8. These formatting instructions provide complete clarity for the CASE code generator or the application programmer.

The internal format is the storage format. Since no editing is required, 9s (nines) and Xs are used to represent numeric and alphanumeric characters, respectively. For numeric fields, the allowable sign values (i.e., $+$ and/or $-$) should be indicated. If a decimal point is involved, its position should be defined.

Note that in the data dictionary, formts can be written in shorthand. For example, 99999999 can be written 9(8).

Figure 7.8   Symbols Used to Format Information Fields

9 = Numeric
A = Alphabetic
X = Alphanumeric (characters or numbers)
Z = Zero suppress a numeric field
* = Lead numeric field for dollar protection
B = Embedded blank in an alphanumeric field
$ = Floating dollar sign if substituted for Z
. = Decimal point
, = Comma (suppressed if preceded by a Z or a $)

Examples of formats followed by sample data.

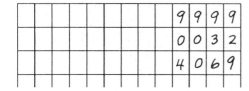

- Four-digit numeric field without zero suppression. Any numeric digit (0-9) may appear in any of the four posistions.

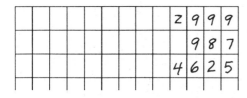

- Four-digit numeric field with zero suppression for the first character only. If the high-order position is zero, it appears as blank; all other digits are printed. The last three positions are printed regardless of value.

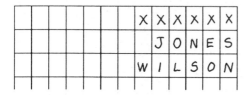

- A six-character alphanumeric field. The numbers 0-9, all letters of the alphabet, and special symbols (i.e., %, *, 7, etc.) can appear in this field.

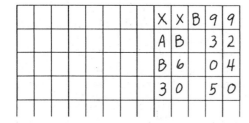

- Five-position field containing two alphanumeric characters and two numeric characters separated by a blank. The numerals are not zero suppressed.

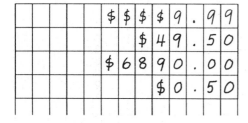

- Eight-position numeric field with zero suppression on the first four positions. Multiple dollar signs indicate a floating dollar sign is to replace the rightmost insignificant zero. All other insignificant zeroes are suppressed. The high-order digit should always be zero to insure the printing of a dollar sign.

Figure 7.8   Continued

```
$ * * * . 9 9
$ * * 8 . 5 0
$ 1 6 8 . 3 5
$ * 8 5 . 0 0
```

• Seven-position field with a fixed dollar sign. The asterisk (*) indicates the use of dollar protection by placing leading asterisks in unused dollar positions.

```
        DATE        ACCT-NO        AMT          Header printed at the
                                                top of every page.
      99/99/99      999999      ZZZ·99          Detail printed for
                                                each item.
```

```
DEPARTMENT    STORE      CORPORATE      Department total printed
   TOTAL      TOTAL        TOTAL        at each change of department.
                                        Store total printed at each
   ZZZ99     ZZZZ99      ZZZZZZ99        change of store. Corporate total
                                        printed at end of report.
```

Source

Data elements can come from one of three places: inputs, tables, or computations. Inputted data elements come directly into the system from an input data stream (e.g., transaction). Table data elements are looked up in predefined tables (e.g., tax-rate tables). Computed data elements are computed from other data elements in the system (e.g., NET-PAY = GROSS-PAY − TAXES).

For a data element to be determined by a table look-up, the table values and the indexes used for the table look-up are defined. This can be accomplished by simply displaying the table. The following example illustrates how a SALARY table in which the table indexes are PAY-GRADE and PAY-STEP may be defined.

| | PAY-STEP | | | | |
PAY-GRADE	1	2	3	4	5
1	500	532	568	612	688
2	645	720	785	822	876
3	832	912	986	1666	1188
4	1112	1186	1290	1380	1440
5	1360	1502	1612	1720	1816
6	1706	1820	1920	2012	2118

PAY-GRADE and PAY-STEP are data elements that must be defined elsewhere in the data-element dictionary. The SALARY table is used by locating the point of intersection for a given set of values for PAY-GRADE and PAY-STEP. For example, a PAY-GRADE of 4 and a PAY-STEP of 3 determine a value of 1290 for SALARY.

Table values may be maintained external to a program (e.g., on disk) and inputted during program execution. Alternatively, they may be coded as part of the program. If table values change frequently, it is preferable to maintain them external to the program so that they can be changed without modifying program codes.

For a data element that is computed, the necessary computation is defined. This can be done in simple algebraic form with one minor variation. The equals sign ($=$) in computer processing means that the variable on the left side of the equation is "set equal to" the computation or variable on the right side of the equation. For example, if $A = 6$ and $B = 3$, the equation $C = A + B + 2$ results in C being set equal to 11, irrespective of C's value prior to the computation.

Each of the data elements used to compute the value for a computed data element must be defined elsewhere in the data-element dictionary. For cross-reference purposes, care must be taken to ensure consistent spelling of the name of each data element.

If a sequence of computations is needed to compute a data element, the complete sequence must be defined. For example, assume the data element to be computed is BALANCE-DUE. The sequence of computations shown here may be used:

Computation:
1. PAYMENT   = CASH DEPOSIT + CHECK DEPOSIT
2. NEW-BALANCE = OLD BALANCE − PAYMENT
3. INTEREST   = NEW BALANCE × INTEREST RATE
4. BALANCE-DUE = NEW BALANCE + INTEREST

There may be logical processes required for inputting or computing data elements. In a simple case, the logical process may be described in the data-element dictionary. For example, consider the following logic associated with the computation of overtime:

IF HOURS WORKED IS GREATER THAN 40, DO COMPUTATION 2
OTHERWISE DO COMPUTATION 1.
1. GROSS-PAY = HOURS-WORKED × PAY-RATE
2. OVERTIME = HOURS-WORKED − 40
   GROSS-PAY = (40 × PAY-RATE) + (OVERTIME × (1.5 × PAY-RATE))

For more complex logical relationships, decision tables (discussed in the following section) should be used. Decision tables are assigned numbers and/or names so they can be referenced from the data-element dictionary as follows:

Validation Rules

For inputted, computed, or data elements looked up in tables, validation is often
required to reduce chances for errors. When known and possible, validation rules
should be defined. For example, for a data element inputted as DATE, the
validation might be as follows:

> VALIDATION RULE: DATE CANNOT BE LATER THAN
> TODAY'S DATE OR MORE THAN 30 DAYS
> EARLIER THAN TODAY'S DATE

For computing OVERTIME, validation rules might be as follows:

> VALIDATION RULE: OVERTIME COMPUTED IS NEVER TO BE
> LESS THAN ONE HOUR OR TO EXCEED 20 HOURS
> FOR ONE WEEK. IF SUCH IS THE CASE, INCLUDE
> TRANSACTION IN ERROR REPORT.

Where Used

A where-used list is created to keep track of each report computation and computer program in which a data element is used. Such a list is needed for control
purposes.

Consider a situation where the data element SALARY is printed on several
reports and displays available throughout the organization. If the organization
(or the government) decides distribution of this salary information should be
restricted, the where-used list pinpoints the reports and displays that should be
reviewed.

In another situation, the definition of a data element may be changed. For
example, assume the data element QUANTITY-ON-HAND is redefined to include back orders. In such a case, it is judicious to review all computations in
which QUANTITY-ON-HAND is used to check for unanticipated effects on
computations. Also, those persons receiving reports affected by this change can
be identified and alerted to the new procedure.

Often during the early and the latter stages of systems design, the specific programs in which a data element is used are unknown. This definition is completed by the programmer during software development.

Entities

This entry defines the entity or entities that the data element is an attribute of, which can be determined from the data model. Note that a data element can be an attribute for more than one entity. For example, in Figure 7.5, CUSTOMER # is an attribute of CUSTOMER, ORDER, ACCOUNT, COMPLAINT, AND CUSTOMER SURVEY. Technically, CUSTOMER # is only an attribute of a CUSTOMER. For all other entities it is a *relationship*. However, since in a relational database CUSTOMER # is physically stored in the ORDER, ACCOUNT, COMPLAINT, and CUSTOMER SURVEY record or table, it can be viewed as an attribute in those cases.

Storage

Storage defines if and where (i.e., in what physical file) a data element is stored. Not all data elements are stored in a file. For example, in processing a weekly payroll for hourly employees, GROSS-PAY is computed by multiplying PAY-RATE times HOURS-WORKED. Since GROSS-PAY is computed each week, it does not have to be stored in the payroll file. In contrast, PAY-RATE changes only when a raise is given. It is relatively stable from pay period to pay period. Consequently, PAY-RATE should be stored so that it is conveniently available for payroll processing. Storage for PAY-RATE is defined as follows:

Storage: Payroll Master File

Storage for GROSS-PAY and HOURS-WORKED is defined as follows:

Storage: Not Applicable (or N/A)

There is generally a one-to-one correspondence between entities and files, which helps in determining which files the data elements belong to (e.g., using Figure 7.5); however, this is not always the case. An entity is a logical structure that may be implemented as a subset of another entity in a physical file. For example, LINE-ITEMS may be treated as a repeating group within an order record.

During the early stages of data-element dictionary development, decisions as to whether certain data elements are to be stored and where they are to be stored may not have been made. In such cases, the storage definitions may remain blank until the decisions are made.

Maintenance

The maintenance information about a data element describes the procedures for maintaining or updating the element. The concept of documenting maintenance is best illustrated by examples:

SEX-CODE:	This is originated when the employee is first put on the payroll. No maintenance is required for this data element unless an error is detected.
BALANCE-DUE:	This is computed during the monthly billing cycle for each customer record in the accounts receivable file.
SALARY:	This is updated after each annual review unless a non-scheduled merit increase is granted.

When to Start the Data Dictionary

Following the creation of the conceptual data model is a good time to begin the data dictionary. Few of the final data elements will have been identified, and there will not be complete data-dictionary information for the data elements, but this is not a problem. The idea is to define as much as can be defined (i.e., make partial dictionary entries) for all data elements currently identified. Getting a head start on this documentation makes subsequent design and data-dictionary documentation easier to manage.

Logic Definitions

A key objective of systems design specifications is to avoid narrative descriptions of what a system is to perform. Narrative descriptions tend to be confusing, incomplete, and unreliable. Therefore, various structured techniques have been developed that allow complex logical relationships to be expressed in nonnarrative form. The primary use of these techniques is to describe

■ Input validation logic

■ Computation logic

■ Output processing and reporting logic

The main techniques used for logic definitions include decision tables, decisions trees, and structured English. It is important to understand that these are different techniques aimed at doing essentially the same thing. The theory behind them is that a structured, well-defined logical process will result in more accurate systems design specifications. Each technique accomplishes this in slightly different ways.

For the purposes of this book, more focus will be placed on decision tables as a structured technique for logic definitions. Once decision tables are understood, it is relatively easy to understand the other techniques and compare their relative value. Each technique is reviewed in the following pages.

Decision Tables

Decision tables are powerful, efficient tools for expressing complex logical relationships in an understandable manner. They may be used in conjunction with the data-element dictionary and output definitions as follows:

1. They may be used in conjunction with the "source" section of the data-element dictionary to define logic used for input validation or computations.

2. They may be used in conjunction with the "output selection logic" section of the output definition to define logic used for selecting records to be included in a report or a terminal display.

Terminology and Structure

A decision table is a matrix containing columns and rows that are used to define relationships. Figure 7.9 illustrates a decision-table format. Ideally, the format should be implemented on a screen, using CASE technology.

There are three major components in a decision table: conditions, courses of action, and decision rules. *Conditions* are events or facts that determine the courses of action to be taken. *Courses of action* are processes or operations to be performed under certain conditions. *Decision rules* express the relationships between combinations of conditions and courses of action.

Decision rules are expressed by making condition entries and course-of-action entries in the matrix provided to the right of condition statements and course-of-action statements (see Figure 7.9). Possible decision-rule entries and their definitions follow:

1. *Condition Entries*
   Y = Yes, condition statement must apply
   N = No, condition statement must not apply
   – = Indifferent, condition statement is irrelevant to the decision and does not have to be checked
2. *Course of Action Entries*
   X = Activate course of action
   • = Do not activate course of action

Decision-Table Example

Figure 7.10 shows a decision table that defines customer billing for the output selection logic of an output definition. Note that a double horizontal line is used to separate conditions from courses of action.

If a customer's balance is less than or equal to zero (decison rule 1, read vertically), no additional conditions have to be checked. A statement is not to be printed for the customer. If a customer's balance is greater than zero, but less than his or her credit limit (decision rule 2), no additional conditions have to be checked. A statement is to be printed for the customer. If the customer's balance exceeds his or her credit limit, then the credit rating is checked. If the credit rating is excellent (decision rule 3), a statement is to be printed for the customer. If the credit rating is not excellent (decision rule 4), a statement is to be printed and a credit warning message is to be added to it.

Constructing Decision Tables

The approach used to construct decision tables varies among systems analysts. The technique discussed below is based on the *progressive-rule development* approach advocated by Keith London. This technique is easy to understand, and it keeps the logic of the problem in sight at all times. The formal rules for constructing a decision table using progressive-rule development are

Figure 7.9 **Format for a Decision Table**

DECISION TABLE FOR _____ APPLICATION
DECISION TABLE NAME _____ REFERENCE NO. _____
PREPARED BY _____ DATE _____

DECISION RULES

CONDITIONS/ COURSES OF ACTION	01	02	03	04	05	06	07	08	09	10	11	12	13	14	15	16	17	18	19	20	21	22	23	24	25	26	27	28	29	30	31	32	33	34	35	36	37	38	39	40	41	42	43	44	45	46	47	48	49	50

Figure 7.10   Completed Decision Table

CONDITIONS/ COURSES OF ACTION	DECISION RULES										
	1	2	3	4	5	6	7	8	9	10	11
BALANCE-DUE $\leq$ 0?	Y	N	N	N							
BALANCE-DUE > 0 AND $\leq$ CREDIT-LIMIT?	–	Y	N	N							
BALANCE-DUE > CREDIT LIMIT	–	–	Y	Y							
CREDIT RATING EXCELLENT?	–	–	Y	N							
DO NOT PRINT STATEMENT	X	•	•	•							
PRINT STATEMENT	•	X	X	X							
ADD CREDIT WARNING TO STATEMENT	•	•	•	X							

1. List conditions and courses of action in the sequence in which they are to be considered.

2. Consider a condition (relevant to the decision) in the positive case (Yes).

3. Consider, in the positive case, all other conditions (relevant to the decision) that must be considered before a course of action can be taken.

4. Use the indifference code ("—") for any conditions that are irrelevant to the decison rule under consideration.

5. Enter the conditions, courses of action, and their respective decision rule entries in the decision table.

6. Start a new decision rule by negating the last positive condition in the current decision rule (leave all other entries the same).

7. Repeat from step 3 onward until the table is complete.

The decision rules in Figure 7.10 were determined using progressive-rule development. Review the rules in Figure 7.10 (sequentially, 1 through 7 in the context of progressive-rule development).

Two additional examples of decision tables are provided in Figures 7.11 and 7.12. Note the use of progressive-rule development in both examples.

Figure 7.12, a decision table for airline reservations, introduces a new concept. The condition "First class available?" is stated as the third and sixth conditions. This repetition is required to adhere to the rule of listing entries in the sequence in which they are considered. In decision rules 5 through 8, the passenger is requesting economy class. Therefore, there is indifference about first-class availability until it is determined that no economy seats are available and an alternative is acceptable. At that point, the next condition to be considered is the availability of a first-class seat.

The repetition of condition statements or course-of-action statements is occasionally necessary to ensure clarity when using progressive-rule development. This type of presentation is helpful to computer programmers because the logic can be translated conveniently into program code in the same sequence as it is presented in the decision table.

Figure 7.11   Decision Table for Payroll Calculations

CONDITIONS/ COURSES OF ACTION	DECISION RULES									
	1	2	3	4	5	6	7	8	9	10
Salaried?	Y	Y	N	N	N					
Hourly?	–	–	Y	Y	Y					
Hours worked < 40?	Y	N	Y	N	N					
Hours worked-40?	–	–	–	Y	N					
Hours worked > 40?	–	–	–	–	Y					
Pay base salary	X	X	•	•	•					
Calculate hourly wage	•	•	X	X	X					
Calculate overtime	•	•	•	•	X					
Produce Absence Report	X	•	X	•	•					

Figure 7.12   Decision Table for Airline Reservations

CONDITIONS/ COURSES OF ACTION	DECISION RULES							
	1	2	3	4	5	6	7	8
First-class request?	Y	Y	Y	Y	N	N	N	N
Economy-class request?	–	–	–	–	Y	Y	Y	Y
First-class available?	Y	N	N	N	–	–	–	–
Economy available?	–	Y	Y	N	Y	N	N	N
Alternative acceptable?	–	Y	N	–	–	Y	Y	N
First-class available?	–	–	–	–	–	Y	N	–
Reduce first-class available	X	•	•	•	•	X	•	•
Issue first-class ticket	X	•	•	•	•	X	•	•
Reduce economy available	•,	X	•	•	X	•	•	•
Issue economy ticket	•	X	•	•	X	•	•	•
Refer to alternate flight	•	•	X	X	•	•	X	X

Y = YES

N = NO

– = Indifference

X = Activate course of action

• = DO NOT activate course of action

Decision-Table Identification

Each decision table should be identified for documentation purposes, preferably in an area provided on the decision table. The minimum points to be documented are illustrated by the following example:

Name of Application or System: Payroll System
Decision Table Name: Pay Calculations
Reference Number: 112
Prepared by: Jill Smith
Date: 5–30–88

Decision Trees and Structured English

Decision trees offer an alternative to structured English or decision tables as a means of defining complex relationships. Figure 7.13 illustrates a decision tree that is functionally equivalent to the decision table in Figure 7–10.

Structured English, or *pseudocode*, is a structured way to describe a process and can usually be translated directly into program code; it resembles COBOL. An example follows:

```
If hours worked less than 40
    Produce absence report
    Compute wages by multiplying hours worked times hourly rate
Else
    Compute wages by multiplying hourly rates times 40, multiplying
    1.5 hourly rate times hours worked over 40, and adding the two
    amounts
Else
    If hours worked less than 40
        Produce absence report
        Pay salary
    Else
        Pay salary
```

Figure 7.13   An Example of a Decision Tree Based upon the Decision Table in Figure 7.10

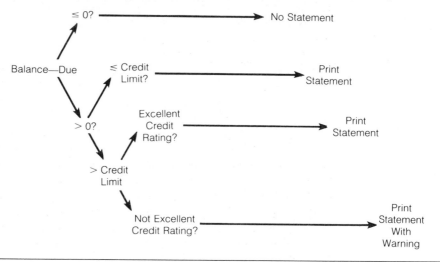

Pros and Cons of Different Logic Definition Techniques

Having reviewed the various logic definition techniques, we can discuss their relative advantages and disadvantages. Structured English has the obvious advantage of being very easily converted to programming language, as it resembles procedural languages such as COBOL, PL/1, and FORTRAN. Its disadvantage is that it is not very easily read by someone who is not a computer programmer and is even more difficult for a nonprogrammer to generate. Also the use of CASE code generators reduces the need for preparing structured English.

Decision trees, on the other hand, are easy for anyone to develop; yet they are more difficult to read than decision tables. Decision tables, can be easily read by programmer and nonprogrammer alike. Programmers can easily convert decision tables directly into program code if necessary. Decision tables, however, are more difficult to develop than are decision trees.

In practice, most designers find decision trees a useful way to conceptualize a problem and the various alternatives to be considered for a set of logic. But since decision trees are more difficult to read, most designers find it a good idea to use a decision tree only as a means of better understanding the problem. Then they will use a decision table to express it in a way that is understandable to users or nonprogramming (as well as programming) personnel. Specifically, constructing a decision tree helps to prepare for step one of progressive-rule development, as used to construct a decision table. The CASE products provide useful support for constructing and documenting logic definitions.

Detail Design Phase

Once the conceptual data model is complete and all entries possible are made in the data dictionary, the detail phases of information requirements and data structuring can begin (review Figure 7.2).

Detail Information Requirements

Detail information requirements involve generating exact specifications for all formal required reports and developing a good enough understanding of ad hoc informal information requests to make sure that the data structure contains the necessary entities and data elements to respond to inquiries.

Formal Reporting Specifications

Examples of a few of the many formal outputs that might ultimately be designed for a system are illustrated in Figures 7.14a through 7.14d. These outputs are based upon the order-processing system for which the conceptual data model was developed earlier in the chapter.

The "Out-of-Stock/Below-Minimum" terminal display (Figure 7.14a) is to notify management of stock-out items. This information not only alerts inventory management to the urgency of obtaining more stock from vendors, it also alerts sales personnel to use caution in delivery commitments for those items. The invoice (Figure 7.14b) is used to report totals and bill customers. The picking list (Figure 7.14c) is used in the warehouse to select items to package for shipment. The packing slip (Figure 7.14d) is used as an insert into a shipping container so that a customer can crosscheck shipment contents when they arrive.

Figure 7.14a   Out-of-Stock/Below-Minimum Report

Date 12-06-88			OUT-OF-STOCK/ BELOW-MINIMUM REPORT				PAGE 1
VENDOR NUMBER	ITEM NUMBER	ITEM DESCRIPTION	STOCK CONDITION	ON HAND	REORDER POINT	QUAN ON ORD	DATE ORDERED
146894	67894220	HAIR DRYER	OUT	∗ ∗ ∗ ∗	60	250	12-02-84
792267	68116772	ELECTRIC SHAVER	BELOW MIN	20	40	100	12-03-84
406843	46214558	GUITAR	OUT	∗ ∗ ∗ ∗	50	175	12-01-84
787743	22116600	LAMP	BELOW MIN	10	40		12-04-84
432687	98987146	CHESS SET	BELOW MIN	42	100	300	12-01-84

Ad Hoc Reporting Ability

Ad hoc inquiries to a database almost exclusively use fourth-generation languages (4GLs) and the relational data model. Examples of popular 4GLs are FOCUS, RAMIS, NOMAD, INFO, NATURAL, DB II, AND SQL. The relational data model logically views data as a set of tables in which column headings describe the data attributes and the rows describe the individual occurrences or values of an entity (record).

For example, Figure 7.15 shows a relation for CUSTOMER in a table. The column headings—CUSTOMER #, CUSTOMER NAME, SHIPPING ZONE, CUSTOMER IMPORTANCE—and other items are the data elements in the table (note that each column heading would have a corresponding entry in the data dictionary). Each row contains the data item values for one customer. The "#" after CUSTOMER indicates that CUSTOMER # is the key or unique identification of the relation. That is, customers are identified by CUSTOMER #.

Relations such as those in the CUSTOMER table are constructed from the data model (Figure 7.5) using the following rules:

1. One relation (or table) is created for each entity.

2. Every attribute creates a column (data item) in an entity.

3. All attributes must be associated with the correct entity.

4. Whenever a one-to-many relation is connected to an entity, the key of the source of the "chicken-foot" symbol must be a column in the "many side"

Figure 7.14b   Invoice Format

INTERMOUNTAIN DISTRIBUTING, INC.

DATE 12-06-88
PAGE 01

SOLD TO

THE TOY STORE, INC.
300 BROWN ST.
P.O. BOX 8275
DENVER, COLORADO 76142

SHIP TO

THE TOY STORE
LORETTO PLAZA
415 MAIN ST.
PORTLAND, OREGON 68172

ORDER NO.   123456
INVOICE NO.   123456
CUSTOMER NO.   123456

ITEM NUMBER	DESCRIPTION	U/M	ORDERED	QUANTITY SHIPPED	B/O	RETAIL PRICE	NET PRICE	EXTENDED PRICE
68246789	JET ROCKET	DZ	12	12	0	12.95	6.95	83.40
42687420	BIONIC MAN	EA	10	5	5	8.50	4.40	22.00
68714321	SKATE BOARD	BX	24	0	24	10.25	8.75	.00
77882169	DOLL HOUSE	EA	1	1	0	22.95	11.45	11.45
30002468	GUITAR	EA	6	6	0	165.00	95.50	510.00

TAX	SHIPPING	QUANTITY DISCOUNT	INVOICE TOTAL	RETAIL TOTAL	NET TOTAL
91.33	53.20	275.60	1,918.00	3,848.00	2,049.07

of the relation. For example, CUSTOMER has a one-to-many relation to orders. Therefore, CUSTOMER # must be included as a column in the ORDER table. This is necessary to identify which orders belong to which customers.

The relations, or table, generated from the preceding procedure can be manipulated for ad hoc information retrieval using powerful nonprocedural operators available in 4GLs. The most commonly used generic operators are SELECT, PROJECT, and JOIN. Applying an operator generates a new relation (though it is not usually stored as such) that can also be manipulated further by relational operators.

The SELECT operator is used to identify which records within a single table are to be retrieved based upon data items and their values. For example, a user might want to retrieve all CUSTOMERS who are located in SHIPPING-ZONE 001, or all in SHIPPING-ZONE 001 who have CUSTOMER-IMPORTANCE equal to 1. The PROJECT operator identifies which data items are to be included in the new table. A sample query might be to SELECT CUSTOMERS with

Figure 7.14c   Picking List

DATE 12-08-88   PICKING LIST   PAGE 01

CUSTOMER NO.	CUSTOMER NAME	ORDER NO.
123456	THE HANDY HUT	123456

BIN LOCATION	ITEM NUMBER	ITEM DESCRIPTION	U/M	SHIP	WEIGHT
A 01-13	46872431	LIPSTICK	BX	5	1.2
A 01-17	61127421	HAIR SPRAY	BX	10	4.0
A 03-04	77864230	HAIR DRYER	EA	20	24.6
B 07-06	46789106	SIDE MOLDING	FT	100	.5
.	.	.	.	.	.
.				.	

TOTAL WEIGHT   121.6

Figure 7.14d   Packing Slip

DATE 12-08-88   PACKING SLIP   PAGE 01

CUSTOMER NO. 123456

SOLD TO

THE HANDY HUT
CENTER BUILDING
P.O. BOX 8172
ATLANTA, GEORGIA 71682

SHIP TO

THE HANDY HUT
CENTER BUILDING
P.O. BOX 8172
ATLANTA, GEORGIA 71682

ORDER NO.   123456
ORDER DATE   12-01-88

ITEM NUMBER	ITEM DESCRIPTION	U/M	SHIPPED	WEIGHT	SUGGESTED RETAIL
46872431	LIPSTICK	BX	5	1.2	1.29
61127421	HAIR SPRAY	BX	10	4.0	1.95
77864230	HAIR DRYER	EA	20	24.6	12.75
46789106	SIDE MOLDING	FT	100	.5	8.95

TOTAL WEIGHT 121.6

Figure 7.15   Customer Relationships

CUSTOMER#	CUSTOMER NAME	SHIPPING ZONE	CUSTOMER IMPORTANCE	OTHER ITEMS
132467	HANDY HUT	001	1	
144789	SHOPPER'S PARADISE	005	1	
149267	DISCOUNT HOUSE	001	2	
184324	KID'S KORNER	020	1	
188765	TEEN'S DREAM	015	3	
198146	PARTY PLACE	001	1	
211472	BEST PRICE	020	3	
223389	DIME STORE	015	2	

SHIPPING-ZONE = 1, CUSTOMER IMPORTANCE = 1, and PROJECT CUSTOMER # and CUSTOMER NAME. Using dBASE III, Ashton-Tate's relational DBMS for microcomputers, the query could be stated

---

List CUSTOMER #, CUSTOMER-NAME for SHIPPING-ZONE = 1 and CUSTOMER IMPORTANCE = 1

(In dBASE III, *list* corresponds to SELECT and *for* corresponds to PROJECT.)

Based upon the data in Figure 7.15, the new relation created by this command would consist of

CUSTOMER #	CUSTOMER NAME
132467	HANDY HUT
198146	PARTY PLACE

This new relation, or table, could be used to create a screen display.

The JOIN operator operates on two tables simultaneously to produce a third table, which combines all entries from the two original tables that satisfy the JOIN condition.

For example, in Figure 7.15, a display containing both CUSTOMER and ORDER information cannot be produced unless the CUSTOMER and ORDER tables are joined. The JOIN condition is that the CUSTOMER # must be the same in both the CUSTOMER and ORDER records. Whenever that condition is met for a pair of records (one from each table), an entry is created in the new table. In dBASE III such a query could be stated as follows:

---

join CUSTOMER with ORDER to TEMP-TABLE for
CUSTOMER# = O-CUST#

As can be seen by these examples, through the use of 4GLs, a properly structured database that contains all the necessary data items, users are in a good position to fulfill their information requirements.

An important concept for creating relational data structure is *normalization* (March, Carlis, and Flory),[1] a technique based upon relational calculus that can be used to ensure the integrity of the database. The concept is quite theoretical and difficult to understand. Fortunately, the systems analyst does not need to fully understand it. By following the data modeling rules presented here, the resulting data model is normalized.

Prototyping and Heuristic Design

So how do we make sure we determine the detail data structure and data items to meet formal and ad hoc information requirements? In Chapter 6, it was pointed out that during information requirements five mistakes were commonly made:

1. Assuming managers know their information requirements
2. Failing to ask the right people
3. Asking them one at a time instead of as a group
4. Asking the wrong questions
5. Not allowing users to refine their requirements through trial and error.

The first four steps have been addressed by techniques covered in Chapter 6 (i.e., JAD and the structural interview). Mistake number 5 is addressed here.

Trial and error (or exponential learning) is an important part of problem solving. For example, people are using trial and error when they

1. Try on clothes before they purchase them
2. Test-drive cars before making a selection
3. Change their college major after a few courses
4. Have several relationships before marriage
5. Rearrange furniture several times when decorating a room.

Trial and error is also a part of determining detail information requirements. Proper analysis prior to the trial-and-error process can substantially reduce the amount of trial-and-error time expended to resolve a solution. For example, in the five preceding situations, the following analysis prior to trial and error could save time:

1. A fashion consultant could narrow the trial-and-error search for a new wardrobe by asking questions about career, lifestyle, budget, and taste, and by observing physical characteristics. Stores and designs could be suggested based upon the answers to these questions, thereby saving search time.
2. A well-trained car salesperson could ask qualifying questions (similar to those of the fashion designer) to better suggest automotive alternatives.

---

1. March, Salvatore T., Carlis, John V., and Flory, André. "Relational Data Base Design and Use for Nontechnical Users." *Managing the Information Center Resource Series*, 93-00-65. Boston: Auerbach Publishers, 1986.

3. A career counselor, based upon interviewing, could suggest majors for students.

4. A marriage counselor could use personality and interest profiles to assist in determining the compatibility of potential marriage partners.

5. Interior decorators could use their analysis techniques to approach a decorating solution that could be refined by trial and error.

As in these examples, the structured interview (Chapter 6) combined with the analytical tools of data modeling provide a good approximation of the system. The approximation reduces the amount of trial-and-error effort necessary to determine detail information requirements. The details can be determined through prototyping or heuristic design.

Origin of Prototyping and Heuristic Design

Information systems technology is now flexible enough to follow more advanced techniques for information requirements determination, using the heuristic design or prototyping. On-line technology makes user interaction with the new system more practical than it would be with batch-produced reports. The ease of use of information-retrieval (query) languages has improved dramatically and is continuing to improve. Database management systems—especially relational-model types—allow significant improvement in data manipulation. More directly in support of management decision making, there is a proliferation of application-oriented models and easy-to-use modeling languages for decision-support systems (discussed in Chapter 14). Although these technological advances do increase support for management decision making, they impose overhead on computer systems; the continuous improvement in hardware cost-performance, however, has reduced the cost of processing, thereby making the use of the additional software practical.

Two labels have emerged for similar but slightly different approaches to systems development that exploit advanced technologies. Heuristic and prototype development represent major departures from traditional systems development by allowing trial and error as part of design. Other labels given to the process are *evolutionary* and *iterative* development.

Prototyping and heuristic-design techniques emerged in the late 1970s as a solution to the problem of major revisions being required by systems after they were allegedly complete. Research on systems that had been developed in organizations revealed that revision costs were staggering, often exceeding original development cost several times. It became clear that it was better to have the trial-and-error activity before rather than after the final system was built.

The notion of prototyping comes from manufacturing, where a prototype product (e.g., a car) is built in a shop (not a factory) and a cycle of testing and modification is repeated several times until it appears the final design is complete. Then the factory can be built with reasonable assurance that an acceptable product has been designed. The idea is avoid building a high-cost factory until the design is right.

The notion of heuristic design comes from the recognition that determining requirements for most things generally involves a lot of trial-and-error (*heuristic*) effort. Though the original procedures for prototyping and heuristic design differed slightly, the two concepts have merged in recent years. A basic model of prototyping or heuristic design is provided in Figure 7.16. As stated at the beginning of the chapter, systems design is *output*-driven. Accordingly, the initial prototype should focus on the output abilities of the system and the data structure

Figure 7.16   **Model of Prototyping or Heuristic Design**

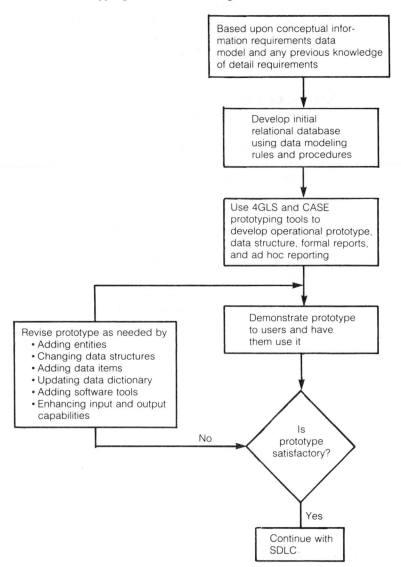

necessary to support them. There is no need during the early stages of the prototype to become preoccupied with the detail specifications or sources of system inputs. Until data structure and data content issues are resolved, concern with the details of inputs is premature.

CASE Tools and Prototyping

Besides the 4GLs and relational DBMSs, CASE products have some powerful tools to facilitate the prototyping process. These tools include menu generators, screen generators, report generators, and code generators.

Menu Generators

Menu generators outline the eventual functions or components of systems (e.g., place orders, check on status of order, cancel order). When linked to other menus, they illustrate how users can branch to the various screen and subscreens to be used for data entry or inquiry. Figure 7.17 illustrates a menu for an order-processing system.

Screen Generators

Screen generators for printed reports are used to format or "paint" the desired layouts and content of a screen without having to center complex formatting information. Using the data dictionary, the screen generator retrieves output formats for data items for editing, and the data item can be easily moved to various screen locations.

Report Generators

Report generators are similar to screen generators in that formats are generated in the same fashion. In addition, they can indicate totals, paging, print edits, sequencing, and control breaks in creating samples of the system reports.

Figure 7.17   Menu for Order-Processing System

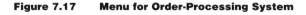

ORDER-PROCESSING SYSTEM
MAIN MENU

1. CUSTOMER PROFILE
2. ORDERS
3. SHIPMENTS
4. INVENTORY
5. VENDORS
6. VENDOR ORDERS
7. COMPETITORS

Code Generators

Code generators allow the analyst to generate modular units of source code from high-level specifications, demonstrate the system, revise specifications, and demonstrate the system again. Code generators play an even bigger role in systems development and are discussed in greater detail in Chapter 9.

Designing the Input System

As users become satisfied with the detail output abilities, the detail structure and content issues are resolved and the input design needs can be completed. Due to the on-line interactive nature of the system, the majority of inputs use the same format as many outputs. That is, there will generally be an output requirement to look at all data about an individual record (i.e., customer, order, inventory item). The format of that output can often serve as the format for inputting those data elements that are to be inputted. For example, consider the customer record display illustrated in Figure 7.18. The data items on the left-hand side of the

Figure 7.18   Customer Record Display

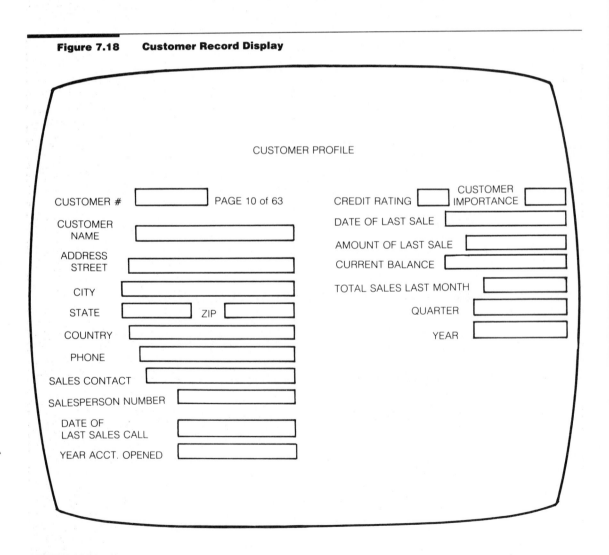

screen would likely be inputted directly for a new record or modified for an existing record. Whether or not data elements are inputted (as opposed to computed) can be confirmed by checking the SOURCE entry in the data dictionary.

The items on the right-hand side of the screen would likely be computed or generated directly from order transaction (verifiable from the data dictionary). For example, credit rating might be based upon evaluation of accounts receivable payment data, whereas customer importance might be based upon certain sales volumes. However, rather than being computed or derived from other data, either could be inputted based upon a judgment made outside the computer. Again, these are issues documented in the data dictionary and approved by users.

At any rate, since the screen format has already been developed for output purposes, it is just as well to use it for input and update purposes. By employing protection features in the system, only certain users should be able to perform inquiry, input, or update functions. For example, salespeople should not be able to change accounts receivable data.

To create a record, a blank screen is provided for the user, as portrayed in Figure 7.18. The "blanks" are filled in and will be (when all software is complete) edited and validated according to the data-dictionary specifications. Data elements such as TOTAL-SALES-LAST-MONTH, QUARTER, AND YEAR are not inputted, but they will be propagated by normal processing as the order-processing transactions occur.

On occasion special input formats will need to be created. For example, in the order-processing system example, the decision was made to let customers place orders directly by using the on-line system. In that case, a customer display might be developed for customer use only. Of course, special care would have to be taken to ensure that customers could not access information other than that appropriate for processing their orders.

Besides on-line inputs, some inputs might come in as completed forms, optical scan sheets, magnetic ink character recognition (MICR, as used on checks), transmitted via data communications from other computers, or received electronically (e.g., on floppy disk). In those cases the format of the document or media would be defined simply by indicating where the data elements were located on the document or media. Remember, data elements and their input formats are defined in the data dictionary.

Documenting Input, Output Functions, and Access

As a design and control consideration, it is important to document the input and output function of a system and which users can access which functions or get copies of which reports. A simple procedure for doing this is to construct matrices as illustrated in Figures 7.19 and 7.20. The matrix in Figure 7.19 defines which users can access what functions from the system menu. For example, based upon Figure 7.19, customers are allowed to view information about their profile, account, and invoices. They can view and update their complaints and customer survey information. They cannot, however, make ad hoc inquiries about other customers. All departments can look at customer complaints, but only customers and sales can place a customer complaint.

Figure 7.19 is conceptual; more detail is needed about exactly which data elements can be retrieved and updated by which user. For each output, letting customers access and update *all* customer profile information may not be desir-

Figure 7.19   **Output-by-User: Matrix**

	Customer						Order					Inventory					Vendors				Shipments			
	Cust. Profile	Cust. Account	Cust. Complaint	Cust. Survey	Invoice	Ad hoc	Order Record	Picking Slip	Packing Slip	Order Summary	Ad hoc	Item Record	Item Analysis	Item by Customer	Below Minimum	Ad hoc	Vendor Record	Item by Vendor	Vendor by Item	Ad hoc	Shipment Record	Shipment Schedule	Order by Shipment	Ad hoc
Customers*	/	/	X	X	/		X			/											/			
Sales	X	/	X	/	/	/	X		/	/		/	/	/	/	/					/	/	/	/
Credit	/	X	/	/	/	/	/							/										
Order Dept.	/	/	/	/	X	/	/			/		X	/		/	/	X	/	/	/		/	/	/
Shipping	/	/	/	/	/	/	/	/	/	/		/									X	X	/	/

*Customers can view only their personal records
/ = View only
X = View and Update functions

Figure 7.20   **Data-Element-by-User Matrix**

	Customer #	Customer Name	Address	City	State	Zip-Code	Phone	Sales Contact	Date of Last Sales Call	Year Acct Opened	Credit Rating	Customer Importance	Date of Last Sale	Amt of Last Sale	Current Balance	Total Sales	Month	Quarter	Year
Customer	/	X	X	X	X	X	X	/											
Sales	X	X	X	X	X	X	/	X	X	/	/	X			/		/	/	/
Credit	/		/	/	/	/	/	/	/	⊗	⊗	/			⊗		/	/	/
Order Dept.	/		/	/	/	/	/	/				/	⊗	⊗			⊗	⊗	⊗
Shipping	/		/	/	/	/	/	/					/	/					

/ = View Privelage Only
X = View and Update and Access Privelage
○ = Automatically updated by Routine Processing but can be corrected by authorized department (designaed by ⊗)

able. For example, Figure 7.20 provides that detail for the customer record. Note that customers can access and update name, address, city, and so forth; they can only access (not update) customer # and sales contact. Information such as credit rating is not available at all. A matrix such as Figure 7.20 would be needed for every output, such as those listed in Figure 7.19.

These matrices should be developed in conjunction with users of the system. Once completed, they provide the documentation needed for detail software development later in the SDLC.

Graphics to Document Overall System

The old saying "Can't see the forest for the trees" is applicable to all the design documentation generated during systems analysis and design. To complete system design and allow the forest to be seen, a straightforward, graphical representation of how the overall system is put together is needed. These graphics are useful for documentation and for presentations to management.

There are different ways to portray systems. Some involve more detail than others. Keeping in mind that the purpose of a system is to support decision making of design centers, a good basic graphical representation is to portray the outputs and function menus available to each decision center. This can be done by translating the outputs listed in the matrix of Figure 7.19 into menus for each decision center. For example, the ordering department might have a menu including

```
ORDER-PROCESSING SYSTEM
1 CUSTOMER
2 ORDER
3 INVENTORY
4 VENDORS
5 SHIPMENTS
```

If the order department selects CUSTOMER, a submenu (also derived from Figure 7.19) would appear displaying:

```
CUSTOMER INFORMATION
1 PROFILE
2 ACCOUNT
3 COMPLAINTS
4 SURVEY
5 INVOICE
6 AD HOC
```

If the ordering department selected CUSTOMER, the system would request CUSTOMER #. The system would then retrieve the profile data for the requested customer (by CUSTOMER #) and display it using a format similar to that of Figure 7.18.

The most conceptual graphic of the new system could then be illustrated as depicted in Figure 7.21. Expanded graphics can be made to provide detail on outputs for each department, as illustrated in Figure 7.22. Graphics for each individual screen display (as in Figure 7.18) can also be included in management/ user presentations. Some on-line outputs also have physical documents. For example, invoices, packing slips, and picking slips are paper documents, as well as

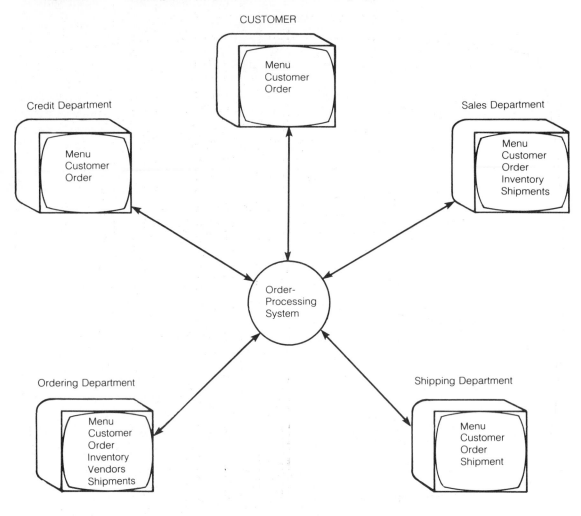

Figure 7.21   Conceptual Graphic of Order-Processing System

CUSTOMER

Menu
Customer
Order

Credit Department

Menu
Customer
Order

Sales Department

Menu
Customer
Order
Inventory
Shipments

Order-
Processing
System

Ordering Department

Menu
Customer
Order
Inventory
Vendors
Shipments

Shipping Department

Menu
Customer
Order
Shipment

screen displays. In the case of the shipping department, these outputs can simply be documented as shown in Figure 7.23.

A more detailed graphic representation of the new design can be provided by creating a DFD for the new system, as described and illustrated in Chapter 6.

System Encyclopedia

Keeping track of all the presentation graphics, data elements, input and output screens, DFDs, logic definitions, and program code (generated from prototyping) is a huge task. To address the problem, many CASE products provide central information repositories, referred to as the *system encyclopedia*. The core of the system encyclopedia is the data dictionary, which links most documentation. Figure 7.24 portrays how the data dictionary can be used to keep track of documentation.

Figure 7.22   Illustration of an Expanded Menu

SALES DEPARTMENT EXPANDED MENU

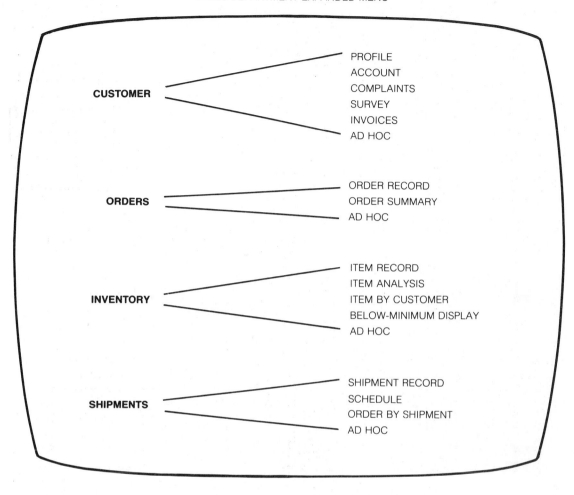

Figure 7.23   Designation of Physical Documents on Screen Displays

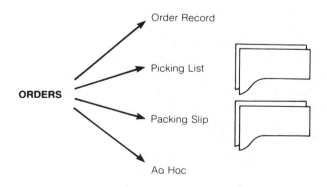

Figure 7.24   Linking Design Specifications Using Unique Data Names

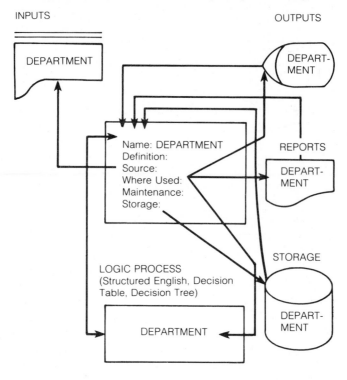

Figure 7.25   System Encyclopedia for Documentation Management

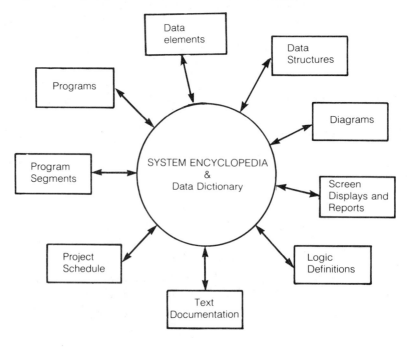

The system encyclopedia expands the functions provided by the data dictionary. It also stores information on

- Screen and report layouts
- Programs
- Special instructions
- Documentation
- Diagrams (flowcharts, data-flow diagrams, data structures)
- Project schedules

Figure 7.25 shows how the encyclopedia coordinates all system documentation. For each item in the encyclopedia, documentation is kept as well as information on who prepared it, the date it was prepared, and other useful instructions.

Summary

Systems design involves both the inclusion and structuring of information for an information system. The deliverables from the design process include output definitions, logical data structure, data dictionaries, logic definitions, input decisions, and presentation graphics.

There are conceptual and detail stages for determining and structuring information requirements. The conceptual stage of information requirements draws from structural interviews, data-flow diagrams (or flowcharts) and previous documentation to determine entities and attributes for the new system. Data modeling is used to construct a conceptual data structure using entity/attribute analysis.

During the detail phase of information requirements determination, the conceptual data model is converted into a working prototype using CASE technology, fourth-generation languages, and relational databases. The prototype is repeatedly demonstrated to users and is revised until formal reports, ad hoc inquiry, and detail data structures are derived.

Presentation graphics are developed for the system. These graphics range from conceptual to detail.

All design documentation for the system is managed through a system encyclopedia, which is an extension of the data dictionary.

Exercises

1. Show how to define a six-character, numeric data field that is to have a decimal point between the hundreds' and tens' positions. How many positions on an output definition does this field require? How would the field appear with a field value of 30? 488? 5069?

2. Construct an output definition field format for the following inventory part codes: **AE91609, CG41113,** and **FR60021.**

3. How is an item in an output definition linked to the data-element dictionary?

4. Create an input definition format for the following input transaction:
   The first field is a nine-digit account number.
   The second field is a fifteen-character field for customer name.
   The third field is a six-digit date field in the format of month, day, and year.
   The fourth field is a two-digit transaction code.
   The fifth field is a six-digit field called **AMOUNT**.

5. Where does a systems analyst define the computations and/or the reports in which a data element is used?

6. Why is it essential to assign unique names to data elements and always use their correct spellings?

7. What are the three possible sources of data elements in an information system?

8. Why is it important to document where data elements are used?

9. What determines whether a data element should be stored in a file?

10. Use decision trees and decision tables to illustrate the logic for the following situations:
    a. ABC Inc. must deduct city tax from employee wages, depending upon where they live and work as follows:
       (1) Employees living in the city must pay the tax.
       (2) Employees living outside the city and working within the city must pay the tax.
       (3) Employees living and working outside the city do not pay the tax.
    b. Salespersons for XYZ Distributing have the possibility of receiving a monthly bonus check. To receive a bonus, a salesperson must achieve the following goals:
       (1) He or she must have sold in excess of ten units or sold orders to a minimum of two new customers.
       (2) He or she must have installed three units or sold an additional five units or sold an additional unit to a new customer.

Selected References

Ackoff, R. L. "Management Misinformation Systems." *Management Science*, December 1967, pp. 147–56.

Alavi, M. "The Evolution of Information Systems Development Approach: Some Field Observations." *Data Base*, 15(3), Spring 1984a, pp. 19–24.

Alavi, M. "An Assessment of the Prototyping Approach to Information Systems Development." *Communications of the ACM*, 27(6), June 1984b, pp. 556–63.

Andrews, W. "Prototyping Information Systems." *Journal of Systems Management*, September 1983, pp. 16–18.

Athey, T. H. "Information Gathering Techniques." *Journal of Systems Management*, January 1980, pp. 11–14.

Bariff, M. L. "Information Requirements Analysis: A Methodological View." Working paper #76–08–02, The Wharton School, January 1977.

Bariff, M. L. and Lusk, E. J. "Cognitive and Personality Tests for the Design of Management Information Systems." *Management Science*, April 1977, pp. 820–29.

Benbasat, I. and Dexter, A. S. "Value and Events Approaches to Accounting: An Experimental Evaluation." *Accounting Review*, October 1977, pp. 735–49.

Berrisford, T. and Wetherbe, J. "Heuristic Development: A Redesign of Systems Design." *MIS Quarterly*, March 1979, pp. 11–19.

Blaylock, B. K. and Rees, L. P. "Cognitive Style and the Usefulness of Information." *Decision Sciences*, 15, 1984, pp. 74–91.

Bostrom, R. P. "Development of Computer-Based Information Systems: A Communications Perspective." *Computer Personnel*, 9(4), August 1984, pp. 17–25.

Burns, R. N. and Dennis, A. R. "Selecting the Appropriate Application Development Methodology." *Data Base,* Fall 1985, 17(1), 19–23.

Carey, T. T. and Mason, R. E. A. "Information Systems Prototyping: Techniques, Tool and Methodologies." *INFOR*, 21(3), August 1983, pp. 177–91.

Cerveny, R. P., Garrity, E. J. and Sanders, G. L. "The Application of Prototyping to Systems Development: A Rationale and Model." Working paper 649, School of Management, State University of New York at Buffalo, October 1985.

Chandler, E. W. and Nador, P. "A Technique for Identification of User's Information Requirements: The First Step in Information Systems Design." *Information Processing 17*, North-Holland Publishing Co., 1972.

Davis, G. B. "Strategies for Information Requirements Determination." *IBM Systems Journal*, 21(1), 1982, pp. 4–30.

DeMarco, Tom. *Structured Analysis and Design Specification*. New York: Yourdon, 1978.

Earl, M. J. "Prototype Systems for Accounting, Information and Control." *Accounting, Organizations, and Society*, 3(2), 1978, pp. 161–70.

Gane, Chris and Trish, Sarson. *Structured Systems Analysis: Tools and Techniques*. Englewood Cliffs, N.J.: Prentice-Hall, 1979.

Grudnitski, G. "A Methodology for Eliciting Information Relevant to Decision Makers," Proceedings of the First International Conference on Information Systems, 1980.

Henderson, J. C. and Nutt, P. C. "On the Design of Planning Information Systems." *Academy of Management Review*, 3(4), October 1978, pp. 774–85.

Holland Systems Corporation. *Strategic Systems Planning*. Ann Arbor, Michigan, document # MO154-048611986.

Howard, R. A. "An Assessment of Decision Analysis." *Operations Research*, 28(1), January–February, 1980, pp. 4–27.

Huber, G. P. "Cognitive Style as a Basis for MIS and DSS Design: Much Ado About Nothing." *Management Science*, May 1983, pp. 567–79.

Hudson, M. H. "Determining Organizational Information Requirements." *Journal of Systems Management*, pp. 6–10.

IBM Corporation. *Business Systems Planning—Information Systems Planning Guide*, Publication # GE20-0527-4, 1975.

Janson, M. A. and Smith, L. D. "Prototyping for Systems Development: A Critical Appraisal." *MIS Quarterly*, December 1985, pp. 305–16.

Jarvenpaa, S. L., Dickson, G. W. and DeSanctis, G. "Methodological Issues in Experimental IS Research: Experiences and Recommendations." *MIS Quarterly*, June 1985, pp. 141–56.

Jenkins, M. A. "Prototyping: A Methodology for the Design and Development of Application Systems." Discussion paper #227, Indiana University, April 1977.

Kennedy, M. E. and Mahapatra, S. "Information Analysis for Effective Planning and Control." *Sloan Management Review*, 16(2), Winter 1975, pp. 71–83.

King, W. R. and Cleland, D. I. "The Design of Management Information Systems: An Information Analysis Approach." *Management Science*, 22(3), November 1975, 286–97.

Kraushaar, J. M. and Shirland, L. E. "A Prototyping Method for Applications Development by End Users and Information Systems Specialists." *MIS Quarterly*, 9(9), September 1985, pp. 189–97.

Langle, G. B., Leitheiser, R. L., and Naumann, J. D. "A Survey of the Application Systems Prototyping in Industry." *Information and Management*, 7, 1984, pp. 273–84.

Lederer, A. L. "Information Requirements Analysis." *Journal of Systems Management*, December 1981, pp. 15–19.

Lederer, A. L. and Mendelow, A. L. "Issues in Information Systems Planning." *Information and Management*, May 1986, pp. 245–54.

Lederer, A. L. and Sethi, V. "The Implementation of Strategic Information Systems Planning Methodologies." Unpublished paper, University of Pittsburgh, 1987.

London, Keith R. *Decision Tables*. Pennsauken, N.J.: Auerbach, 1972.

Lucas, Henry C. *The Analysis, Design, and Implementation of Information Systems*. New York: McGraw-Hill, 1976.

March, Salvatore T., Carlis, John V., and Flory, André. "Relational Data Base Design and Use for Nontechnical Users." *Managing the Information Center Resource Series*, 93-00-65. Boston: Auerbach Publishers, 1986.

Martin, J. *Strategic Data-Planning Methodologies*, Englewood Cliffs, N.J.: Prentice-Hall, 1982.

McKeen, J. D., Naumann, D. J., and Davis, G. B. "Development of a Selection Model for Information Requirements Determination," Working paper MISRC-WP-79-06, Graduate School of Business, University of Minnesota, June 1979.

Montalbano, Michael. *Decision Tables*. Palo Alto: SRA, 1974.

Montazemi, A. R. and Conrath, D. W. "The Use of Cognitive Mapping for Information Requirements Analysis." *MIS Quarterly*, 10(1), 1986.

Munro, M. C. "Determining the Manager's Information Needs." *Journal of Systems Management*, June 1978, pp. 34–39.

Munro, M. C. and Davis, G. B. "Determining Management Information Needs: A Comparison of Methods." *MIS Quarterly*, June 1977, pp. 55–67.

Munro, M. C. and Wand, Y. "A Systems Theory Approach to Understanding Information Requirements Analysis." Working paper WP-05-81, Faculty of Management, University of Calgary, September 1981.

National Cash Register Company. *Accurately Defined Systems*. Dayton, Ohio, 1968.

Naumann, J. D. and Jenkins, M. A. "Prototyping: The New Paradigm for Systems Development." *MIS Quarterly*, September 1983, pp. 29–44.

Newell, A. and Simon, H. A. *Human Problem Solving*. Englewood Cliffs, N.J.: Prentice-Hall, 1972.

Nolan, Richard L. "Systems Analysis for Computer Based Information Systems Design." *Data Base*, Winter 1971, pp. 1–10.

Nutt, P. C. "Evaluating MIS Design Principles." *MIS Quarterly*, June 1986, pp. 138–56.

Pollack, S. L., Hicks, H. and Harlon, W. "A Decision Table Approach to Systems Analysis." *Data Base*, Spring 1970.

Rockart, J. F. "Chief Executives Define Their Own Data Needs." *Harvard Business Review*, March–April 1979, pp. 215–29.

Schussel, George. "The Role of the Data Dictionary." *Datamation*, June 1977.

Segall, M. J. "The Use of Prototyping to Aid Implementation of an On-Line System." *Systems, Objectives, and Solutions*, 4, 1984, pp. 141–56.

Semprivivo, Phillip C. *Systems Analysis: Definition, Process, and Design*. Palo Alto: SRA, 1976.

Sethi, V. and Teng, J. M. C. "Choice of an Information Requirements Analysis Method: An Integrated Approach." *INFOR* (forthcoming), 1987.

Shevlin, Jeffery L. "Evaluating Alternative Methods of Systems Analysis." *Data Management* 21 (1983): 22–25.

Slovic, P., Fleissner, D. and Bauman, S. "Analyzing the Use of Information in Investment Decisionmaking: A Methodological Proposal." *Journal of Business*, April 1972.

Specht, P. H. "Job Characteristics as Indicants of CBIS Data Requirements." *MIS Quarterly*, September 1986, pp. 270–87.

Taggart, W. M. Jr., and Tharp, M. O. "Dimensions of Information Requirements Analysis." *Data Base*, 7(1), Summer 1975, pp. 5–13.

Taggart, W. M. Jr., and Tharp, M. O. "A Survey of Information Requirements Analysis Techniques." *Computing Surveys*, 9(4), December 1977, pp. 273–90.

Teng, J. T. C. and Galletta, D. F. "MIS Researchers' Perspective on Methodological and Content Issues of MIS Research." Unpublished paper, University of Pittsburgh, 1987.

Valusek, J. R. and Fryback, D. G. "Information Requirements Determination: Obstacles within, among, and between Participants." Proceedings of the SIGCPR ACM Conference on End User Computing, May 2–3, Minneapolis, 1985.

Yadav, S. B. "Determining an Organization's Information Requirements: A State of the Art Survey." *Data Base*, Spring 1983, pp. 3–20.

Zani, W. M. "Blueprint for MIS." *Harvard Business Review*, November–December 1970, pp. 95–100.

Wetherbe, James C. *Systems Analysis for Computer-Based Information Systems*. St. Paul, Minn.: West Publishing, 1979.

_____. *Executive's Guide to Computer-Based Information Systems*. Englewood Cliffs, N.J.: Prentice-Hall, 1983.

_____. "Traditional Approaches to Systems Development." *Auerbach Information Management Series*, December 1982.

_____. "Advanced Approaches to Systems Development." *Auerbach Information Management Series*, December 1982.

_____. "System Development: Heuristic or Prototyping?" *Computerworld*, April 1982.

_____. "Taking the System for a Test Drive." *Corporate Report*, April 1982, pp. 38.

_____. "Evolution in Systems Analysis and Design." *Proceedings of the 1982 Annual Conference of the Association for System Management*, Kansas City, May 1982.

_____. "A Systems Specification Model for Instruction in Systems Analysis and Design." *Journal of Data Education*, July 1978, pp. 4–10.

_____. "Development and Application of Industry-Based Cases in Systems Analysis and Design." *Journal of Data Education*, October 1978, pp. 5–7.

Wetherbe, James C. and Berrisford, Thomas R. "Heuristic Development: A Redesign of Systems Design." *MIS Quarterly*, March 1979, pp. 11–19.

Wetherbe, James C. and Davis, Gordon B. "Information Requirements Determination Tools: Borrowing from the Best." *AIDS 1982 Proceedings*, San Francisco, November 1982.

Wetherbe, James C. and Dickson, Gary W. *Management of Information Systems*. New York: McGraw-Hill, 1984.

Wetherbe, James C., Bowman, Brent, and Davis, Gordon B. "Three Stage Model of MIS Planning." *Information and Management* 6 (1983): 11–25.

Yourdon, Edward and Constantine, Larry L. *Structured Design: Fundamentals of a Discipline of Computer Program and System Design*. New York: Yourdon, 1978.

Chapter

8

Selecting Technology and Personnel

In the preceding two chapters the information to be included in an information system and the structuring of that information were discussed. This chapter discusses concepts and techniques for determining and evaluating the technology and personnel required to support an information system. These procedures are necessary for the major decision of whether to use the existing technology and personnel in an organization or to replace or expand it.

Determining and evaluating technology and personnel is a process in which both management and systems analysts generally play an active role. This chapter fits into the information systems framework and relates to the preceding two chapters as follows:

	Inclusion	Structure
INFORMATION	Chapter 6	Chapter 7
TECHNOLOGY	Chapter 8	
PERSONNEL	Chapter 8	

This chapter is organized into three major sections: "Inclusion," "Technology," and "Personnel." The first section discusses general concepts of selection applicable to both technology and personnel. The next two sections address unique selection issues as they pertain to technology and personnel.

Systems Approach

The systems approach discussed in Chapter 2 provides a framework for determining what technology and personnel to include in a system. To review, this framework involves the following steps:

1. Establishing systems objectives
2. Determining and selecting entities or components possessing the necessary attributes to accomplish the objectives
3. Structuring or relating the attributes of the entities included in the system such that the objectives are obtained

The first two steps are relevant to this chapter; the third step is of primary interest in the next chapter.

The systems objectives are based upon the capabilities required to support an organization's information requirements. For example, an objective might be to provide a two-second response time to inquiries about the balances of customers' savings accounts. Given this objective, alternative combinations of both technology and personnel capable of achieving it should be determined. The most cost effective combination should be selected for inclusion in the information system. Personnel, in this context, pertains to both development and operational personnel. Development personnel are primarily involved with creating the application programs needed in the system. Operational personnel are primarily involved with using the completed system to provide information.

Decision Making

The selection of technology and personnel for inclusion in an information system is a decision-making process. In Chapter 6 (see Table 6.3), decision making was defined as a process involving intelligence, design, and choice.[1] It is helpful to consider inclusion decisions in the context of intelligence, design, and choice.

Intelligence

During the intelligence phase, those persons involved in the decision-making process learn about the technology and personnel available to support the desired information system. Selection and evaluation criteria should be based upon the attributes required of the technology and the personnel. For example, important attributes for a computer programmer may include CASE expertise, accounting background, database experience, and on-line experience. The importance of these attributes should, of course, be related to the technology to be used. In this example, the technology is likely to include an on-line computer system with CASE technology. This could be used for an accounting application in conjunction with a database management system.

Design

With an adequate level of intelligence about both the problem and the possible means (technology and personnel) to address it, various combinations of technology and personnel should be considered. These combinations may be alternative means to achieving the solutions defined during systems design (as discussed in Chapter 7). For example, rather than hire programmers, an alternative may be to contract with another company to write the required programs. Selection criteria may be further defined as a better understanding of alternatives is achieved.

The best alternative can be selected only if that alternative is recognized and considered. It is important to do more than merely look for an alternative that will work. The decision maker should attempt to identify all reasonable alternatives that deserve consideration in order to select the very best.

Choice

Choice involves selecting the alternative that best satisfies or approximates the selection criteria established during the intelligence and design phases. A combination of both economic and subjective analyses is usually required. For example, the difference in cost between two pieces of equipment can be analyzed, but there are also subjective considerations such as the reputations of the equipment suppliers and the capabilities of their technician. Such subjective consid-

---

1. Herbert Simon, *The New Science of Management Decision* (New York: Harper & Brothers, 1960), p. 54.

erations can usually be quantified; however, it must be remembered that the numbers do represent subjectivity.

Technology

The systems approach of considering objectives and technology with attributes that can contribute to achieving objectives is an important dimension to selecting technology. In a technologically dynamic field such as computing, it is unfortunately common for those responsible for technology decisions to continue to use hardware and software with which they are familiar, rather than to thoroughly investigate and consider other technological alternatives. To illustrate, minicomputers were a cost-effective alternative for on-line processing several years before most organizations considered implementing them as an alternative to large, centralized computers. Even computer professionals are occasionally reluctant to change to technologies that differ from those with which they are familiar. In other words, they tend to invest their efforts into structuring old technology in a new way rather than considering whether new technology might be more effective and efficient.

The reluctance of a computer professional to capitalize on new technologies is not dissimilar to that of an accountant who resists changing from a manual accounting system to a computerized one. No amount of structuring of the manual system can provide the performance that is possible with computer technology. There exists somewhat of a paradox when computer professionals, who have brought so much change on others, are reluctant to accept change themselves.

One of the greatest contributions of the systems approach to problem solving is that inclusion is considered prior to proceeding with structuring. It is far better to find the technology that can best perform the tasks required than to fit the task to a more familiar but less suitable technology.

Intelligence

To properly consider technology alternatives, the decision maker(s) responsible for technology selection must know about the state-of-the-art hardware and software technology. Research into existing technology is therefore required. The primary sources of intelligence about technology are

- In-house expertise
- Vendor representatives
- Consultants
- Literature
- Other organizations

In-House Expertise
Computer professionals within the organization are a logical starting point when aiming to acquire information about technology that may be included in an information system. In very simple or obvious cases, they may provide the only

expertise required. For example, if an organization is implementing a new personnel information system in conjunction with an existing payroll information system, simply using the existing computing equipment may be a clear-cut decision. However, for more complex and sophisticated situations, such as redesigning several major information systems to use new technology, additional expertise may be desirable.

A limitation of using in-house staff is that they are not likely to be aware of the latest technological developments available from all hardware and software suppliers. They may also be biased toward hardware and software with which they are familiar.

Vendor Representatives

Representatives from the various manufacturers of hardware and/or software (e.g., IBM, Unisys, NCR, Honeywell, CDC, DEC, Hewlett-Packard, Apple, etc.) are a major source of information on technology. Vendor representatives can be contacted and requested to give presentations and demonstrations of their company's available or forthcoming hardware and/or software relevant to the information system under consideration.

The major drawback of vendor representatives as an information source is that they tend to be biased toward the hardware and/or software they are marketing. It is important to acquire information from several different vendor representatives to ensure an objective perspective of alternative technologies.

Consultants

Consultants can be contracted to provide information on technology and to make recommendations as to selection. Their insight can be helpful because they generally have a wide variety of experience with different organizations and technologies. This additional perspective can be highly beneficial.

On the negative side, consultants may be expensive, are subject to hardware/ software biases (as is any computer professional), and do not have to live with the results of the recommendations they make.

Literature

The various professional and trade journals unique to the computing industry are an excellent source of information on technology as well as on other aspects of computing and information systems. It is worthwhile to subscribe to several of these publications. Some of the common ones are

1. *Auerbach Communications*
2. *Computer Decisions*
3. *Computerworld*
4. *Data Base*
5. *Data Communications*
6. *Data Management*
7. *Datamation*
8. *Data Pro*
9. *Information WEEK*
10. *Infosystems*
11. *Mini-Micro Systems*

12. *MIS Quarterly*

13. *MIS Week*

Two very useful sources of information on technology are available through industry research services provided by DataPro and Auerbach. Both companies market subscription services that provide information on hardware and software supplied by different companies. These sources can be used to determine vendors that should be contacted for additional information.

The shortcomings of literature as a source of intelligence are as follows:

1. Articles in publications are generally a result of an author's (or authors') decision to make a contribution in a certain area. Therefore, there may or may not be articles that specifically discuss technology pertinent to the information system under consideration. The articles that are available may contain author biases.

2. Hardware/software information services, though extremely comprehensive, are, of necessity, generalized. They do not specifically discuss hardware or software as it applies to a particular organization's problems.

Other Organizations

Other organizations that have developed or are developing similar computer-based information systems can provide considerable insight into the advantages and disadvantages of technology. A common way of locating such organizations is through vendor representatives. It is quite appropriate to request a list of organizations to which the vendor has sold similar hardware and/or software. In situations where there is some question as to the capability of a vendor's hardware and/or software, visiting with some of their other customers is highly desirable.

The disadvantages of information based on visits to other installations are the following:

1. Other installations are not identical, so care must be taken not to inaccurately generalize.

2. The reasons for another organization's technology decisions may no longer apply due to new technological developments that provide better alternatives.

3. If the other oroganization is a competitor, the organization may not be willing to share all or even any of the information about its environment.

General Selection Criteria

The intelligence phase provides a general understanding of what information systems capabilities are possible with existing technology. Determination can be made of the technological and economic feasibility of supporting the information system as desired. For example, in an on-line retail information system, are there combinations of retail terminals and computers that can perform the necessary processing functions at a price that is realistic for the organization? If the technology does not yet exist or is cost prohibitive, it may be necessary to compromise the desired information system to strive for one that is realistic.

Such intelligence about current technology and its cost allows the organization to establish general selection criteria for the type of system desired. The following are examples of selection criteria questions:

1. Will centralized or distributed processing be used?

2. What size and speed processor will be needed?

3. What storage capacity will be needed?

4. What should the price range be?

Design

During the design phase, the information about technology is structured into specific alternatives. The objective of this phase is to design the most economical alternative for properly supporting the information system. There is generally a wide variety of alternatives. It is easy and also tempting to get overly involved with the rather exciting technological elegance of them. For example, it is interesting to compare processor instruction sets and speeds, disk access techniques, and operating software architecture. But alternatives should be compared in terms of *what they do,* not *what they are.*[2] This perspective can best be illustrated by analogy. In purchasing a television set, the type of technology used to produce a quality picture, ensure reliability, and so forth, may be interesting. But the selection of a television set should be based on performance, or what the television does, not purely on its technology. The value of technology is based on the results it produces, not in how interesting or complex it is.

RFQ and RFP

The end result to be accomplished during the design phase is the creation of systems specifications that can be used to request hardware and/or software proposals from vendors. These specifications are generally expressed in either a *request for quotation* (RFQ) or a *request for proposal* (RFP).

An RFQ is used when the hardware and/or software to be procured is somewhat predetermined. This usually applies in situations where an organization has existing computing facilities that need only be expanded to accommodate a new information system. In such cases, an additional disk drive or additional computer memory may be all that is required. Therefore, an RFQ that specifies the exact hardware and/or software required is prepared. The vendor(s) respond with specific prices. Figure 8.1 illustrates an RFQ for personal computers. Some technical terms in Figure 8.1 may be unfamiliar to the reader. Familiarity with the terms pertinent to an RFQ is acquired during the intelligence phase of an actual selection exercise. For the purposes of this discussion, such familiarity is not required.

An RFP is used for more complex procurement activities. Generally, there is uncertainty as to the best hardware and/or software to procure when an organization is installing its first computer system or replacing all or most of an existing one. An RFP is less restrictive than an RFQ in that an RFP indicates the functional capabilities required but allows latitude to vendors in configuring and proposing different hardware/software alternatives. In fact, the RFP approach allows vendors to participate in the design of alternatives to meet an organization's requirements. For example, one vendor may propose both centralized and distributed technology for an on-line system. The organization can take advantage of the technological creativity available from a wide variety of computer professionals working for different hardware and/or software suppliers.

---

2. Erik M. Timmreck, "Computer Selection Methodology," *Computing Surveys*, V (1973): 201.

Figure 8.1   **An RFQ for Personal Computers**

REQUEST FOR QUOTATION

Six Personal Computers

Specifications:

     1. IBM PS/2 Model 60

     2. 80286 Microprocessor, 10 MHz

     3. 128 KB ROM Memory

     4. 1 MB RAM Memory

     5. Video Graphics Array, IBM PS/2 Color Display 8513

     6. Auxiliary Storage: 1.44 MB 3 1/2 inch Diskette Drive
                              70 MB Fixed Disk Drive

     7. Operating System: IBM DOS 3.3

     8. IBM Quietwriter III

A detailed discussion of RFPs is beyond the scope of this text. Figure 8.2 summarizes the contents common to an RFP in outline form.

Mandatory and Desirable Specifications

Specifications for RFQs and RFPs are generally expressed in terms of mandatory and desirable specifications.

Mandatory specifications indicate hardware and software features and capabilities that must be available within an alternative in order for it to be considered. For example, a mandatory requirement for an on-line system might be the ability to concurrently support at least thirty terminals with an average response time of less than two seconds.

Figure 8.2   **An Outline for an RFP**

RFP

FOR

ACME MANUFACTURING

I.   INFORMATION

      A.   Purpose of Request for Proposal
      B.   Summary of Requirements
      C.   Background Information

II.   GENERAL INFORMATION AND REQUIREMENTS

      A.   Schedule of Selection Process
      B.   Proposal Deadline
      C.   Vendor Contact with ACME Manufacturing
      D.   Alternative Proposals
      E.   Proposal Evaluation Criteria
      F.   Caveats

III.   APPLICATIONS TO BE SUPPORTED

      A.   Production Scheduling
      B.   Material Requirements
      C.   Financial Systems

IV.   SPECIFICATIONS

      A.   Representative Hardware Configuration
      B.   Representative Hardware Listing
      C.   Attributes and Support Requirements

V.   CONTRACTUAL EXPECTATIONS

VI.   EVALUATION OF PROPOSALS

      A.   General Approach
      B.   Hard-Dollar Evaluation
      C.   Soft-Dollar Evaluation

VII.   SPECIFICATIONS FOR VENDOR PROPOSALS

      A.   Vendor Reply Format
      B.   Vendor Cost Schedule

Desirable specifications indicate hardware and software capabilities that are of value but are not essential to the operation of the system. Desirable attributes pertain to enhancements to mandatory specifications or to new features or capabilities not discussed under mandatory specifications. For example, it may be desirable, but not necessary, for an on-line system to support more than thirty terminals.

Care must be taken in specifying both mandatory and desirable specifications. The writer must not use specifications that are too "vendor-oriented" or unfairly restrictive. For example, it would be unfairly restrictive to specify that disk access time must be exactly thirty milliseconds if it were known that only one vendor's disk access time was exactly that speed. To avoid such unfairness, specifications should be expressed in terms of minimal acceptable levels. An example follows:

Disk access time must not exceed 40 milliseconds

Hard and Soft Dollars

The criteria by which RFQs and RFPs will be evaluated should be defined in the systems specifications. Obviously, a major selection criterion is the cost, or *hard dollars,* required to purchase the hardware and/or software. Any response to an RFQ or RFP that meets all mandatory specifications should be given a hard-dollar evaluation. Those that do not meet the mandatory specifications should be considered nonresponsive to the specifications and eliminated from further consideration.

Systems specifications that are only desirable may be assigned *soft-dollar* values. A soft-dollar value is a subjective assessment of the economic worth of a desirable feature or capability of a hardware or software component. Soft dollars are a mechanism to reward vendors for features or capabilities provided above and beyond mandatory requirements. For example, vendors who can deliver hardware in advance of the mandatory delivery date might be given a soft-dollar credit against the hard-dollar cost of their equipment.

Vendors should be notified in advance as to what soft-dollar evaluations will be made of their quotations or proposals. Providing this information allows vendors to assess the importance and, therefore, the need to address desirable specifications.

Figure 8.3 illustrates how mandatory and desirable systems specifications can be expressed. Note that vendors may be awarded different amounts of soft dollars, depending on the levels at which they satisfy a desirable specification.

Choice

The responses to RFQs and RFPs provide the information required to proceed with choosing an alternative. The objective is to select the most cost effective alternative.

Cost Analysis

In the simplest case, an RFQ or RFP will contain only mandatory requirements. In such a case, all that is necessary is a hard dollar evaluation. The RFQ or RFP with the lowest cost is awarded the order.

In cases where desirable specifications are used, both hard-dollar and soft-dollar evaluations are required. The hard-dollar evaluation is conducted as in the preceding paragraph to determine the *actual cost* of each alternative. The soft-dollar evaluation is then conducted as follows:

1. For each desirable attribute provided by a vendor, the associated soft-dollar value is *deducted* from the vendor's hard-dollar cost.

2. The vendor with the lowest cost after soft-dollar adjustments is awarded the bid. If the system is to be rented or leased (rather than purchased), soft-dollar deductions are prorated over the anticipated life of the rental agreement. For example, if a soft-dollar adjustment of $60,000 is prorated for a system leasing for $20,000 a month for five years, the adjusted price is $19,000 a month.

Figure 8.3   **Mandatory and Desirable RFP Specifications**

MANDATORY SPECIFICATIONS

I.   Main Storage (Memory)

    A. System must have minimum storage capacity of 32 million bytes.

    B. Fetch cycle for one word should not exceed 100 nanoseconds.

    C. Store cycle for one word should not exceed 200 nanoseconds.

    D. Memory should exhibit physical modularity to enable functional isolation for trouble-shooting and fault detection.

    E. Memory should permit the use of parity bits for error detection.

DESIRABLE SPECIFICATIONS

I.   Delivery Date

Vendors will be awarded "soft-dollar" credits toward the cost of their proposals for bettering the mandatory delivery date. The schedule of credits follows:

Delivery Date	Credit
May 1990	$60,000
June 1990	40,000
July 1990	20,000
August 1990	-0-

In some cases, desirable specifications may have hard-dollar costs. For example, additional hardware may be required to support additional (desirable) terminals. This can be adjusted for by conducting separate hard-dollar evaluations for vendors capable of supplying the particular feature or capability.

Proposal Validation

A vendor's response to a customer's proposal is a written document describing a system or systems that the vendor considers responsive to the customer's requirements. However, in an attempt to provide an attractive price, a vendor may propose a system that is inadequate; it may not be able to properly process the work load. Therefore, the customer should check whether each proposed system is indeed capable of handling the work load before a final choice is made.

The most accepted and comprehensive means of validating a proposal is the use of a benchmark.[3] A *benchmark* is a point of reference from which performance measurements of hardware and software can be made. Benchmarks are most commonly conducted by running computer programs and/or mixes of computer programs that are the actual work, or are representative of the actual work, to be performed by the new system. By measuring the time required to process such work loads, projections can be made as to the ability of the proposed system to process the total work load.

Benchmarks can be used to validate and compare the performance of various systems or just to validate the performance of the most economical proposal. Comparative benchmarks are advantageous in situations where the costs of different proposals are very close. In such cases, a slight performance advantage may be the deciding factor. If one proposal is substantially more economical than the others, a single benchmark may be conducted. If the most economical system performs satisfactorily, it may be selected. Otherwise, the next lowest proposal should be benchmarked. The benchmarking is continued until a system that performs satisfactorily is identified. Because each benchmark generally requires time and effort on the part of both customer and vendor, the single benchmark approach should be used when appropriate.

Personnel

As discussed at the beginning of this chapter, both development and operational personnel are required to support an information system. The need to make personnel selection decisions is invoked by the need to replace existing personnel (i.e., personnel who have been promoted, transferred, or terminated, or have resigned) or to expand the number of personnel.

The system analyst is seldom involved in the selection of managers and clerical personnel to use and operate the system. However, systems analysts are often called upon to make recommendations about, and/or to participate in, the training of development and operational personnel.

The area of personnel selection in which the systems analyst usually participates is the selection of information systems personnel such as programmers, other systems analysts, and managers for such groups. Therefore, the following discussion of personnel selection is oriented toward the selection of personnel who work directly in the data-processing or information systems area. The framework is applicable to the selection of personnel for other positions as well.

Intelligence

In personnel selection, intelligence pertains to determining the capabilities required of personnel who are to participate in the development and maintenance

---

3. Eric M. Timmreck, "Performance Measurement: Vendor Specifications and Benchmarks," *The Information Systems Handbook* (Homewood, Ill.: Dow Jones-Erwin, 1975), pp. 365–67.

of information systems. These capabilities requirements are based upon the characteristics of the information processed and the technology used to process it. For example, if an information system involves on-line database technology for production scheduling, it is desirable to select personnel with related training and capabilities.

A careful review of the information systems and their respective technology provide the basic framework for selection criteria. Such criteria can be further defined during the design phase.

Design

The recruitment and selection of qualified personnel has been a difficult task for most organizations. The number of computer specialists available has not been sufficient to meet the demand for them. Title VII of the Civil Rights Act of 1964 and the guidelines set up within affirmative action programs have further complicated the task of personnel selection from a procedural perspective.

Now management's task is two-fold. First, a good personnel selection must be made. Second, the selection of one candidate over others must be justifiable if audited by an affirmative action officer of the Equal Employment Opportunity Commission (EEOC). Such justification can be particularly difficult if, as a result of national advertising for a position, several hundred applicants respond.

Detailed Descriptions

The common approach to personnel selection and affirmative action issues is to ensure that job descriptions and specifications for positions are detailed and current. Well-defined position requirements can substantiate the selection of one candidate over others.

The use of position requirements has been fairly successful, particularly when selecting personnel for well-defined and relatively stable positions—for example, data-entry clerks, control clerks, and computer operators. For more technologically diversified positions in programming, systems analysis, and management, job descriptions often lack the precision needed to serve as selection criteria and/or the flexibility needed for satisfactory selection of such personnel.

For example, in filling a programmer position, management may be looking for an individual with extensive payroll experience, familiarity with a certain programming language, and experience with a particular teleprocessing monitor on a certain hardware configuration using a certain operating system. In filling another programmer position in the same company, management may be interested in an individual with order-entry experience, the ability to utilize distributed processing with a specific minicomputer, and experience in using a programming language with specific intelligent terminals. A single programmer job description is inadequate for both.

Certainly, job descriptions can play a meaningful role in an organization. However, unless many job descriptions are maintained for many job classes (an unwieldy and continual task), such descriptions do not provide a comprehensive basis for personnel selection. The job descriptions for most data-processing or information systems positions tend to be quite general and often fail to articulate the specifics required. Further, in most cases they do not provide sufficient documentation to support a given personnel selection from an affirmative action or EEOC perspective if necessary.

Selection Technique

As an alternative to using job descriptions as selection criteria, a structured selection technique has been developed. It provides both a clear definition of the selection criteria and comprehensive documentation for EEOC audits.[4] Research on the use of this technique indicates that it results in higher acceptance of offers made to prospective employees and in reduced turnover. The six steps used in the technique are explained below. They provide a framework for the design phase of the decision-making process.

1. *Formulate selection committee:* The selection committee should be a group of individuals collectively competent to evaluate the necessary attributes required for a candidate to successfully fill a given position, preferably with a vested interest in the successful filling of the position. An essential committee member and likely chairperson is the immediate supervisor of the position to be filled. Other potential members of the committee include members of the candidates' potential peer group, individuals in subordinate positions to the position being filled, and members of departments who use services of the information systems department. A diversified committee is apt to follow a rigorous evaluation process. Also, a new employee will likely gain more acceptance from the organization.

2. *Atttribute determination:* The committee should collectively agree upon the attributes that are meaningful to the selection of a candidate for the position. Attributes should be categorized—education, work experience, test scores (if applicable), management background (when relevant), and so on. A sample profile of attributes for a programmer position is given in Figure 8.4.

3. *Attribute weighting:* At this point in the process, the committee assigns a weight to the value of each attribute as it relates to the position under consideration. One way to do so is by employing a scale of 1 through 10. This is a critical step; it prevents overemphasizing or overlooking an attribute or a category for a particular position. For example, technical attributes are weighted higher for a programmer than for a manager. Conversely, communication skills may be weighted higher for a manager than for a programmer. Sample weights for attributes are also shown in Figure 8.4.

4. *Score applicants:* The scoring of applicants is accomplished by having each committee member independently rank each applicant by attribute on a scale of 1 through 5. Alternatively, the scoring of applicants may be performed collectively by the committee.

The scoring of applicants is particularly significant. This step provides the means for determining and also justifying the selection of one applicant over another. Therefore, it is important that the scoring criteria be fair and justifiable. For example, for most information systems positions, it would be appropriate to score an applicant with a degree in computer science and/or information systems higher than an applicant with a degree in general business. However, the applicant with a degree in general business would likely be scored higher than an applicant with a degree in, say, history.

If anonymity of applicants is desirable, name and demographic data may be deleted from resumes by someone not on the selection committee, prior to the scoring of applicants. If few applicants are involved personal interviews may also be considered. This does, of course, eliminate anonymity of the applicants.

---

4. James C. Wetherbe and V. Thomas Dock, "Breaking the Description Dilemma: Personnel Selection by Group Analysis," *Data Management*, December 1976, pp. 16–18.

Attribute	Weight

Education

Technical School	8
Bachelor's	10
Master's	5
	23

Technical Training

COBOL	10
FOCUS or similar 4th Generation Language	6
Database	8
Data Communication	8
	32

Work Experience

Banking Applications	10
Supervisory	5
	15

Interpersonal Skills

	7
	7

After the attribute ranking has been completed, a total weighted attribute score may be computed for each applicant. The weighted attributes to be added are the products of attribute weights and attribute ranks.

5. *Score analysis:* At this step, various forms of analysis may be employed, depending on the preference of the committee. One approach is

a. Exclude the highest and lowest scores for each attribute. Then compute an average by attribute, using the remaining scores. For example, if scores of 18, 28, 30, 32, and 38 are stated for an attribute by committee members, the scores 18 and 38 are excluded when averaging. The result is an average score of 30 obtained by the computation (28 + 30 + 32)/3.

b. Compute a composite score for each applicant. This is accomplished by totaling the weighted attribute average scores.

c. Rank applicants in descending order, based on their composite score.

6. *Verification:* The committee should now evaluate certain applicants, preferably the top three through five, to confirm that the scoring technique has provided the quality of candidates desired. If the attribute identification and weighting have been accomplished properly, the committee will generally achieve consensus on the top three through five candidates. If not, reevaluation of the attributes and weighting should be considered.

Choice

After review of the resumes, the committee should select the candidate(s) to be personally interviewed. In cases where travel expenses are a constraint, the committee should interview the highest-scoring individual first. If the committee agrees that this person is satisfactory and the candidate accepts the position offered, the process is concluded. Otherwise, the remaining candidates should be interviewed in descending order until a suitable candidate has been determined who accepts the position.

An illustration of the selection technique, involving six applicants for a programmer position, is given in Figure 8.5. It uses the criteria shown in Figure 8.4. The scoring grid in Figure 8.5 consists of a matrix correlating each of the six applicants by attribute score, weighted score (W.S.), total weighted score (T.W.S.) by category, composite score, and ranking.

The ranking of applicants by composite score indicates that Applicant A is the number-one candidate (quantitatively speaking), with a score of 349. Applicant B, with a score of 314, has scored high enough to warrant consideration also. The remaining applicants' scores are much lower. The committee should interview Applicant A and Applicant B. The interview should consist of a one- or two-day visit to the MIS department, depending upon the importance and the level of the positions being filled. The total time interval between candidates' visits should be as short as possible to ensure a continuity of evaluation of all candidates.

Each candidate should meet with the various members of the selection group on a one-on-one basis and then meet with the entire group. The one-on-one meetings serve to establish rapport with the candidate and to let him or her ask most of the questions. The group interview is an excellent forum for the group

Figure 8.5   Scoring Grid for Applicants for Computer Programmer for a Bank

Attributes	Education							Technical Training									Work Experience					Interpersonal Skills			Composite Score
	Technical School or Equivalent	W.S.	Bachelor's	W.S.	Master's	W.S.	T.W.S.	COBOL	W.S.	FOCUS or equivalent	W.S.	Data Base	W.S.	Data Communications	W.S.	T.W.S.	Banking Applications	W.S.	Supervisory	W.S.	T.W.S.	Ability to Communicate	W.S.	T.W.S.	Composite Score
Weight	8		10		5			10		6		8		8			10		5			7			
Applicant A	4	32	5	50	3	15	97	5	50	5	30	5	40	4	32	152	5	50	3	15	65	5	35	35	349
Applicant B	5	40	4	40	0	0	80	5	50	5	30	5	40	3	24	144	5	40	3	15	55	5	35	35	314
Applicant C	5	40	5	50	0	0	90	4	40	3	18	4	32	3	24	114	4	40	2	10	50	4	28	28	282
Applicant D	3	24	3	30	4	20	74	3	30	4	24	4	32	4	32	118	4	40	1	10	50	3	21	21	263
Applicant E	2	16	3	30	2	10	56	3	30	0	0	3	24	2	16	70	5	50	0	0	50	3	21	21	197
Applicant F	2	16	2	20	0	0	36	5	50	0	0	3	24	4	32	106	4	40	0	0	40	2	14	14	196

W. S. - Weighted Score
T.W.S. - Total Weighted Score

members to ask challenging questions of the candidate where the entire group can observe the candidate's responses.

Quickly after the meeting with the candidate, the group should meet to discuss their perceptions of the candidate and reevaluate the impact of those perceptions on the scores of the applicant. That is, the group may want to adjust the original scores that were based upon application or resume evaluations after they have the advantage of the greater insight gained from a personal interview.

Select a Candidate

The final decision on a candidate should be based upon the group analysis, but it should ultimately reside with the immediate supervisor. Of course, it is unwise for a supervisor to significantly deviate from the recommendation of the group, just as it is unwise for the group to insist on one of two comparably qualified applicants in its recommendation to the supervisor.

Use of Group Analysis

One might expect that using group analysis for staffing decisions in the manner just described would result in (1) better evaluation of candidates and (2) convincing evidence to applicants that (a) they have been thoroughly evaluated, (b) that they are qualified to handle the job if it is offered to them, and (c) that they will have widespread support if they will accept the job. Moreover, using group analysis for staffing decisions might be expected to produce a higher acceptance rate by candidates, less turnover, and minimum problems with affirmative action complaints.

A study conducted at the University of Houston confirmed these expectations. It indicated that an organization could experience almost twice the acceptance rate and half the turnover rate by employing group-analysis techniques.[5] Also, group-analysis documentation has proved quite effective in the event of affirmative action questions and hearings.

Summary

Technology and personnel are components required to support an information system. The selection of technology and personnel should be based on their ability to contribute to the objectives of the information system. This selection is a decision-making process involving the three phases of intelligence, design, and choice.

Intelligence pertaining to technology is available from in-house expertise, vendor representatives, consultants, literature, and other organizations. This intelligence is used to design alternatives through RFQs or RFPs. The specifications defined in RFQs and RFPs are categorized as mandatory and desirable, and assigned hard-dollar and soft-dollar values, respectively. The selection criteria for technology to be included in an information system should be based upon both cost and effectiveness.

---

5. Albert H. Napier, "Peer Evaluation Selects Professionals," *Journal of Systems Management*, January 1980, pp. 6–9.

Intelligence about the personnel required in an information system is based primarily on the technology used and the type of information processed. The design of alternatives and the actual selection of personnel can be accomplished using a quasi-quantitative selection technique in which selection attributes are defined, weighted, and scored for each applicant. This selection technique provides an effective, efficient means for selection decisions and comprehensive documentation for affirmative action and/or EEOC audits.

Exercises

1. Define and discuss potential sources of intelligence about hardware and software.
2. What is the difference between an RFQ and an RFP? When should each be used?
3. What are the possible consequences of having mandatory specifications that are really not necessary (i.e., that should be categorized as desirable specifications)?
4. What are soft-dollars and why are they used?
5. Discuss the limitations of job descriptions as personnel selection criteria.
6. Within the personnel selection technique discussed in this chapter, what is the importance of weighting attributes for selection purposes?

Selected References

Bassler, Richard A., and Demoody, Harold C. *Computer System Evaluation and Selection—An Annotated Bibliography and Keyword Index*. Arlington, Va.: College Readings, 1971.

Cuthert, Norman H., and Paterson, Janis M., "Job Evaluation: Some Recent Thinking and Its Place in Investigation." *Personnel Management*, September 1966, pp. 156–62.

Equal Employment Opportunity Commission. *Fifth Annual Report*. Washington, D.C., 1971.

Ferrari, Domenico. "Workload Characterization and Selection in Computer Performance Measurement." *Computer* V (1972): 18–24.

Fife, Dennis W. *Alternatives in Evaluation of Computer Systems*. (MTR-413). Bedford, Mass.: Mitre Corp., 1968.

Galbraith, J. *Organizational Design*. Reading, Mass.: Addison-Wesley, 1977.

Goff, Norris F. "The Case for Benchmarking." *Computers and Automation*, May 1973, pp. 23–25.

Gray, Stuart. "1982 Salary Survey." *Datamation*, vol. 28, no. 11, October 1982, pp. 114–31.

Hillegass, John R. "Standardized Benchmark Problems Measure Computer Performance." *Computers and Automation*, January 1966, pp. 16–19.

Howard, Phillip C. ed. "Measuring System Performance with Benchmarks," *EDP Performance Review* 50 (1973): 1–7.

Joslin, Edward O. ed. *Analysis, Design, and Selection of Computing Systems*. Arlington, Va.: College Readings, 1971.

Joslin, Edward O. *Computer Selection*. Reading, Mass.: Addison-Wesley, 1968.

Longenecker, Justin G. *Principles of Management and Organizational Behavior.* Columbus, Ohio: Merrill, 1973.

McFarland, F. Warren; Nolan, Richard L.; and Norton, David P. *Information Systems Administration.* New York: Holt, Rinehart & Winston, 1973.

Miner, John B., and Miner, Mary G. *Personnel and Industrial Relations, A Managerial Approach.* London: Collier-Macmillian, 1973.

Mintzberg, H. *The Structuring of Organizations: A Synthesis of the Research.* Englewood Cliffs, N.J.: Prentice-Hall, 1979.

Napier, H. Albert. "Peer Evaluation Selects Professionals." *Journal of Systems Management,* January 1980, pp. 6–9.

Norman, A. K. *Industrial and Organization Psychology.* Englewood Cliffs, N.J.: Prentice-Hall, 1971.

Owens, W. A., and Jewell, D. O. "Personnel Selection." *Annual Review of Psychology,* 1969, pp. 26–29.

Paul, L. "Programmer Aptitude Tests Worth the Risk." *Computerworld,* November 29, 1982, p. 1.

———. "Bank Coins Program to Identify, Retain Best DP Trainees." *Computerworld,* Feb. 7, 1983.

Simon, Herbert. *The New Science of Management Decision.* New York: Harper & Brothers, 1960.

Timmreck, Eric M. "Computer Selection Methodology." *Computing Surveys* 5 (1973): pp. 199–222.

Timmreck, Eric M. "Performance Measurement: Vendor Specifications and Benchmarks." *The Information Systems Handbook.* Homewood, Ill.: Dow Jones-Irwin, 1975, pp. 365–67.

Wetherbe, James C. *Systems Analysis for Computer-Based Information Systems.* St. Paul, Minn. West Publishing, 1979.

———. *Executive's Guide to Computer-Based Information Systems.* Englewood Cliffs, N.J.: Prentice-Hall, 1983.

Wetherbe, James C. and Berrisford, Thomas R. "Heuristic Development: A Redesign of Systems Design." *MIS Quarterly,* March 1979, pp. 11–19.

Wetherbe, James C. and Davis, Charles. "A Decision Support System for the Planning and Control of a Chargeout System in a Large-Scale Computing Environment." *Data Base,* June 1980, pp. 13–20.

Wetherbe, James C. and Dickson, Gary W. *Management of Information Systems,* New York: McGraw-Hill, 1984.

Wetherbe, James C., and Dock, V. Thomas "Breaking the Description Dilemma: Personnel Selection by Group Analysis." *Data Management* 14 (1976): 16–19.

Wetherbe, James C., and Dock, V. Thomas "Breaking the Description Dilemma: Personnel Selection by Group Analysis." *Datamation,* December 1978, pp. 16–42.

Zmud, Robert W. *Information Systems in Organizations.* Glenview, Ill.: Scott, Foresman, 1983.

Chapter 9

Systems Development

Introduction

In this chapter, systems development is discussed. During this stage, the systems design specifications created in Chapter 7 are converted into operable software. Specific topics include software development, quality assurance, testing, and user-developed systems. These four processes pertain to the structuring of the technology and personnel included in information systems.

Systems development is a process in which both programmers and systems analysts are actively involved. The systems analyst may or may not become involved in software development, depending on the orientation of the organization and the qualifications of the programmer and the systems analyst. The objective of this chapter is to provide a solid understanding of software development, but it is beyond the scope of this book to provide operational skills in all the techniques discussed. Such training comes from books and courses in computer programming. However, whether systems analysts become involved in programming or not, the material in this chapter is important for them to coordinate and oversee software development.

This chapter relates to the preceding three chapters as shown in the following chart:

	Inclusion	Structure
INFORMATION	Chapter 6	Chapter 7
TECHNOLOGY	Chapter 8	Chapter 9
PERSONNEL	Chapter 8	Chapter 9

Development

Systems development refers to structuring hardware and software to achieve effective and efficient processing of an information system. To a great extent, the structure of the hardware included in the system is determined as a result of the selection process discussed in Chapter 8. Hardware installation, which is usually performed by the vendor, may be all that is required. Therefore, for the systems analyst, development primarily encompasses the structuring or programming of application software to support the information system. Accordingly, the following discussion is directed at concepts and techniques of software development.

In the early days of data processing, computer programming was not subjected to traditional management measurements of quality and quantity. Rather, the mystique and glamour of computer technology often caused management to take a "hands-off" approach to the activity of computer programming. The standards and quality control so essential to uniformity and reliability were often not enforced. Programs developed in such environments are difficult to understand, test, debug, and modify.

The computing industry is plagued with unfortunate stories of information systems that have become crippled or useless with the turnover of critical programmers. In many of these cases, the program code was so complex and confusing that it was easier to write new programs than to figure out how the old ones worked.

Fortunately, over the years considerable progress has been made in the area of computer programming standards and control. The major concepts are as follows:

- Structured programming
- Top-down development
- Structured software design techniques
- CASE and the programmers' workbench
- Chief-programmer teams

These concepts are discussed in the following subsections.

Structured Programming

Surprisingly, the programming of complex computations does not represent the major cause of software problems. Rather, the major contributor to confusing and error-prone program codes is the nonstandard and undisciplined use of control (branching) logic. Often the logic flow erratically jumps about in the program code so that it is extremely difficult to follow.

The main culprit of undisciplined coding is the unrestricted use of unconditional branches (e.g., GO TO statements or their equivalents). The application of structured programming standards to control logic can resolve this problem. Programs can be written in a straightforward and readable fashion. The complexity and confusion involved in tracing undisciplined branching can be eliminated.

Structured programming is based on a solid theoretical foundation. In the 1960's Bohm and Jacopini formally proved that only three basic control structures are necessary for computer programming.[1] These structures are

1. The SEQUENTIAL structure—Statements are executed one after the other.

2. The DO WHILE structure—One or more statements are repeated as long as a condition is true (looping process).

3. The IF THEN ELSE structure—One of two statements is branched to, based on whether a given condition is true or false (Boolean relation).

The logic within these three structures is presented in Figure 9.1. The three structures can be combined with each other to define program logic from beginning to end. Though unconditional branches are not necessary, on occasion they may be used for purposes of programming practicality. However, such cases should be made on an exception basis and carefully documented and justified.

Note that each of the structures shown has a single entry and a single exit. This single entry/exit property can be maintained as the structures are combined to build program modules or segments. For example, Figure 9.2 shows a logic flow that contains all of the structures and has the single entry/exit property.

---

1. C. Bohm and G. Jacopini, "Flow Diagrams, Turing Machines, and Languages with Only Two Formation Rules," *Communications of the Association for Computing Machinery* (1966): 366–71.

Figure 9.1 **Basic Control Structures of Structured Programming**

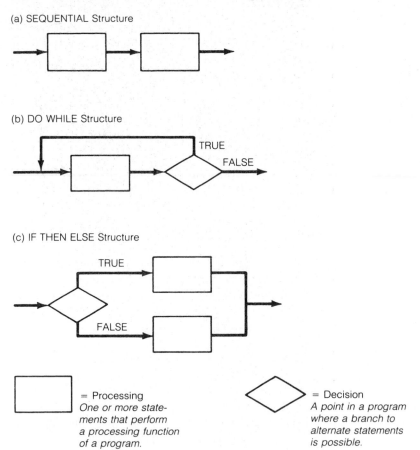

(a) SEQUENTIAL Structure

(b) DO WHILE Structure

(c) IF THEN ELSE Structure

☐ = Processing
*One or more state-
ments that perform
a processing function
of a program.*

◇ = Decision
*A point in a program
where a branch to
alternate statements
is possible.*

The use of the single entry/exit structure in organizing logical structures allows the clustering of program functions into single entry/exit modules or sub-systems. Assigning program functions to specific modules allows the modules of a complex program to be independently programmed by separate programmers. Because each module has only one point of entry and one point of exit, the interfacing of these modules into one program or system is simplified. Since each module's function is clearly defined before it is programmed, the modules fit together to form a complete program.

In addition to allowing program functions to be programmed independently, modularization simplifies program testing, debugging, and modification. Since program functions are defined in specific modules, they can usually be tested, debugged, and modified as separate modules of program code. This reduces confusion caused by program statements branching into or out of modules at points other than where modules are designed to be entered or exited.

Top-Down Development

Program development originally used a *bottom-up* procedure. The lowest-level functions such as READ and WRITE modules, were coded first. They were then

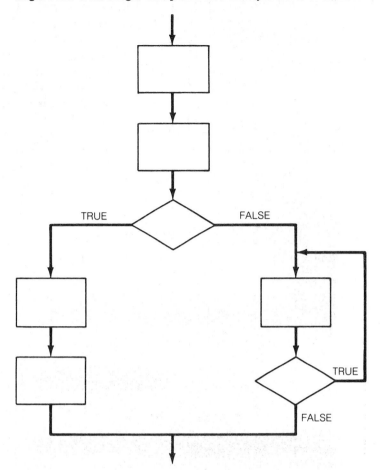

tested and prepared for integration with other program functions.[2] This approach often results in data definition and interface problems during integration of the functions. Consequently, integration is delayed until these problems are resolved. Another major drawback of the bottom-up approach pertains to debugging program malfunctions after the functions have been integrated. It may be extremely difficult to determine which of the many program modules combined during integration is the source of a particular problem.

A bottom-up approach philosophically contradicts the manner in which a system evolves. The systems development process evolves from a broad and conceptual level to an increasingly detailed and operational level. Much of the low-level detail required for a bottom-up approach is not available until late in the systems development cycle.

Top-down program development is a procedure in which program requirements are factored from the top to the bottom. Top-down development is sometimes referred to as an *outside/inside* approach. The program to be developed is

---

2. IBM, "Improved Programming," p. 3.

first considered in its broadest sense. This involves considering the program in terms of its inputs and outputs (i.e., what is to happen *outside* the program). The program is then factored into the progressively smaller functional units required to achieve the desired transformation of inputs to outputs (i.e., what is to happen *inside* the program).

Structured design is a means for conveniently factoring large complex programs into manageable single entry/exit modules. The factoring continues until modules are defined at a level directly transferable to program code. This factoring process is illustrated in Figure 9.3. Each module (A through O in Figure 9.3) can be independently programmed, tested, debugged, and, if necessary, modified.

During the testing and debugging of a program module, other modules to be integrated may not yet exist. Dummy units called *program stubs* are used to simulate such modules. Program stubs do not usually perform meaningful computations. They may send acceptable responses back to another module so that it can continue processing as if the program stubs had in fact executed as real modules. The eventual integration of the various program modules is not a problem because well-defined data definitions and interfaces are inherent to the top-down approach.

Three guidelines have proven useful for top-down program development:

1. Use indentation in source code to highlight the DO WHILE and IF THEN ELSE structures.

2. Keep program modules to fifty or fewer statements.

3. A program module should be coded only after the higher-level module that invokes it has been coded and tested. (For example, in Figure 9.3, Module A should be coded and tested before Modules C and D.)

Structured Software Design Techniques

A variety of commercially developed products are available that further refine and operationalize the concepts of structured programming and top-down development. Though these products often have cosmetic differences, there is a great deal of similarity in how they work and how they are used. We will review some of the more common techniques. This review is designed to provide a general understanding of the various techniques, but it is beyond the scope of this book to provide skills training in each of the techniques. With the conceptual understanding provided in the preceding discussions and from the review in this chapter, however, it should become easy for you to grasp the specifics of any of the following structured software design techniques:

1. HIPO—hierarchy plus input process output

2. SADT—structured analysis and design techniques

3. W/O—Warnier/Orr technique

4. SDD—structured decomposition diagram

HIPO

IBM has developed a system of procedures and forms called HIPO (hierarchy plus input process output) that is designed to facilitate effective and efficient top-down program development. HIPO provides excellent discipline and structure

Figure 9.3   Top-Down Approach to Developing a Computer Program

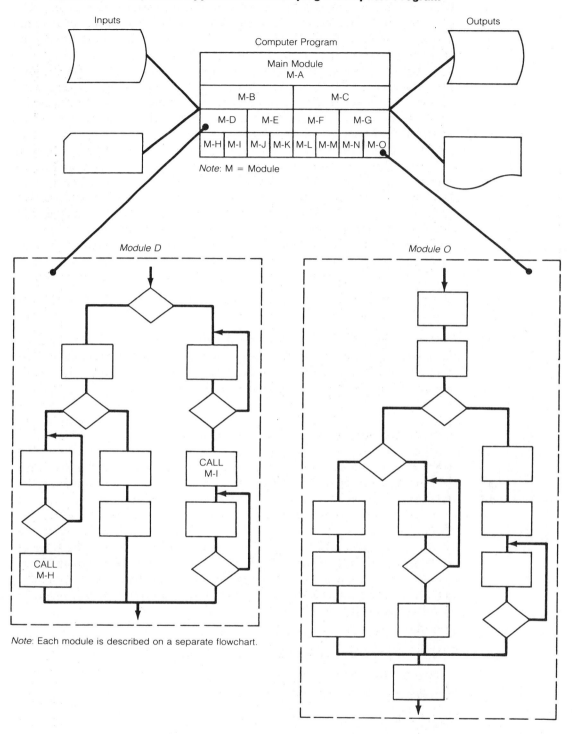

Note: Each module is described on a separate flowchart.

to factor complex programs into individual single entry/exit modules. Each module is defined in terms of its inputs, processes, and outputs.

HIPO provides a visual representation of software that serves as both a design tool and documentation. Completed HIPO forms (or charts) provide a blueprint for program code and reduce the need for program flowcharts. The HIPO charts partition all program modules, define the hierarchy and organization of the modules, and define the communication between modules. HIPO has been criticized because it does not usually define the sequence of execution (i.e., conditional branching or iterations) as flowcharts do. However, when necessary, HIPO specifications can be augmented with flowcharts.

The actual HIPO chart looks much like an organizational chart, with the hierarchy showing how functions or activities are composed of more elementary functions (see Figure 9.4). The process of factoring, or functional decomposition, continues until a function has been reduced to its most basic activities. Each function is placed within a box and located at the appropriate level on the hierarchy chart and labeled with a *noun-verb* action statement. For example, the function of "compute gross pay" may be factored or decomposed into "compute straight-time wages" and "compute overtime." Each box or function within the chart is numerically labeled referring to the next higher function so that organization can be maintained.

The noun-verb action statement labeling each function in a HIPO chart is further described through IPO (input, process, output) charts (see Figure 9.5). IPO charts consist of three boxes showing the input, process, and output. For higher-level functions, the process may be the control statements for one or more subfunctions. As lower-level functions are described in an IPO chart, the process becomes very close to the actual code to be used in programming the function.

HIPO is generally regarded as a good technique for limited analysis work. It is criticized, however, for lacking the characteristics that would allow it to be useful in complex system structures. Specifically, due to processes being analyzed separately, the relationships between functions or data elements that are critical to any system are not evident with HIPO. Also, the consistent use of the technique and its iterations are not controlled.

SADT

SADT is a product of Soft Tec Incorporated and was developed as a complete methodology for the use and control of the tools used in analyzing and designing a system. It provides a graphical representation of the hierarchy functions and data within a system (see Figure 9.6). As illustrated in Figure 9.6, SADT displays the inputs, processes, outputs, controls, mechanisms, and relationships between the functions and data in the system.

Note that in Figure 9.6 there is a macro diagram at the top of the figure which is then (as with HIPO) factored down into more detailed boxes conveying a specific operation to be performed to achieve processing. The arrows connecting the vertical faces of functions in a SADT diagram represent data interfaces and are labeled appropriately. The arrows connecting the horizontal faces of the SADT diagram represent the external constraints on the function and the mechanism used to perform the function.

Beyond specifying the tools involved in the SADT technique, SADT also includes comprehensive procedures and rules for using SADT.

Figure 9.4   HIPO Hierarchy Chart

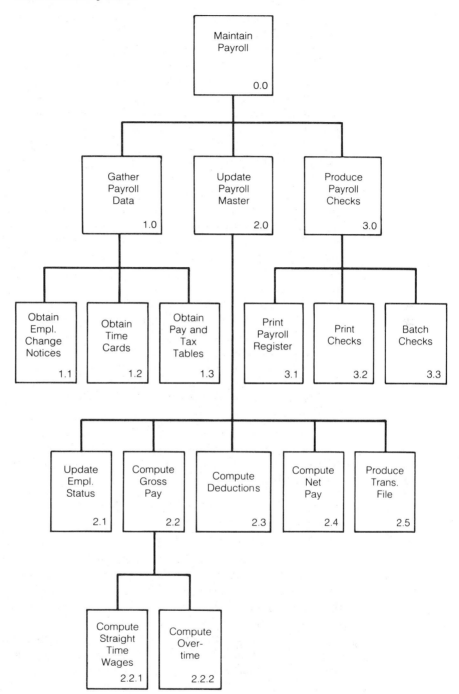

Figure 9.5   **HIPO Input, Process, Output**

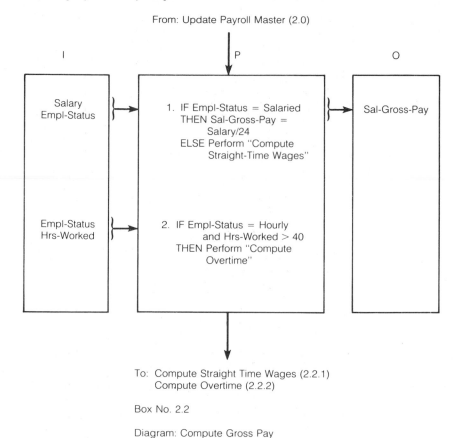

Box No. 2.2

Diagram: Compute Gross Pay

Implementation of SADT requires a strong management commitment due to its high cost and extensive training requirements; due to the high volume of work, automation is often required for complex systems.

Warnier/Orr Technique

The Warnier/Orr technique (W/O) is a systems design and program design technique that is unique in its use of graphical displays consisting of a hierarchy of brackets to portray activities or data elements. Figure 9.7 illustrates a W/O diagram and defines the symbols used.

The W/O technique consists of several steps that ultimately end in the programming of the system. The first step of the technique involves the diagramming of system outputs. System outputs, including reports and files, are then decomposed, and a hierarchy is shown by brackets that enclose data or functions supported to identify outside the brackets. Parallels are used to show how the data or function is controlled—sequentially, logically, or repetitively.

W/O diagrams can be used to schedule the times when data will be required in order to complete the desired outputs on time. For example, the reporting cycle (yearly, quarterly, monthly) can be shown with its scheduled output. This ensures that the data needed to produce the output is available on time.

Figure 9.6   **SADT Diagram**

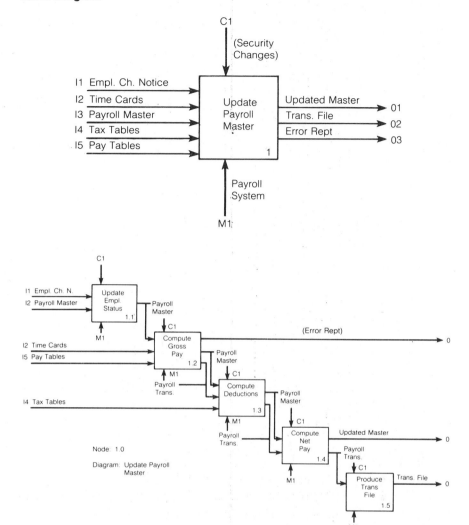

Another feature of the W/O diagram that is particularly useful is called *change analysis*. During a change analysis, all occurrences that might cause a change in a data element, function, or reporting cycle are investigated. If the design cannot handle the occurrence, appropriate changes are made. This allows the design to encompass predictable changes in the environment.

The final stage in the W/O technique consists of translating diagrams into program code. Each bracket in the W/O diagram becomes a module in a computer program. (Note that each bracket in a W/O diagram is similar to a decomposed box in an IPO or SADT chart.) The arrows in a W/O diagram represent the control statements while the functions become the working statements. The W/O diagrams are duplicated to show the working statements as derived from the functions. This program diagram can then be pseudocoded (i.e., put into structured English) showing the control statements as derived from the arrows. To

Figure 9.7   W/O Diagram

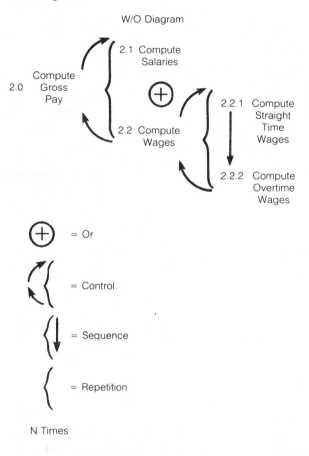

complete the technique, pseudocode can then be translated directly into program code and structured walk-throughs at the control points within the entire process.

W/O diagrams are very useful for data-structuring techniques, as the diagrams come very close to programming code. However, they are criticized for not providing adequate functional analysis. In other words, while the W/O diagrams clearly show composition, they fail to show the links between the various functions.

SDD

SDD is a hybrid structured analysis technique developed by CTEC Incorporated. It is derived from the previously discussed SADT and the W/O techniques and attempts to borrow the best from both techniques. SADT diagrams are used for analyzing functional hierarchy, and W/O diagrams are used to describe the function. Figure 9.8 illustrates an SDD version of SADT. Note that the data interfaces (arrows) of SADT are circumvented, but the decomposition of the data remains. Mechanisms arrows are also not used in this SDD version of SADT. Diagramming is completed by showing the data interfaces in a modified W/O diagram, as illustrated in Figure 9.9. This hybrid version of SADT and W/O technique decomposes both functional activities and data elements without the voluminous diagramming required in the SADT technique.

Figure 9.8   SDD Truncation of SADT

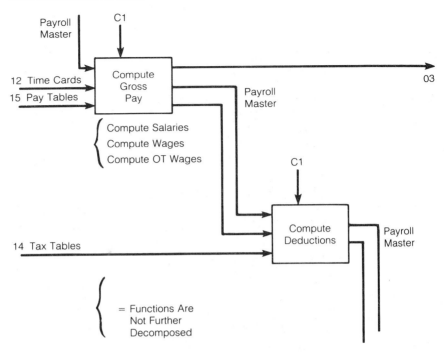

The analysis phase of SDD is completed with a data/function matrix as illustrated in Figure 9.10. This is also a hybrid of the decomposition of data elements and functions. The data/function matrix depicts the role, if any, that each data element plays within each function. Note that in Figure 9.10, each intersection of a data element and a function is labeled with a "I" for input and/ or "O" for output or "C" for constraint, if appropriate. The result of this matrix is a complete diagram showing the links between data elements and functions.

SDD has three advantages over the SADT or W/O technique used separately. First, the combining of W/O technique with SADT provides "shortcut" notation without disturbing the flow of SADT diagramming. Second, the voluminous diagram required in SADT can be circumvented without losing the hierarchical

Figure 9.9   SDD Modified W/O Diagram

			Inputs	Outputs
	1.1 Update Empl. Status		Empl. Ch. Notice Payroll Master	Payroll Master
	1.2 Compute Gross Pay	1.2.1 Comp. Sal. 1.2.2 Comp. Wages 1.2.3 Comp. OT	Payroll Master Time Cards Pay Tables	Payroll Master Payroll Trans
Update 1.0 Payroll Master	1.3 Compute Deductions		Payroll Master Payroll Trans Tax Tables	Payroll Master Payroll Trans
	1.4 Compute Net Pay		Payroll Master Payroll Trans	Updated Master Payroll Trans
	1.5 Produce Trans. File		Payroll Trans	Trans. File

Figure 9.10   SDD Data/Function Matrix

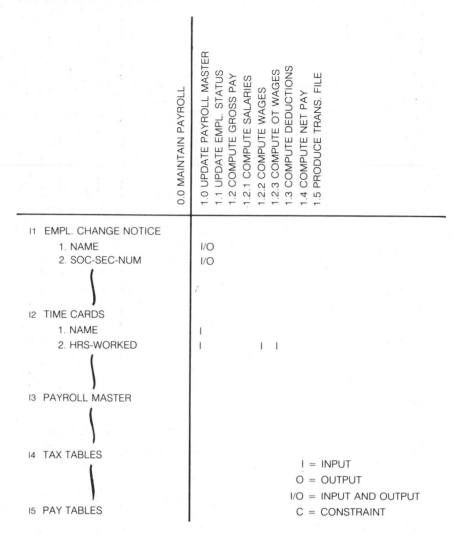

structure. Third, the data function matrix provides a link between data elements and functions that is not found in either SADT or W/O diagrams.

Review of Structured Software Design Techniques

There is a great deal of similarity between the various structured techniques discussed above—all are graphical representations of functional and data hierarchies. They start at a high conceptual level and factor to detail, providing enough definition to be readily translated into computer programming instructions.

There is no research indicating which techniques work the best. Rather, there is only anecdotal information from people who have worked with different techniques.

As pointed out at the beginning, the purpose of this discussion was to provide familiarity with the variety of techniques available, recognizing that different techniques have different characteristics and different strengths and weaknesses. Again, keep in mind that the theory behind the techniques is to provide a more closed/stable/mechanistic, disciplined approach to doing software development. Also, note that all techniques draw extensively from the theoretical framework of general systems theory discussed in Chapter 2.

CASE and the Programmers' Workbench

The CASE (computer-aided systems engineering) tools discussed in Chapters 5, 6, and 7 have a special subset of tools generally referred to as the *programmers' workbench*. The old adage—the shoemaker's children are the last to get shoes—is as applicable for the programmer as it is for the systems analyst. The previous techniques discussed in this chapter, though undoubtedly helpful, are manual, labor-intensive techniques. Outside of the emergence of report generation and fourth-generation languages, there have been few truly powerful strides made to provide computer-based support for computer programming.

Fortunately, in recent years there has been at least an evolution, if not a revolution, in the approaches to software development. Products such as Excelerator, PACBASE, CASE 2000 Designaid, Information Engineering Workbench, and Teamwork are helping to improve software quality and productivity, and are also improving working conditions for programmers.

These products are in an evolving state and many features will be added and fine-tuned during the next several years. However, a certain common functionality is emerging among these types of products.

Figure 9.11 provides a model for the programmers' workbench. A discussion of major tools and their functions follows.

Diagramming Tools

Data-structure diagrams provide computer-based graphics support for generation of program structure charts, such as those discussed in the preceding section of this chapter. This support reduces the tedious effort to generate and modify these charts.

An important feature is the ability to verify the accuracy and completeness of the information that is entered for creation of a chart. Referred to as a *syntax checker,* this feature is very helpful for reducing errors. Although the variety in syntax used to graphically portray charts limits the kind and type of syntax checking that is done, it is hoped that in the future more standardization of graphics will eliminate this problem.

Prototyping

The software necessary to support prototyping and heuristic design (procedures discussed in Chapter 7) is part of the programmers' workbench. Most CASE products include menu generators, screen generators, and report generators.

Code Generator

Code generators allow the programmer to generate modular units of source code from high-level design specifications. The code generators translate *icons* (used to indicate various program functions) and data-defined flows of information into structured, top-down development computer programs.

Although generators have difficulty generating software for complex systems specifications, they can at least create prototypes that can be refined into complex

Figure 9.11   **Model of the Programmers' Workbench.**

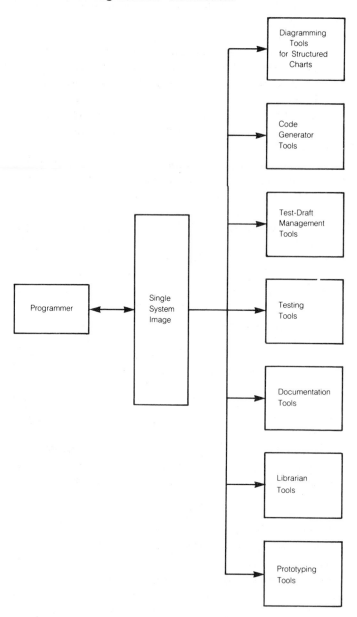

systems. Code generators are also useful for generating reusable code in those instances where systems have standard modules in common.

Test-Data Management Tools

Test-data management tools support the creating, sorting, and retrieving of test data. Besides managing the data, these tools help determine what test cases need to be created. They allow programmers to use original design documentation and data design to determine test paths to be followed through the database. From these paths, test cases can be generated that test those functions of the system.

The test cases can be stored and reused for repeated or future tests. This process of generating reusable data is similar to generating reusable code.

Testing Tools

Testing tools include test-execution control and test-results verification. *Test-execution control* allows the programmer to capture intermediate results of the test. Thus, in the event of an abnormal end (*abort*) to a program execution, this tool allows the system to capture critical data such as the program statement being executed at the time of abort and the record being processed at time of abort. Consequently, processing can continue after recovery from the abort.

Test-results verification allows comparison of results from one test program to results of other programs and projected results. The tool notes differences among the different results.

Documentation Tools Documentation is perhaps the most dreaded task of the programmer. Workbenches reduce the difficulty of producing documentation by providing convenient access to text-editing and data-dictionary facilities.

Program Librarian Tools A major inhibitor of programmer productivity are the clerical activities peripheral to software development, such as filing and maintaining source code and creating, updating, and backing up program libraries. Also, multiple versions of a program must be kept track of as enhancement and maintenance are being incorporated and listed.

If these activities are not interfering with software development, they are usually being neglected—which is worse. Failure to perform these tasks jeopardizes the integrity and operation of an information system. Newly hired people have great difficulty understanding and taking responsibility for an undocumented system.

The librarian tool provides facilities for maintaining multiple versions and records of changes to the same program or systems. It also allows testing of system changes without affecting the production version if one is in use.

Chief-Programmer Teams

Because of the increasing size and complexity of major programming efforts, a team effort is often required to complete a system on schedule. The coordinating and interfacing of the efforts of several computer programmers has traditionally proved to be a cumbersome task. The advent of structured programming, top-down design, and programmers' workbench has provided the necessary framework for a team approach to program development. Coordination and integration of activities are simplified by the modularity of tasks that these interrelated techniques afford.

The use of a team of programmers headed by a chief programmer has proved to be highly productive. It is also a means of achieving high-quality program development. The team members perform specific functions, as follows:

1. *Chief programmer:* The chief programmer is a highly proficient programmer who has overall responsibility for the program design and coding. He or she defines and assigns program modules to be coded by other team members. He or she reviews all code and supervises the testing and debugging of code. The chief programmer informs management of project status and arranges for additional team members when necessary.

2. Backup programmer: The backup programmer is an experienced programmer who works closely enough with the chief programmer to assume his or her responsibilities, if necessary. He or she also participates in coding of program modules.

3. Other: Additional team members (primarily programmers) are assigned to the team as necessary. These team members generally have specialized skills (e.g., coding speed, application knowledge, special tools or knowledge of unique coding techniques).

Chief-programmer teams provide increased control and integrity to the development process. They also provide for professional growth and technical excellence in programming. Inexperienced programmers are exposed to senior-level programmers in an environment that prepares them for leadership roles on future teams.

Quality Assurance

Quality, or the lack of it, has become a major issue worldwide. Other than increasing productivity, perhaps no other general management issue has received so much attention. And, as many quality experts are quick to point out, poor quality is a major cause of lost productivity. Poor quality can cause production delays, cost overruns, and unacceptable products.

The development of application software has traditionally been one of the most quality-negligent activities in industry. The software industry is laced with stories of million-dollar errors on checks printed by computers, of lost customer accounts, of on-line systems being down for hours or even days. There has been so much emphasis on meeting deadlines that there is often little time for quality assurance. It seems, as so often happens in MIS, "urgency drives out importance." No recipient of an information system would pressure for a deadline to be met if he or she knew that meeting that deadline would result in unacceptable quality. But deadlines can become so urgent that the importance of quality is overlooked.

Recently, the more "street-wise" managers have placed greater emphasis on quality in software development. Starting with quality-oriented management attitudes, MIS organizations have been able to significantly increase quality in software development by using technologies including design overviews, detail reviews, and code inspections. In addition to increasing quality, these techniques provide a potent educational process. Some attitudes and techniques necessary to assure quality are discussed next.

Management Attitude

The right place to start with quality is with a management attitude that insists on *economic* quality. That is, quality that is cost effective. Quality that costs more than it is worth is not economic.

For example, an on-line system that is up 98 percent of the time may be acceptable for a wholesale distributors' on-line order-entry system. It would not make economic sense to add a second backup computer to provide 99.8 percent uptime. Note that such reliability, however, would make economic sense for an on-line airline reservation system.

Management must establish quality standards that are reflective of their organization's requirements, communicate these attitudes, and then support and reward quality performance.

A particularly popular process for establishing an attitude of quality is the *quality circle*. In a quality circle, key organizational participants in a process are freed from their regular duties to have collective discussions on ways to improve quality. This allows "importance" to get some leverage over "urgency"; it shows that management places enough importance on quality to take time away from production. Also, the synergy that is facilitated by group discussions involving individuals with different perspectives often generates outstanding ideas.

Quality circles can be established to focus on programming techniques, documentation, use of new technology, and so forth. They often have a strong training orientation more skilled personnel share their techniques with those less skilled. Quality circles complement the chief-programmer team concept discussed earlier.

Design Overview

At a more specific but still macro level of quality assurance is the design overview, which should be incorporated into the MIS function's systems development methodology. It consists of a peer evaluation of the designer(s) overall approach to software development concerning one or more programs in an information system.

In a design overview, the peers critique the design but generally do not offer suggestions for improvement (since that would put the reviewers into the role of designer). The designer(s) respond to the criticism with explanations or suggested improvements. If satisfactory responses to criticism cannot be provided at the first meeting, subsequent meetings are scheduled. This process continues until the peers are willing to "sign off" on the software design.

Detailed Review

Once design overviews are complete, systems designers begin detail design of software. Upon completion of this effort, the next level of quality assurance is the *detail review*, or what IBM calls a *structured walk-through*.

During this stage of quality assurance, systems designers step through the logic of their software in a presentation to qualified peers. Peers check for use of well-structured techniques and top-down design, as well as specific logic. As with a design overview, critiques are made, and if necessary an iterative process of explanations and corrections is made until the peers agree to sign off on the detail design.

Code Inspection

The final stage of quality assurance prior to a system testing is code inspection. During code inspection, a programmer other than the original coder inspects line by line the instructions in a clearly compiled computer program to see if everything seems correct and if coding conventions have been followed.

During program development, computer programs and program modules are tested. As the programs are integrated into the total information system, overall testing of the system is possible. When individual modules or programs are tested independently it is called a *unit test*. When all programs are tested together it is called a *systems* or an *integrative test*.

Testing is a critical stage of the systems development cycle. It is the major checkpoint prior to the actual implementation of the system. The consequences of system malfunctions are minimized when they are discovered and resolved during testing rather than during implementation. The testing tools and test-data management tools provided by CASE and the programmers' workbench greatly enhance the ability to do high-quality testing in an efficient manner.

Validation

System problems are discovered by subjecting the system to extensive validation. The major system processes to be validated are:

1. *Clerical processing:* The data-collection and preparation procedures must be performed correctly.
2. *Input processing:* Transactions must be properly checked for errors and must be applied to the right records in the right files.
3. *Computational processing:* The proper variables must be computed using the proper arithmetic.
4. *Logic processing:* Decision rules must be executed using the correct sequencing and branching.
5. *File accessing:* Records must be stored in, and retrieved from, the right locations.
6. *Output processing:* The correct variables must be printed in the right places on printouts and displays.

Validation of these processes can be accomplished through the processing of real or fabricated transactions that represent normal and abnormal conditions. The outputs resulting from transaction processing can be validated by checking to see if they are correct. For example, the amount due on a loan can be computed on a calculator and compared to a corresponding output provided by the computer system.

An information system must be able to handle exceptions (i.e., abnormal transactions). To validate this aspect of the system, transactions that deviate from normal transaction form should be prepared and processed. For example, payroll transactions with invalid department codes, missing social security numbers, pay rates below the minimum wage rate, or hours-worked that exceed the allowable limit should be prepared. A properly developed information system should detect such errors and report them on error reports.

Debugging

The process of correcting system malfunctions is called *debugging*. The source of an incorrect output can exist anywhere in an information system. The problem can occur during data collection or preparation, during input processing, in the computational or logical processes of one or more computer programs, or at any other point. Therefore, to debug a malfunction, the source of its occurrence must be isolated before the problem can be resolved.

Malfunctions occur somewhere between the initial entry of data to the system and the final output. In other words, if the data entering the system are correct, but the information leaving it is incorrect, then the malfunction resides somewhere between entry and exit. The specific location of the malfunction can be isolated by working backward from the output to the input and/or forward from the input to the output. Each point in the system where the data are used can be checked to determine where the malfunction occurred. This is, the data can be displayed before and after each system process until the specific process causing the problem is isolated. In many cases, data are not reported or displayed before and/or after a process (e.g., before and after a sort). In such cases, a special program may have to be written or a simple dump of the file taken for debugging purposes. Figure 9.12 illustrates a system with identified checkpoints for isolating a system malfunction.

Once isolated, a system malfunction can be resolved, say, by a change in program code, data-collection procedure, or data-entry procedure. If the malfunction exists in a computer program, the use of structured programming facilitates debugging. Since the function of each module is defined, each module that uses the data in question can be determined and checked for malfunction. Once isolated, the program module(s) can be corrected and integrated back into the system.

User-Developed Systems

One thing is clear. Given the insatiable demand for more and better information to support decision making and operations of organizations, not enough computer programmers can be graduated from our colleges and universities to handle the work load.

This situation is similar to the one faced by the phone company years ago. Had the phone company not automated as it did, everyone in the work force would have been a telephone operator today in order to handle the telephone work load of today. By automating, the phone company *did* make all of us telephone operators but in a simplified fashion. In the early days of phone use, we had to go through a telephone operator to do everything, including making a local call. Gradually, via automation, the phone company shifted more and more of the work load to us. First we could make local calls by ourselves, then long distance, then credit-card calls, and so forth. Today we get assistance from a telephone operator only when we can't make the call on our own, and often that assistance comes from a computerized operator. Note that there are still telephone operators, but they are able to provide more service per operator than in pre-automation days, and costs per transaction have gone down.

The evolution of phone use is analogous to what is happening with computer use. In the early use of computers, managers or users of computer services were totally dependent upon the computer technician for everything. The gradual evolution of *user-friendly languages* is allowing managers to do more and more on their own. The computer technician's role, like the telephone operator's, is becoming one of helping users to help themselves.

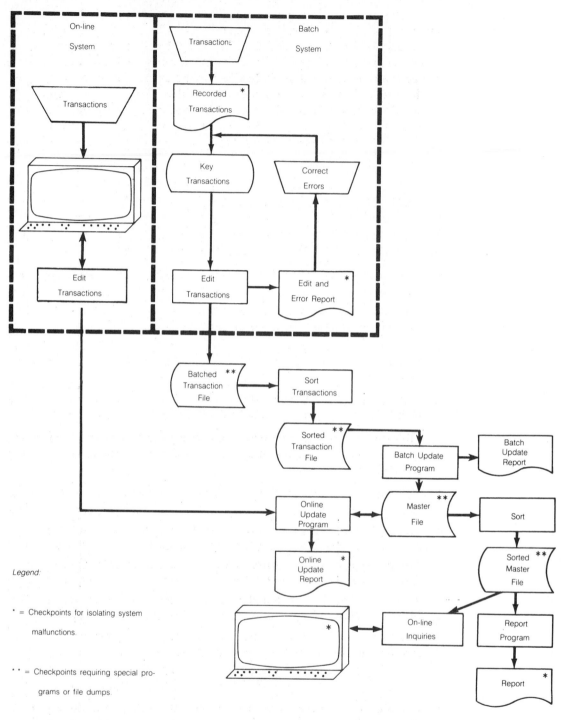

Legend:

* = Checkpoints for isolating system malfunctions.

* * = Checkpoints requiring special programs or file dumps.

With user-friendly, high-level query languages (as discussed in Chapter 4), users can develop in a day or two enough skills to satisfy most of their immediate and ad hoc information requirements, provided that properly collected and organized data are stored on the computer.

Computer technicians and information systems professionals will still be needed to manage and control the increasing complexity or hardware, software, and databases needed to support this new environment. But since managers will not have to explain to a technician everything they need done, they will be able to have direct access to the computer, and overall productivity will significantly increase. Programmers will be involved in programming complex editing and updating software, and users will program more of the ad hoc output programs.

Managing User-Developed Systems

The trend toward users programmers is a good one. However, if not managed properly, it could have the following negative consequences:

- Systems may be developed that cannot be integrated with other systems or corporate databases.
- Users may purchase mini- or microcomputer systems that are incapable of handling the job they were bought for.
- Users may develop programs that are not subjected to adequate quality-assurance checks to ensure that accurate, reliable, processing has been performed.
- Systems may be developed without adequate documentation, thereby rendering the system useless when its developer leaves the organization.

To avoid the preceding problems, as well as to provide an environment that is supportive and encouraging of user-developed systems, the following actions are recommended:

- The information systems function should set up a user assistance or information center to support end-user computing and programming.
- The philosophy of this group should be service oriented. The group should propose solutions to user problems and then help users to help themselves.
- The group should keep current on new technological developments in the areas of micrographics, personal computers, software packages, and so forth.
- A uniform commitment should be made by the group to provide consulting support in the use of certain types of micro- and minicomputers.
- The group should maintain proficiency in the use of all major user facilities and provide one-to-one user consulting when needed.
- A user hotline should be established to provide instant response to users having questions about, or difficulty in, the use of end-user computing.
- Procedures should be established for quality assurance and documentation checks on systems developed by users that will be available for general use and/or will become an integral part of any department's operation.

Medtronics Inc. is a progressive company in the use of managers as programmers. To encourage programming by managers, this company provides a financial bonus to managers who develop proficiency in the use of programming languages pertinent to their areas of responsibility.

Regarding the use of managers as programmers, Tom Morin, VP of Information Systems, says: "We're convinced that the 'carrot' approach is the way to go. We are providing the best consulting service we can to our user-managers, and top management has been supportive enough to reward those managers who develop the skills."

Chapter 15, "End-User-Computing," covers the management issues of end-user computing in greater detail.

Summary

Software development has been plagued by missed deadlines, cost overruns, and errors. Techniques that have improved the development of software include: structured programming, top-down development, structured software design techniques, CASE programmers' workbench, and chief-programmer teams.

Beyond improving the process of developing software, quality-assurance strategies have proven very effective. Quality assurance starts with a quality-oriented management attitude and employs techniques such as design overviews, detailed reviews, and code inspections.

Testing of software follows quality assurance. Testing includes validation and debugging of software. The problems and magnitude of testing have been reduced with the advent of better software development and quality-assurance techniques.

User-developed systems are a major trend in software development. The advances in programming languages have made it possible for users to develop much of their own software. Good management of user-developed software is critical to protect both the users and the organization.

Exercises

1. Discuss the complications caused by unconditional branches in a computer program.

2. Flowchart the following sequences and define which of the three basic control structures from Figure 9.1 apply:
   a. (1) $A = B + C$
      (2) $D = A + 5$
      (3) $E = D \times A$
   b. (1) $A = A + 1$
      (2) Repeat the preceding step until $A = 10$, then add A to B, yielding C.
   c. (1) $A = B + C$
      (2) If A is negative, multiply it by -1; otherwise, divide it by 5.

3. Using structured programming control logic, flowchart the following program module with a single entry/exit and no unconditional branching.

Problem Description

This program module is to read transactions, each of which contains a customer account number (C-ACCT) and a charge (CHARGE) made by that customer. More than one transaction may be entered for a customer. The transactions are in account number order.

As the program module reads transactions, it is to add charges to individual customer totals (C-TOT) and to a grand total (G-TOT). Each individual customer's total is to be printed after it is tallied. When the end of the transaction file is reached, the grand total is to be printed. The program module has then completed processing. HINTS: All totals must be initialized to zero. Conditional branches can branch to other conditional branches, and two or more branches can branch to the same place.

4. Discuss the disadvantages of bottom-up program development. How are these disadvantages overcome by the top-down approach?

5. Discuss the work of the backup programmer in a chief-programmer team.

6. Why should testing of a system include abnormal as well as normal transactions?

7. An accounts receivable system is to receive batch transaction on magnetic disk. The transaction validation rules for the computer program in the accounts receivable system are as follows:
   a. Account number—Located in record position 1–6. Must be numeric.
   b. Customer name—Located in record position 7–27. Can contain alphabetic characters or blanks.
   c. Store number—Located in record position 28–30. Must be a number between 99 and 201.
   d. Amount charged—Located in record position 31–37. Must be all numeric except for a decimal point in column 35. Cannot exceed 1,000 or be negative.

   Create at least ten transactions that can be used to validate the input-error-detection module(s).

8. Why is structured programming helpful for both preventing program malfunctions and isolating them if they occur?

9. Discuss the purpose and functions of the programmers' workbench. If possible, work with or observe a demonstration of one or more CASE tools.

10. What are the steps involved in quality assurance? Briefly describe each step.

11. Discuss how to debug a system containing several programs, reports, and sorts.

12. Discuss the pros and cons of user-developed systems.

Selected References

Alavi, Maryam. "Software Development Alternatives." From *The Handbook of MIS Management*, Supplement 1, Edited by R. E. Umbaugh. Boston: Auerbach Publishers, 1986, pp. 907–14.

"The Analysis of User Needs." *EDP Analyzer*, January 1979, pp. 3–12.

Baker, F. T. "Chief Programmer Team, Management of Production Programming." *IBM Systems Journal*, 11 (1972): 56–73.

Baker, F. T. "System Quality Through Structured Programming." Fall Joint Computer Conference, 1972.

Barsoff, H., Henderson, V. D., and Siegel, S. G. *Software Configuration Management: An Investment in Product Integrity.* Englewood Cliffs, N. J.: Prentice-Hall, 1981.

Bohm, C. and Jacopini, G. "Flow Diagrams, Turing Machines, and Languages with Only Two Formation Rules." *Communications of the ACM 9* (1966): 366–71.

Dahl, O. J.; Dijkstra, E. W.; and Hoare, C. A. *Structured Programming.* London: Academic Press, 1972.

Dijkstra, E. W. *Structured Programming.* New York: Academic Press, 1972.

Elspas, B.; Levitt, K. N.; Waldinger, R. J.; and Waksman, A. "An Assessment of Techniques for Proving Program Correctness." *ACM Computing Surveys* 4 (1972); 97–147.

Gane, Chris and Sarson, Trish *Structured Systems Analysis: Tools and Techniques.* Englewood Cliffs, N. J.: Prentice-Hall, 1979, pp. 9–60.

Highsmith, J. "Data Structured Systems Development." From *Systems Development Management.* Boston: Auerbach Publishers, Inc., 1983. (Article 31–02–05)

Huling, Jim. "Key Elements of CASE Kits: Prototyping, Code Generators." *Computerworld*, April 20, 1987, p. 74.

Huling, Jim. "Tools of the Trade: Is CASE Really a Cure-all?" *Computerworld*, April 20, 1987, pp. 73–86.

IBM. "Improved Programming Technologies—An Overview." *Installation Management.* White Plains, N. Y.: IBM, 1974, p. 3.

Jones, Martha Nyvall. "HIPO for Developing Specifications." *Datamation*, March 1976, pp. 112–14.

Knutsen, K. Eric. "Business Systems Analysis: Program Design." from *The Information Systems Handbook*, Edited by Warren McFarlan and Richard Nolan. Homewood, Ill.: Dow-Jones-Irwin, 1975, pp. 539–67.

Knutsen, K. Eric, and Nolan, Richard L. "On Cost/Benefit of Computer-Based Systems." *Managing the Data Resource Function.* Edited by Richard Nolan. St. Paul, Minn.: West Publishing, 1974, pp. 253–76.

Martin, James. *Applications Development Without Programmers.* Englewood Cliffs, N. J.: Prentice-Hall, 1982.

McGowan, Clement L., and Kelly, John R. *Top-Down Structured Programming Techniques.* New York: Mason/Charter Publishers, 1975.

Mendes, Kathleen S. "Structured Systems Analysis: A Technique to Define Business Requirements." *Sloan Management Review,* Summer 1980, pp. 54–61.

Mills, Harlan D. *Mathematical Foundations of Structured Programming.* IBM Report Number FSC 72-6012. 1972.

Mills, Harlan D. "Top-Down Programming in Large Systems." In *Debugging Techniques in Large Systems,* Edited by Randall Rustin. Courant Computer Science Symposium 1, NYU: 1971, pp. 41–56.

Mills, Harlan D., and Naughton, John J. "Programming Standards and Control." From *The Information Systems Handbook*, Edited by Warren McFarlan and Richard Nolan. Homewood, Ill.: Dow-Jones-Irwin, 1975, pp. 568–91.

Nolan, Richard L., and Seward, Henry H. "Measuring User Satisfaction to Evaluate Information Systems." In *Managing the Data Resource Function*, Edited by Nolan. St. Paul, Minn.: West, 1974, pp. 253–76.

Parnas, D. L. "On the Criteria to be Used in Decomposing Systems into Modules," *Communications of the ACM* 15 (1972): 1053–58.

"Program Design Techniques," *EDP Analyzer*, March 1976, pp. 6–7.

Ross, Douglas T. and Brackett, John W. "An Approach to Structured Analysis." *Computer Decisions*, September 1976, p. 44.

Rudkin, Ralph I. and Shere, Kenneth D. "Structured Decomposition Diagramd: A New Technique for Systems Analysis." *Datamation*, October 1979, pp. 133–40.

Ryan, Hugh W. "The Designers' Work Bench." From *The Handbook of MIS Management*, Supplement 1, Edited by R. E. Umbaugh. Boston: Auerbach Publishers, 1986, 887–98.

Ryan, Hugh W. "The Programmers' Work Bench." From *The Handbook of MIS Management*, Supplement 1, Edited by R. E. Umbaugh. Boston: Auerbach Publishers, 1986, pp. 879–85.

Sammet, J. C. "Perspective on Methods of Improving Software Development." *Software Engineering*, Vol. 1, 1970.

Weinberg, Gerald M. *The Psychology of Computer Programming*. New York: Van Nostrand Reinhold Co., 1971.

Wetherbe, James C. *Systems Analysis for Computer-Based Information Systems*. St. Paul, Minn.: West Publishing, 1979.

———. *Executive's Guide to Computer-Based Information Systems*. Englewood Cliffs, N. J.: Prentice-Hall, 1983.

Wetherbe, James C. and Dickson, Gary W. *Management of Information Systems*. New York: McGraw-Hill, 1984.

———. *Cases in Management of Information Systems*. New York: McGraw-Hill, 1984.

Chapter

10 Implementation and Evaluation

Introduction

In this chapter, the final stages of the systems development life cycle (SDLC) are explained. The last two steps include the following: (1) implementation of the information system and (2) evaluation of how well it has met user requirements. Both of these stages, particularly implementation, involve a great deal of behavioral consideration to ensure success of the system. In Chapter 1 it was pointed out that interpersonal or behavioral skills are very important to effective systems analysis. Important insights into behavioral factors that can positively or adversely affect information systems development are provided in this chapter. As is shown in these discussions, behavioral factors are by no means limited to the final stages of information systems development.

Implementation is covered in the first part of the chapter, followed by a discussion of evaluation of information systems.

Implementation

After development and testing has been completed, implementation of the information system can begin. During systems implementation, the project team (described in Chapter 5) should be brought back to full strength. During the software development stage, project teams tend to play a passive role as the technical steps of program development and testing evolve. However, broad organizational representation, accomplished through the project team, is required to complete the systems development cycle.

It may be appropriate to reorganize and reorient the project team to complete what has become a much more refined and defined information system. More insight into the actual use and operation of the system is now possible and is likely to affect the organizing and planning for implementation.

Another important reason for revitalizing the project team is that the interaction of the team members with their respective departments is important to facilitate communications concerning forthcoming changes. Members of user departments tend to be more comfortable and cooperative if they participate, or feel they are adequately represented, during the implementation of a new information system.

Unfortunately, one of the least understood and most overlooked issues of information systems is implementation. Often, the biggest error made is that planning for implementation is usually postponed until it is time to switch from the old system to the new. This is a big mistake.

In Chapter 5, the importance of conducting technical, economic, and operational evaluations (feasibility studies) before developing information systems was discussed. Technical evaluations determine if a system is technically feasible; economic evaluations determine if a system is cost justified; operational evaluations determine if a system will work in the context of a particular organization using particular employees.

Remember, operational feasibility determines if a system can be implemented. To postpone a thorough assessment of the feasibility of implementing a system until after it is designed and developed can cause unfortunate consequences.

Case Studies in Implementation

Consider the following case studies:

A major retail chain developed a new sales entry system using retail terminals. To save cost, the company selected a system that used keyboard entry of sales price and stock-keeping unit (SKU) numbers instead of selecting the more expensive alternative of optical scanning of such information.

Technical and economic evaluations had been supportive of the less expensive system. There were, however, implementation problems. The stores have high transactions volumes. Thus, when the terminals were installed, long lines resulted due to the extra time required to key in *sku* numbers. People waiting in lines became hostile and vocally critical to the salesclerks. In frustration, salesclerks began to index any set of *sku* keys just to expedite customer orders. This caused many problems for the retail chain.

In another case, an equipment rental company had its personnel provided with a decision-support system that took into consideration eighty different variables in preparing customer quotes. The system was developed by technical people and was quite sophisticated. Unfortunately, it was never used. Why? The people working at the rental counters explained that the system was too complex and difficult to learn for personnel who typically had limited computer exposure. Also, turnover was high in these positions, and therefore managers had little incentive to put forth the effort necessary to teach employees the system.

Implementation Parallels Systems Development

Though many examples can be provided similar to the preceding two, one theme always emerges. Implementation is not something to consider as the last step of systems development. It must be considered every step of the way. In other words, implementation is an activity that must parallel the systems development processes, as illustrated in Figure 10.1.

For example, in the case of the retail chain, the company could have installed a few of the new retail terminals in a store on a pilot basis before any computer software or systems were developed. Then salesclerks could have tried out the pilot system. In the case of the equipment rental company, some rental personnel could have been allowed to work with the earliest version of the decision-support system to see if they had any difficulty with it.

Allowing users of a system to experience what a new system will actually be like prior to conversion helps implementation problems to be uncovered and resolved before the system becomes set in concrete. (Note the compatibility of

Figure 10.1 **Parallel Relationship between Implementation and Systems Development Life Cycles**

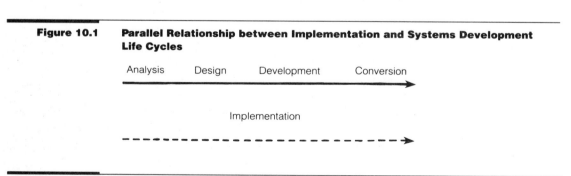

this concept with heuristic development and prototyping as discussed in Chapter 7.)

Given that implementation issues should be continuously reviewed during system development, the next issue is preparing for, and actually converting to, the new system. There are two key dimensions to this: *user training* and the *conversion* strategy.

User Training

Thorough user training is critical to the successful implementation of an information system. User-training requirements for a new information system can be categorized as either clerical or managerial:

1. *Clerical:* Must be instructed in the processing of transactions. This may involve training in the operation of computer terminals and/or the filling out of documents that will be converted to machine-readable form.

2. *Managerial:* Must be informed as to the format and content of reports and terminal displays and how to request reports or make terminal inquiries. Managers also need to understand how programmed decisions are made and how to use any decision-support capabilities.

It is often advantageous to have representatives (possibly project team members) from the different departments conduct training sessions within their respective departments. Departmental representatives should also participate in the development of user documentation (e.g., user procedures manuals) to be used in operating the system. Clerical and managerial personnel are often less intimidated by computer technology if a new information system is explained to them and documented for them by someone from their own department rather than by a computer technician.

If an information system affects several departments and a large number of employees, it may be advisable to formalize user-training requirements in a user-training checklist to ensure complete coverage. An example of a user-training checklist is given in Figure 10.2. Note that the specifics of training, responsibility, and time frame are clearly defined.

Frequently, user training overlaps program development. Much of the required training can be accomplished without the use of an operational system. Also, systems tend to be implemented in a modular fashion (e.g., the inventory module of an order-processing system may be operational before the order-entry module). This allows operational training to be staggered in the sequence in which modules are operationalized. If an early start on user training is possible, it is wise to proceed.

Conversion

Conversion is the process of changing from the old system to the new system. It requires careful planning to establish the basic approach to be used in advance of the actual cutover. Two major planning decisions that establish the basic conversion methodology pertain to the amount of parallel processing to be used and the sequence of conversion activities.

Figure 10.2   Illustration of a Training Checklist

Training Checklist
For New Order-Processing System

Inventory Subsystem

Description of Activity	Individual Responsible	Department Responsible	Start Date	Completion Date	Approved By
Management Training					
Overview	J. Johnson	Plant Mgr.	9/1	9/1	
Information Available	B. Benton	Inventory	9/5	9/7	
Programmed Decisions	B. Benton	Inventory	9/5	9/7	
Decision-Support Capabilities	H. Hendrix	Info. Systems	9/8	9/8	
Feedback Procedures	R. Scammel	Info. Systems	9/8	9/8	
Clerical Training					
Overview	J. Johnson	Plant Mgr.	9/2	9/2	
Document Preparation & Handling	C. Carnes	Inventory	9/5	9/10	
Terminal Operation	C. Carnes	Inventory	9/10	9/30	

Accounts Receivable Subsystem

Description of Activity	Individual Responsible	Department Responsible	Start Date	Completion Date	Approved By
Management Training					
Overview	B. Barnes	Finance	9/20	9/20	
Information Available	B. Black	Accounting	9/25	9/26	
Programmed Decisions	S. Smith	Credit	9/27	9/27	
Decision-Support Capabilities	B. Black	Accounting	9/28	9/28	
Feedback Procedures	R. Scammel	Info. Systems	9/28	9/28	
Clerical Training					
Overview	B. Barnes	Finance	9/21	9/21	
Document Preparation & Handling	S. Sharp	Credit	9/25	9/27	
Terminal Operation	S. Sharp	Credit	10/1	10/15	

Order-Entry Subsystem

Description of Activity	Individual Responsible	Department Responsible	Start Date	Completion Date	Approved By
Management Training					
Overview	G. Gordon	Sales	10/1	10/1	
Information Available	J. Jackson	Order Processing	10/2	10/5	
Programmed Decisions	J. Jackson	Order Processing	10/6	10/6	
Decision-Support Capabilities	G. Gordon	Sales	10/10	10/15	
Feedback Procedures	R. Scammel	Info. Systems	10/15	10/15	
Clerical Training					
Overview	J. Jackson	Order Processing	10/2	10/2	
Document Preparation & Handling	W. Wilson	Order Processing	10/5	10/7	
Terminal Operation	W. Wilson	Order Processing	10/8	10/20	

Parallel and Cutover Conversion

Conversion methodology can be defined in terms of a continuum ranging from parallel to cutover (cold turkey). In a parallel conversion, both the old and new systems are concurrently processed until the new system stabilizes. This ensures the reliability of the new system prior to abandoning the old one. In a cutover conversion, the use of the new system immediately terminates the use of the old system.

The conversion approach may fall somewhere between parallel and cutover. For example, transactions processed on the old system can be collected for a week and then reprocessed on the new system over the weekend. In this example, parallel processing applies to the transactions processed, but not to the time frame in which they are processed.

Parallel processing reduces the risk of implementing a new system. However, it has disadvantages. First, parallel processing is expensive. Often, additional personnel (or overtime) and/or equipment are required. For example, assume a wholesale distributor is converting from batch to on-line processing of customer orders. Parallel processing during the implementation requires preparing the transaction documents for the old system in addition to entering the transactions for the new system. Personnel will be required to operate both the old and the new systems. If the computer that supports the on-line processing is to replace an old computer, both the new and the old computers are required until conversion is completed and parallel processing is terminated.

For some systems, it is virtually impossible to conduct parallel processing. For example, it is not plausible to parallel process two on-line airline reservation systems. In cutover conversion, rigorous testing of the new system is critical prior to implementation. Where parallel processing is possible, value judgments must be made to assess the trade-offs between risks and costs.

Another popular strategy for conversion is the *pilot approach*, which one unit of the business is converted first to allow for a "shake down" before other units are converted. For example, the retail chain discussed earlier could convert a single store, evaluate results, and make adjustments before converting all stores.

Finally, a *phased-in approach* can be used. This is an incremental conversion whereby different business units are brought on-line in a stepped sequence. For example, the inventory module of an order-processing system can be phased in, followed by accounts receivable, and, finally, order entry.

Conversion Sequence

In implementing a new information system, it is important to determine the sequence in which system segments are to be converted. The ordering of the conversion sequence is generally a function of the following variables:

- *Organizational functions:* The different departments, branches, plants, or other logical breakdowns
- *Organizational cycles:* The normal cycle of organizational activities within organizational functions (e.g., accounting or billing cycles)

It is generally advisable to begin conversion with those segments of the system that appear to offer the least difficulty. For example, the conversion may start with the best-managed and best-organized functions with the shortest cycles. This enhances the probability of initial successes, which in turn enhances the credibility and momentum of the new system and its implementation.

During the conversion, management should attempt to minimize any additional internal or external requirements on the department undergoing conversion. A department is generally disrupted enough during conversion without having to handle the complications of additional disturbances.

A checklist of conversion activities and their sequence should be made to ensure that all bases are covered. Such a document defines responsibilities and schedule dates for conversion activities in much the same manner as the user-training checklist defines training requirements (look again at Figure 10.2).

The final task of the conversion sequence is to define what will signify the complete implementation of the new system. This definition may vary. Whatever the definition of implementation, it should be agreed upon in advance by organizational members. When the defined implementation has been achieved, the system should be formally approved and accepted, and implementation should be considered completed.

Behavioral Considerations

The major orientation of the material in this book is toward the technical area of information systems. However, serious problems in the development and operation of information systems may result from the behavioral reactions of those affected by changes brought about by these information systems. An understanding of the nature and causes of such behavior is helpful to both management and systems analysts. They are in a better position to properly address resultant problems.

Behavioral considerations are by no means limited to the implementation stage of the SDLC. Rather, they are a consideration throughout the cycle. But given the orientation that implementation parallels the SDLC, this seems the appropriate place for discussing behavioral considerations. Research by Gary W. Dickson and John K. Simmon provides a useful framework for understanding and minimizing dysfunctional behavioral reactions that may be brought on by information systems development and operation.[1] The following discussion is based upon their research findings.

Behavioral Groups
Three basic behavioral groups are involved and/or affected by information systems development and operation. These groups are as follows:

1. *Operating personnel:* Includes all nonmanagerial personnel in the organization except those classified as technical staff. Operating personnel are further subdivided into nonclerical and clerical. Nonclerical personnel provide the inputs to the system (sales, orders, time cards, production order, etc.). Clerical personnel process the inputs and convert them into forms suitable for processing.

2. *Operating management:* Includes personnel in positions ranging from the first-line supervisor up to and including middle management.

3. *Top management:* Includes the chief executive officers of the organization. The president and vice-presidents are in this group.

---

1. Gary W. Dickson and John K. Simmon. "The Behavioral Side of MIS, Some Aspects of the People Problem," *Business Horizons*, 1970.

Dysfunctional Reactions

The three most common forms of dysfunctional reaction to information systems are explained as follows:

1. *Aggression:* This behavior represents attacks (either physical or nonphysical) intended to disrupt or destroy the system. The most dramatic aggression behavior is sabotage. Less dramatic (and more common) behavior occurs in the form of attempts to "beat the system" or to disrupt it by outsmarting it.

2. *Projection:* This behavior represents attempts to blame the system for difficulties or problems that are, in fact, not caused by the system. When a problem occurs, personnel simply claim that the problem is a computer error or problem.

3. *Avoidance:* This behavior represents defensive reactions by personnel who withdraw from, or ignore, the system because they are intimidated, frustrated, or simply reluctant to familiarize themselves with the new system and any changes it has brought about.

During the development and operation of a new information system, all three of the behavioral groups usually experience varying degrees of uncertainty, unfamiliarity, and increased complexity. Because each group has a somewhat different relationship with information systems, the groups are also affected in different ways by the new system. Usually, they exhibit different behavioral reactions. These reactions are discussed next and portrayed in Figure 10.3.

Operating Personnel

Clerical operating personnel are particularly affected by new information systems. Their jobs may be changed considerably or even eliminated. Work and interpersonal relations are usually modified. There may be increased work rigidity and time pressure. The usual response to such effects is projection behavior: all of the undesirable effects and any other ensuing problems are blamed on the computer and its technicians. Though clerical personnel may want to indulge in avoidance behavior, their low-level positions in the organization tend to make that impractical.

Nonclerical operating personnel are usually required to provide more input when a new system is installed (e.g. information on competitors, allocation of staff time). They tend to demonstrate aggression behavior because of the increased inconvenience imposed on their daily work activities to fulfill these information requirements.

Operating Management

Operating management is usually highly affected by new information systems. As a result of the additional information available from new systems, top management is usually more informed about the effectiveness and efficiency of operating managers. This often leads to closer evaluation and control of their activities.

Figure 10.3   **Relationship of Behavioral Groups to Behavioral Patterns**

	Aggression	Projection	Avoidance
Top Management			X
Operating Management	X	X	X
Operating Personnel			
Clerical		X	
Nonclerical	X		

Many of the decisions made by operating managers may be displaced by programmed decision making performed by the new information system. Programmed decisions usually create free time for managers. This increased free time allows operating managers to engage in other, more creative and constructive efforts. However, the overall effect as perceived by the managers is a change in their job content. Operating management, therefore, has reason to feel their status or power is threatened. They often experience feelings of insecurity. These feelings may be strong. If so, reactions of aggression, projection, and avoidance are apt to occur.

Top Management

Top management is usually less affected by, and concerned with, new information systems than other groups are. They may feel some insecurity because a technology they do not understand plays a larger role in their organization. However, their high-level positions provide them the option of avoiding new systems. They have little reason to exhibit projection or aggression behavior.

Top management is usually more involved with information systems that do more than transaction processing, information providing, and programmed decision making. In other words, they are more affected by systems that provide decision-support capabilities.

Guidelines for Minimizing Dysfunctional Behavior

The recognition of the nature and causes of dysfunctional behavior as it relates to new information systems does not, by itself, solve any problems. The following discussion pertains to steps or procedures that can be used to minimize dysfunctional behavior. There is no single best solution. Accordingly, the following guidelines are plausible solutions, but they are not absolute answers.

Operating Personnel

Nonclerical operating personnel generally react to new information systems with aggression behavior caused by what are usually increased requirements for input preparation. For example, salespersons may have to fill out new reports describing the factors that caused the loss of customer or client. To the salesperson, these reports may appear to be needless, additional work. However, information generated from the reports may be used to recognize patterns in the competitiveness of the company's product line. Such information can then be used to make decisions about product modifications or new product developments. These will, in turn, enhance the salespersons' ability to regain lost customers and acquire new ones. If salespersons understand and believe this strategy, they are less likely to react negatively to the additional input requirements.

The key issue is that if information is valuable, those providing it should be sold on the importance of collecting and processing it. This is best accomplished when top management communicates its support of the information system and its use. Top management is usually more effective than technical staff in gaining support for a new information system from operating personnel

Clerical operating personnel's common dysfunctional reaction of projection is caused by fears that their jobs may be changed adversely or even eliminated. This reaction is often brought on because clerical personnel are the last to know what is going on as new systems are being developed. The almost secretive atmosphere that surrounds the new systems creates concerns that something bad is about to happen. Such concern can be minimized if top management provides an atmosphere of openness and security about forthcoming changes. When clerical personnel are made aware of what is taking place and are assured of their job security, projection behavior can be reduced. It is virtually impossible for the systems analyst to create such an atmosphere without top management's involvement and support.

Operating Management

As computer technology first began to be used by organizations, much was said and written about the displacement of operating management. It appeared to many that the ability of the computer to process large volumes of data and generate solutions would reduce or eliminate the need for operating management. Understandably, operating managers were, and often still are, threatened by the computer. It is not too surprising that operating managers often exhibit all three forms of dysfunctional behavior: aggression, projection, and avoidance.

The computer can (and does) make a large number of decisions previously made by operating managers. It is quite efficient at making programmed decisions for highly defined and structured problems. However, the computer has not displaced operating managers. Rather, these managers have been released to address much more complicated and less structured problems created by more complicated organizational processes. This pattern is apt to continue. Therefore, the need for operating managers, with even greater skills, still exists.

The fact remains, however, that operating managers continue to experience anxiety as new information systems are developed and put into operation. They are often required to accept changes in job content as the computer takes over routine tasks. To minimize dysfunctional behavior resulting from these circumstances, top management should be open about forthcoming changes and include operating management in the design of new information systems. Only top management can convincingly stress that operating managers will play a more challenging and important role in the organization even though some of their decision-making activities will be relinquished to the computer.

Top Management

A foremost guideline for overcoming dysfunctional behavior toward new information systems is to obtain top-management involvement and support. In spite of the importance of doing so, getting top management involved can be difficult. As discussed earlier, top management's likely reaction to new information systems is avoidance. The systems analyst must take care to describe information systems problems in nontechnical terms when communicating with top management and soliciting their support. Top-level managers, like everyone, are much more comfortable with things they understand and, if they understand, are much more likely to get involved.

An additional means of gaining top-management support is to stress that information systems projects are consistently more successful from an operational, economical, and technical perspective when top management is involved and supports them. Indeed, eliminating avoidance behavior in top management can significantly contribute to eliminating dysfunctional behavior in the remainder of the organization.

General Guidelines

The preceding discussion addresses guidelines for minimizing dysfunctional behavior within each behavioral group. A few guidelines for minimizing dysfunctional behavior apply to all behavioral groups:

1. Information systems should be thoroughly tested and validated prior to implementation. A malfunctioning system can generate undesirable behavior on the part of all personnel.

2. The development and implementation schedule should be realistic. If personnel are rushed into a new system, problem and negative reactions are likely to occur.

3. The system should be kept as simple as possible, adequate training should be provided, and systems outputs should be designed to fit the needs of users. Otherwise, personnel tend to avoid using the system and will keep their own (private) manual information systems because they are afraid to use or depend on the computer-based information systems.

4. Any changes in job content resulting from new information systems should be recognized and adjusted for. In particular, job performance evaluations and the accompanying reward systems should be modified to reflect job changes. For example, if salespersons are required to fill out new competitive-analysis forms, the manner in which they perform this task should be evaluated and properly considered as part of their overall job performance. Alternatively, if a part of an employee's job is eliminated, he or she should no longer be evaluated on that basis. For example, if the computer begins to make inventory reorder decisions, the inventory manager should not be held responsible for inventory decisions that he or she cannot control.

As a final note, the importance of allowing those affected by change to participate in bringing about that change cannot be overemphasized. It is often thought that most people are simply reluctant to accept change. However, this belief is contradicted by the often enthusiastic manner in which people accept changes in such things as clothes and hair fashion. People enjoy driving the latest car or having the latest gadgetry (e.g., electronic calculators and microwave ovens). The reason for this apparent contradiction is that people, as consumers, have a great deal of control over these changes. They can decide by their purchasing behavior if they want to accept or reject new products. If these products were forced on people, they would likely react quite differently. By allowing organizational personnel to participate in or, even better, control organizational changes, acceptance and even enthusiasm is much more probable.

Evaluation

The final step (and, one of the most important steps) of the systems development cycle is evaluation. Evaluation provides the feedback necessary to assess the value of the information and the performance of the personnel and technology included in the information system. This feedback serves two functions. First, it provides information as to what adjustments to the information system may be necessary. Second, it provides information as to what adjustments should be made in approaching future information systems development projects.

There are two basic dimensions of information systems that should be evaluated. One dimension is concerned with whether the system was developed and is operating properly. The other dimension is concerned with whether the proper information system is provided.

Development Evaluation

Evaluation of the development process is primarily concerned with whether the system was developed on schedule and within budget. This is a rather straightforward evaluation. However, it requires that schedules and budgets be established in advance and that records be kept of actual performance and costs. Procedures for scheduling and budgeting information systems development are discussed in Chapter 11 under project planning.

Few information systems have been developed on schedule and within budget. In fact, many information systems have been developed without clearly defined schedules or budgets. The mystique and uncertainty associated with information systems development has often resulted in its not being subjected to traditional management-control procedures. Fortunately, an increasing experience base and better understanding of systems development by both technicians and management have resulted in a greater emphasis on planning and control of information systems development.

Operation Evaluation

The evaluation of the information system's operation pertains to whether the hardware, software, and personnel are capable and do, in fact, perform as they are supposed to. Operation evaluation answers such questions as

1. Are all transactions processed on time?
2. Are all values computed accurately?
3. Is the system easy to work with and understand?
4. Is terminal response time within acceptable limits?
5. Are reports processed on time?
6. Is there adequate storage capacity for data?

Operational evaluation (like development evaluation) tends to be relatively straightforward if evaluation criteria are established in advance. For example, assume a system is supposed to be capable of supporting one hundred terminals with less than a two-second response time. Evaluation of this aspect of system operation is easily done after the system is operational.

Information Evaluation

An information system may perform exceptionally. Computations may be correct; transactions may be efficiently processed; and reports may be distributed on time. However, a system that merely functions properly may not be a system that is proper for the needs of an organization. An information system should also be evaluated in terms of the information it provides. This aspect of information system evaluation is difficult. It cannot be conducted in the straightforward quantitative manner that is possible for development and operation evaluations.

The objective of an information system is to provide information to support the organizational decision system. Therefore, the extent to which information is relative or not relative to decision making is the area of concern in evaluating the information provided by an information system. However, it is a practical impossibility to directly evaluate an information system's support of decisions made in an organization. It must be measured indirectly.

A viable approach for indirectly measuring and evaluating the information provided by an information system has been proposed by Richard L. Nolan and Henry H. Seward.[2] Their approach is based on the concept that the more frequently a decision maker's information needs are met by an information system, the more satisfied he or she tends to be with the information system. Conversely, the more frequently necessary information is not available, the greater the effort (and frustration) required to obtain the necessary information, and hence, the greater the dissatisfaction with the information system. This concept is illustrated in Figure 10.4. Since satisfaction with an information system correlates with the ability of the information system to support decision making, satisfaction can be used as a surrogate to evaluate the information provided by an information system.

Measurement of user satisfaction can be accomplished using the interview and questionnaire techniques discussed in Chapter 6. If management is generally satisfied with an information system, it is reasonable to assume that the system is meeting the organization's requirements. If management is not satisfied, modifications ranging from minor adjustments to complete redesign may be required.

Maintenance

Most information systems require at least some modifications after development. This need arises from a failure to anticipate all requirements during systems design and/or from changing organizational requirements. The second of these, changing requirements, continues to impact most information systems as long as they are in operation. Consequently, periodic systems maintenance is a requirement of most information systems. Systems maintenance involves adding new data elements, modifying reports, adding new reports, changing calculations (e.g., payroll tax tables), and the like.

Maintenance can be categorized as either scheduled maintenance or rescue maintenance. *Scheduled maintenance* is anticipated and can be planned for. For example, the implementation of a new inventory coding scheme can be planned in advance. *Rescue maintenance* refers to previously undetected malfunctions that are not anticipated but require immediate resolution. A system that is properly developed and tested should have few occasions of rescue maintenance.

One problem that occurs in systems development and maintenance is that as more systems are developed, a greater proportion of systems analyst and programmer time is spent on maintenance. Figure 10.5 illustrates this relationship. As an author, I experience this same problem. Having written twelve books that require revisions every five years, I must spend most of my authoring time

---

2. Richard L. Nolan and Henry H. Seward, "Measuring User Satisfaction to Evaluate Information Systems," *Managing the Data Resource Function*, 1974, pp. 253–76.

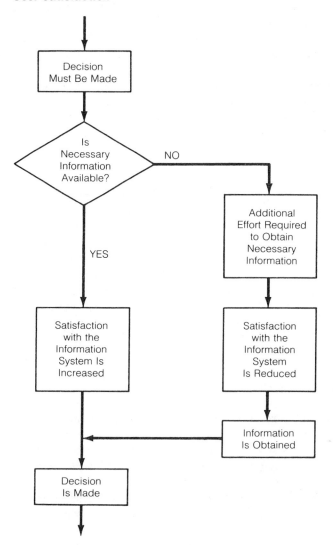

revising books as opposed to authoring new books. Fortunately, word-processing technology dramatically increases my productivity. Otherwise I could not keep up with the work.

Information systems departments can increase their capacity by

1. Adding more staff
2. Using more powerful production technology (e.g., fourth-generation languages)
3. Encouraging end-user computing (see Chapter 15)
4. Reducing maintenance by developing the right systems the right way

Step 4, "Reducing maintenance," is accomplished by properly conducting the systems development life cycle discussed in Chapters 5 through 10. In particular, note these points from the following chapters:

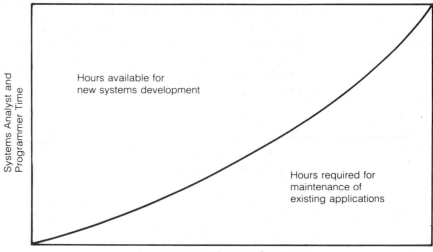

Chapter 5: Make sure the problem or opportunity is properly identified.

Chapter 6: Don't change it until you understand it; use structured interview with groups of users (JAD) to get conceptual information requirements right

Chapter 7: Use prototyping/heuristic design to work out the detail information requirements *before* the system is developed.

Chapter 9: Use good software engineering principles and CASE/workbench tools.

An information system may remain in an operational and maintenance mode for several years. The system should be evaluated periodically to ensure that it is operating properly and is still right for the organization. When a system becomes a problem (i.e., no longer satisfies the organization's needs) or new opportunities are available (e.g., new technology), an information system may be replaced by a new information system generated from a new information systems development cycle.

EDP Audit

Beyond the routine evaluation of an information system, a more formalized type of evaluation is an EDP (electronic data processing) audit. EDP audits can be conducted for more than one information systems concurrently. There are basically two types: The first type is the *financial audit*. This is concerned with whether the computer-based systems that support the organization's financial and accounting processing are conducted with adequate control, security, and asset protection. The second type is the *performance audit*. This is concerned with how effectively and efficiently the information system function is being managed.

For either a financial or a performance EDP audit, the age-old problem of auditing in general still exists. That is, to conduct a good audit, the auditor has to have as much, or more, expertise than those being audited. Nowhere is this problem more of an Achilles' heel than it is for EDP auditors. This is because the rate of technological development in the field of information systems requires continual updating of auditing methodology. Consequently, it is very difficult to truly control and audit computer systems.

Management must trust that EDP auditing will keep the honest, responsible people honest and responsible. However, a dishonest technical giant is in a powerful position to exploit the organization. This is not a problem that is going to be resolved in the short run. There is and will continue to be a shortage of technical giants, and those that exist are not usually interested in auditing as a career.

Financial Audit

Financial EDP audits are a technological evolution of manual audits. They tend to take one of three forms:

- Auditing *around* the computer
- Auditing *through* the computer
- Auditing *with* the computer

Auditing around the computer was the typical approach used in the early days of EDP auditing due to the lack of technical expertise of those conducting audits. The strategy was to treat the computer and computer programs as "black boxes" that accounting information went into and came out of. To audit a system, transactions were processed through the systems and cross-checked with manual processing to validate the accuracy.

Auditing around the computer is not likely to uncover evidence of fraudulent or mischievous actions hidden in computer systems. For example, consider the employee who put an instruction into a computer's payroll program telling the computer to destroy the entire payroll file in the event that he was terminated (i.e., in the event that the programmer no longer had an active record in the payroll file). The employee was promoted to another division and forgot to change the program. The payroll file was destroyed and considerable expense incurred to reconstruct it.

Auditing around the computer would not likely reveal such a problem in advance of it happening. To detect problems such as this requires actually going through the various instructions contained in the computer programs, or auditing *through* the computer, and requires a higher level of technical expertise on the part of the EDP auditor. Of course, the more sophisticated the technical staff, the better their ability to disguise any of their wrongdoings from the auditor.

Auditing *with* the computer means using the computer to select records for auditing and running various cross-checking totals. To do this type of auditing requires computer software capable of performing various audit functions and an auditor proficient at using the software.

One variation of auditing with the computer involves establishing a fictitious mini-company that is processed along with the organization's normal processing. The mini-company has its special set of accounts that can be updated and processed as the real company's accounts are processed. Audits can then be made to see if the computer system processes accounts properly on the mini-company separately from the real company.

As with auditing through the computer, auditing with the computer can be outsmarted by the superior technician. Both techniques, however, are more rigorous than auditing around the computer.

Performance Audit

As stated earlier, a performance audit is concerned with how effectively and efficiently the information system function is operating. A performance audit collects and evaluates information to answer such questions as

- Is the staff adequately trained?
- Is state-of-the-art technology and methodology being used?
- Are computing and staff resources being adequately utilized?
- Is there an information system plan?
- Are users satisfied with services?
- Does adequate documentation exist for all systems?
- Do information systems have an adequate backup and recovery plan?

Ostensibly, a performance audit tends to be an extensive question-and-answer session in which the auditors use an exhaustive checklist to evaluate the management and procedures used by an information systems function. Performance audits are often done because general management does not feel qualified to do them, or does not want to take the time to become familiar enough with information systems to evaluate their performance.

Performance audits have also become a fashionable way to terminate an information systems manager. In such situations, top management calls for a performance audit, explaining to the auditor that it is dissatisfied with the information system and would like an audit to tell it what is wrong. Information systems managers seldom fare well under such conditions. Not too surprisingly, the prudent information systems manager always checks the source and the motivation of a performance audit.

Who Performs an Audit

Financial audits are typically performed by accounting firms such as a "big eight" firm. Performance audits are often performed by accounting firms who have expanded their services to include EDP performance audits. Performance audits are also performed by consulting firms or individuals who specialize in information systems consulting.

One interesting conflict of interest that seems to be ignored in selecting auditors is that organizations will often use the same accounting or consulting firm for both EDP consulting and EDP auditing. This creates a situation where an auditing firm is auditing its own recommendations—hardly an advisable situation.

As management is becoming more knowledgeable and comfortable with information systems, they have become more willing to become active in performance evaluations of information systems. This is a positive trend. Information systems should be evaluated like any other organizational function. Outside, independent evaluations are useful, but they are no substitute for an internal manager's judgment. Outside audits should assist management, not do its job for it.

Summary

Implementation, though following development and testing, is a process that should parallel the systems development process. It begins with considering operational feasibility and should be considered during every step of the SDLC.

Behavioral considerations associated with systems development are extremely important but frequently overlooked. Dysfunctional behavior resulting from poorly managed implementation can be disruptive and even prevent the success of an information system. The four behavioral groups affected by an information system include operating personnel, operating management, technical staff, and top management. People in these groups react to new information systems by engaging in behavioral reactions that include aggression, projection, and avoidance. Dysfunctional behavior can be minimized by properly considering why it occurs and taking appropriate actions.

A key part of implementation is management and clerical training, followed by conversion from the old system to the new. The conversion process may or may not employ parallel processing. In either case, the conversion sequence should be clearly defined.

After the system is implemented, it should be evaluated. Evaluation is concerned with determining whether the system works right (i.e., was developed and operates properly) and whether it is the right system (i.e., provides the proper information). Once evaluated, an information system remains in maintenance mode until it is replaced by a new information system. Audits of a system involve financial and performance audits.

Exercises

1. Discuss the importance of reactivating the project team prior to implementation.
2. Discuss the behavioral groups of top management, middle management, and operating personnel as they relate to dysfunctional behavior during systems development.
3. Discuss some of the guidelines for minimizing dysfunctional behavior.
4. When is parallel processing not practical? Give examples.
5. Discuss how management can evaluate the value of the information provided by an information system.
6. Discuss how a system can work right but not be the right system, and vice versa.
7. Discuss how maintenance costs can be reduced.
8. Why is it so difficult to audit EDP?
9. What are the differences between a financial audit and a performance audit?

Selected References

Alavi, M. and Henderson, J. "Evolutionary Strategy for Implementing a Decision Support System" *Management Science* 27 (1981); 1309–25.

Alter, Steven. "How Effective Managers Use Information Systems." *Harvard Business Review*, November–December 1976.

Alter, Steven. "Implementation Risk Analysis." Working paper series, University of Southern California, Graduate School of Business Administration, Los Angeles, Calif., 1976.

Anderson, John C., and Narasimhan, Ram. "Assessing Project Implementation Risk: A Methodological Approach." *Management Science*, June 1979.

Anderson, N. B.; Hedberg, B.; Mercer, B.; Mumford, E.; and Sole, A. *The Impact of Systems Change in Organizations.* Germantown, Maryland: Sisthoff and Noordhoft, 1979.

Bardoch, Eugene. *The Implementation Game: What Happens after a Bill Becomes Law.* Cambridge, Mass.: MIT Press, 1977.

Bass, Bernard M., and Leavitt, Harold J. "Some Experiments in Planning and Change." *Management Science*, September 1963.

Bavelas, A. "Communication Patterns in Task-Oriented Groups." *Journal of the Accoustical Society of America*, November 1950.

Bennis, Warren G., Benne, Kenneth O., and Chin, Robert. *The Planning of Change.* New York: Holt, Rinehart, and Winston, 1963.

Blanning, Robert. "The Decision to Adopt Strategic Planning Models." The Wharton Working Papers, The Wharton School, University of Pennsylvania, Philadelphia, Penn., 1979.

Churchman, C. West. "Managerial Acceptance of Scientific Recommendations." *California Management Review*, Fall 1964.

DeSanctis, Gerardine. "An Examination of Expectancy Theory Model of Decision Support System Use." *Proceedings of the Third International Conference on Information Systems*, Ann Arbor, Mich.: 1982.

Dickson, Gary W. and Simmons, John K. "The Behavioral Side of MIS." *Business Horizons*, August 1970, pp. 59–71.

Dickson, G. W., Simmons, J. K., and Anderson, J. C. "Behavioral Reactions to the Introduction of a Management Information System at the U.S. Post Office: Some Empirical Observations." In *Computers and Management*, 2d ed., edited by D. Sanders. New York: McGraw-Hill, 1974.

Doktor, Robert, and Hamilton, William F. "Cognitive Style and the Acceptance of Management Science Recommendations." *Management Science*, April 1973.

Ginzberg, Michael. "Steps Toward More Effective Implementation of MS and MIS." *Interfaces* 8 (1978): 57–63.

Ginzberg, Michael J. "A Process Approach to Conducting Management Science Implementation." Doctoral dissertation, MIT, Cambridge, Mass., 1975.

Ginzberg, Michael J. "Key Recurrent Issues in MIS Implementation Process." *MIS Quarterly*, June 1981.

Ginzberg, Michael J. "Steps Towards More Effective Implementation of MS and MIS." *Interfaces*, May 1978.

Janion, J. "Post Office Shows How Not to Develop Information System." *Computerworld*, August 4, 1971.

Keen, Peter, G. W. "Information Systems and Organizational Change." *Communications of the ACM* 24 (1981): 24–33.

Keen, Peter G. W. "Information Systems and Organizational Change." *Communications of the ACM*, January 1981.

Kolb, David, and Frohman, Alan L. "An Organization Development Approach to Consulting." *Sloan Management Review*, Fall 1970.

Lewin, Kurt. "Group Decision and Social Change." In *Readings in Social Psychology.* Edited by T. Newcomb and E. Hartley. New York: Holt & Co., 1952.

Little, John D. "Models and Managers: The Concept of a Decision Calculus." *Management Science*, April 1970.

Lucas, Henry C., Jr. *Implementation: The Key to Successful Information Systems.* New York: Columbia University Press, 1981.

McKenney, James L., and Keen, Peter G. W. "How Managers' Minds Work." *Harvard Business Review*, May–June 1974.

Mumford, Enid, Mercer, E., Mills, D., and Weir, M. "Problems of Computer Introduction." *Management Decision* 1 (1972).

Nolan, Richard L., and Seward, Henry H. "Measuring User Satisfaction to Evaluate Information Systems." In *Managing the Data Resource Function*, Edited by Richard Nolan. St. Paul, Minn.: West Publishing. 1974, pp. 253–76.

Pettigrew, A. *The Politics of Organizational Decision Making*. London: Travistock, 1976.

Powers, Richard and Dickson, Gary W. "MIS Project Management: Myths, Opinions, and Reality." *California Management Review,* Spring 1973.

Radnor, Michael, and Neal, Rodney. "The Progress of Management Science Activities in Large U.S. Industrial Organizations." *Operations Research,* March–April 1973.

Radnor, Michael and Rubenstein, A. H. "Implementation in Operations Research and R&D Organizations." *Operations Research,* November–December 1970.

Roberts, Edward B. "Strategies for Effective Implementation of Complex Corporate Models." *Interfaces,* November 1977.

Robey, Daniel. "User Attitudes and MIS Use." *Academy of Management Journal,* September 1979.

Rudelius, C. W., Dickson, G. W., and Harley, S. W. "The Little Model That Couldn't: How a Decision Support System Found Limbo." *Systems, Objectives, and Solutions,* August 1982.

Schein, Edgar. "Management Development as a Process of Influence." *Industrial Management Review,* May 1961.

Schein, Edgar H. *Process Consultation: Its Role in Organizational Development*. Reading, Mass.: Addison-Wesley, 1969.

Schein, Edgar. *A Theory of Group Structures, vol. 1: Basic Theory*. New York: Gordon and Breach, 1976.

Schultz, Randall L., and Sleven, Dennis P., (eds.) *Implementing Operations Research Management Science*. New York: American Elsevier, 1975.

Schultz, Randall and Slevin, Dennis P., eds. *Implementing Operations Research/ Management Science*. New York: American Elsevier Publishing Co., 1975.

Vroom, Victor H. *Work and Motivation*. New York: Wiley, 1964.

Wetherbe, James C. *Systems Analysis for Computer-Based Information Systems*. 1st ed. St. Paul, Minn.: West Publishing, 1979.

_____. *Executive's Guide to Computer-Based Information Systems*. Englewood Cliffs, N. J.: Prentice-Hall, 1983.

Wetherbe, James C. and Dickson, Gary W. *Management of Information Systems*. New York: McGraw-Hill, 1984.

_____. *Cases in Management of Information Systems*. New York: McGraw-Hill, forthcoming.

Zmud, Robert W. and Cox, James F. "The Implementation Process: A Change Approach." *Management Information Systems Quarterly* 3 (1979); 35–43.

Zmud, Robert W. "Locus of Control, Ambiguity Tolerance, and Information Design Alternatives: Correlates of Decision Behavior." *Proceedings of the American Institute for Decision Sciences* 1 (1979).

Zmud, Robert W. and Cox, James F. "The Implementation Process: A Change Approach." *MIS Quarterly,* June 1979.

Chapter

11

Project Management for Systems Development

Introduction

Project management is a systems approach to management that uses personnel from different organizational functions to complete specific tasks. These are special, one-time tasks that cannot be effectively handled by traditional organizational structures or departments.

As pointed out in Chapters 4, 5, and 6, most information systems transcend departmental boundaries, and their development cannot be the exclusive responsibility of any one department. Rather, a joint, team effort is required. Project management is an organizational structure designed to minimize disruption to the regular organizational tasks while allowing special projects to get completed. As a working definition, *project management* involves planning, scheduling, directing, and controlling human, financial, and technological resources for a fixed-term task that will result in the completion of specific goals and objectives.

Project management was originally developed for large-scale military projects (e.g., Polaris submarine). It became popular for space exploration (NASA) and complex industrial projects. Also as many organizations struggled with information system projects that were continually over budget, late, and generally unsatisfactory, project management techniques were found to be helpful and gained popularity for systems development. Project management, though a necessarily complex variation from traditional departmental structures for doing tasks, provides better control, better customer relations, shorter development time, lower costs, and improved quality and reliability.

Project management skills include knowledge of both scheduling/quantitative tools and organizational/behavioral skills. This chapter covers both categories of skills respectively.

Project Management Scheduling and Quantitative Tools

Project management provides an overall framework with which the system development life cycle (discussed in Chapters 5–10) can be planned, scheduled, and controlled. The fundamental tools used in project management include milestones, program evaluation and review technique (PERT), and Gantt charts.

Milestones

Milestone planning techniques allow projects to evolve as they are developed. Rather than try to fully predict all project requirements and problems in advance, management allows the project to progress at its own pace. *Milestones,* or checkpoints, are established to allow periodic review of that progress so that management can determine whether a project merits further commitment of resources, whether it requires adjustments, or whether it should be discontinued.

Milestones can be based upon time, budget, or deliverables. For example, a project's progress might be evaluated weekly, monthly, or quarterly (i.e., time). It might be evaluated after a certain amount of resources are used, such as $50,000 (i.e., budget). Or it might be evaluated after certain deliverables are completed or events occur such as a preliminary study. Figure 11.1 illustrates the simplicity

Figure 11.1 **Illustration of Milestones**

of a milestone chart. Remember, each milestone can be based upon time, budget, and/or deliverables.

PERT

A commonly used planning technique is PERT (*program evaluation and review technique*). A PERT plan diagrammatically represents the network of tasks required to complete a project (hence, it is sometimes referred to as a *network chart*). It explicitly establishes sequential dependencies and relationships among the tasks. A PERT diagram consists of both activities and events. Activities are defined as time- and resource-consuming efforts required to complete a segment of the total project. Events represent the completion of segments, or parts of segments, of the project. Activities are represented by solid lines with directional arrows; events are represented by circles. Dotted lines are used to represent sequential dependencies where they exist, but no task has to be performed to progress from the first event to the second one. Activities and events are coded or described to designate their functions in the overall project.

Figure 11.2 shows a PERT chart and an accompanying project table that defines the responsible personnel, and estimated and actual times and costs, for each task in the project. By comparing the actual times and costs with the planned times and costs in the PERT chart, management can monitor and control project performance.

A final advantage of PERT is that total time required to complete the project can be determined by locating the longest path (in terms of time) in the chart. This path is referred to as the *critical path*. For scheduling purposes, any delay of tasks in the critical path results in an equivalent delay on the overall project.

Gantt Charts

Gantt charts are a planning technique that, like PERT, provide definitions of tasks to be performed and when they are to start and finish (Figure 11.3). A Gantt chart does not show sequential dependencies, as a PERT chart does. For example, in Figure 11.3, the chart does not show that analysis must be done before design begins. Consequently, a Gantt chart does not have as much information as a PERT chart, but it is much easier to prepare.

A particularly nice feature of Gantt charts is that a macro-Gantt chart can be factored into one or more levels of micro-charts. For example, in Figure 11.3, the line labeled "Analysis" could be made into a separate Gantt chart consisting of the subtasks that constitute analysis (e.g., feasibility study, interview, flow-charting). Such micro-charts provide more detail, allow for specific assignment of responsibility, and facilitate better estimating of time requirements.

Figure 11.2   A PERT Chart and Accompanying Project Table

PERT Chart

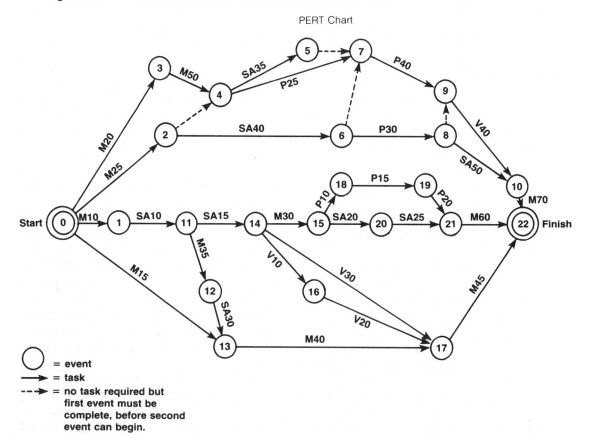

○ = event

→ = task

--→ = no task required but first event must be complete, before second event can begin.

Project Table

Task	Responsibility	Planned Start	Planned Completion	Actual Start	Actual Completion	Planned Costs	Actual Costs
M10	JCW	1/10	1/15	1/10	1/15	N/A	N/A
M15	SCM	1/10	1/30	1/10	1/15	N/A	N/A
M20	JCW	1/15	1/30	1/15	1/30	N/A	N/A
M25	JCW	1/15	1/30	1/15	2/02	N/A	N/A
SA10	NJW	1/15	1/30	1/15	1/30	$ 800	$ 670
M50	SCM	1/30	2/05	1/30	2/04	N/A	N/A
SA15	NJW	1/30	2/07	1/30	2/06	$1,200	$1,250
SA40	BWB	1/30	2/25	2/02	2/26	$3,600	$3,200
SA35	RER	2/05	2/20	2/08	2/20	$ 500	$ 750
P20	JLP	10/01	10/25	10/05	10/25	$1,100	$1,400
SA50	NJW	10/25	11/10	10/20	11/10	$1,500	$1,500
V40	IBM	10/25	11/10			$5,000	
M60	JCW	10/25	11/20			$2,000	
M45	SCM	10/25	11/20			$1,000	
M70	JCW	11/10	11/20			$2,500	

Legend:
M = Managerial   SA = Systems Analysis   P = Programming   V = Vendor

Figure 11.3   Illustration of a Gantt Chart

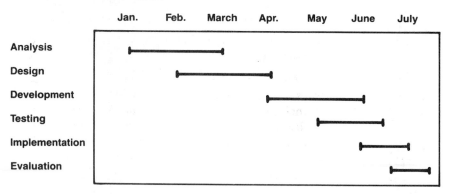

Project Planning

Perhaps the greatest mistake in project planning is to establish budget and schedule first. For example, management decides to put in a new order-entry system in nine months and is willing to spend $500,000. This leaves one important issue undefined—what exactly the new system will do. Setting budget and time frames prior to defining the system constrains design, often without the people involved realizing it. By default, management has defined the limits of the new system.

The proper sequence for managing a project is to *first* get a good functional definition of what the system is to do and then have people with experience and expertise in information system project management develop a budget and schedule. If management cannot live with the schedule or budget, then reductions in capability of the new system can be made to improve the schedule and/or budget.

In developing a budget, schedule, and specifications for a system, several properties of projects and of project management should be understood and considered. The project properties that most significantly influence the overall nature of a project are the following:

1. *Predefined structure:* The more predefined structure a project has, the more easily it can be planned and controlled. For example, transaction-processing applications inherently have a great deal of predefined structure. Their structure reduces the difficulty of designing computer applications to process them. Conversely, decision-support systems (e.g., a market forecast system) are not usually well-structured. They require considerable definition and structuring before they can be computerized.

2. *Stability of technology:* The greater the experience with a given technology to be used for a new system, the more predictable the systems development process is. On the other hand, when a new information system is to use new and unproven hardware and/or software, many unforeseen problems that impede the development process may arise.

3. *Size:* There is an inverse relationship between project size in terms of person-years and costs, and the ability to accurately plan the number of person-years and the costs that will be required to complete the project. That is, the larger the project, the more difficult it is to estimate the resources required to complete it.

4. *User proficiency:* The more knowledgeable and experienced user-managers are with their functional area, and the more knowledge and experience they have

in developing systems, the higher their "user proficiency" will be and the easier it is to develop a system for them. Less knowledge and experience among users results in greater difficulty in developing systems.

5. *Developer proficiency:* The more knowledge and experience the systems analyst assigned to a project has, the easier will go the project, and vice versa.

Any given project can possess any variation of each of the preceding properties. For example, a project can have predefined structure but use unstable technology or it can be a massive undertaking but have low user and developer proficiencies (e.g., most initial on-line airline reservation systems could be described in this fashion).

The properties of a very straightforward project are the following: (1) predetermined structure, (2) use of stable technology, (3) small size of project, and (4) high user and developer proficiency. The more any of these properties deviates in the opposite direction, the less straightforward and therefore more difficult it is to plan the project.

Techniques for Project Management

The techniques for managing information systems discussed earlier in this chapter can be categorized as informal (milestone), formal (PERT) and in between (Gantt). The technique appropriate for a given project is contingent upon the properties of the project. Informal project management techniques are appropriate for projects that are neither straightforward or predictable. Formal techniques are better suited for projects that are relatively straightforward and predictable.

Guidelines

The inappropriate application of formal or informal project management techniques can have unfortunate consequences, including project failure. The use of informal techniques for a straightforward project needlessly forgoes planning definition and control. For example, PERT and Gantt planning techniques provide structure for time and cost estimates. To forgo such planning is unfair to the organization and to the systems developers, whose performance cannot be evaluated as accurately as it could be if more formal techniques were used.

The use of formal techniques for projects that are not straightforward or predictable generally results in dysfunctional constraints on what needs to be a relatively innovative and creative process. Systems developers are often forced to cut corners and stifle innovative processes in order to keep the project on schedule and within budget. When approaching new areas of systems development, there must be sufficient slack allowed to nurture innovation.

The techniques used for projects can be combinations of formal and informal techniques. For example, if the specific project tasks and their sequential dependencies are known, but it is not known how long they will take or how much they will cost, a PERT chart can be constructed without time and cost estimates. Such an approach provides more definition than a milestone approach, without unduly constraining the project time or producing cost estimates that may be unrealistic. Figure 11.4 illustrates the relationship between project characteristics and the selection of project planning techniques.

Prototyping and Project Planning

In Chapter 7 the notion of constructing a prototype and heuristic design was discussed. Ideally, a project team should not commit to any specific budget or

Figure 11.4   Effect of Properties of Project on Selection of Project Planning Technique

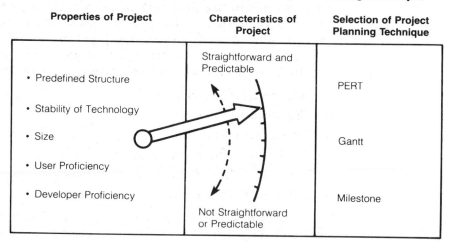

schedule until the users are satisfied with a working prototype. Until the users agree on the prototype, it is impossible to determine what they want from the project. Once this "deliverable" is accurately defined, both budgets and time estimates will be more realistic. This was done in the project proposed in Chapter 5 (Figure 5.2), where estimates were very rough until a prototype was completed. Prototyping was estimated at two to four weeks.

Consider the analogy of buying a car. A family might decide it wants a new car. "New car" is not a precise definition. Cost, availability, and functionality will influence the final decision. Before the family determines exactly how much it wants to spend or exactly what it wants in a car, it typically goes through a trial-and-error process of test-driving, reading consumer reports, and comparison shopping. It then makes a decision, often adjusting what it is willing to spend and how long it is willing to wait so that its functional requirements are best met.

This is a simplistic example. However, strange as it may seem, when it comes to information systems, management often sets budget and schedule first with little idea of the limitations or extravagance impact of those decisions. New systems should be taken for a "test-drive." A good strategy for the project team is to get management to determine their requirements on a milestone basis through the creation of a working prototype. Desired functionality can be accurately resolved for the "target" system, including trade-offs of cost versus benefit for certain features. Prototyping does not take long—a crude prototype can be created in a day or two. Refinement might take days to weeks.

Once the target system is resolved, techniques such as Gantt and PERT can be extremely useful for determining the cost and time required to deliver the target system.

Detailed Project Planning

Detailed project planning requires nailing down cost, time, and deliverables for a project. The most precise way of doing this is through a PERT chart. However, to develop to a PERT chart a planner must start at a conceptual milestone level, add the detail necessary to construct a Gantt chart, and then construct a PERT chart. Accordingly, the framework provided in Figure 11.4 indicates for some projects it will be easier and quicker to establish PERT charts than for others. That is, ill-defined, and therefore not straightforward, projects must stay in the milestone mode of operation longer than straightforward projects.

An accurate PERT chart cannot be constructed until the project team knows

- Tasks needed to be completed
- Time estimates for each task
- Cost estimates for each task
- Which task(s) immediately precedes which task
- Which task(s) immediately follow which task
- Which tasks can be performed concurrently

Coming up with the preceding information is easier for familiar projects than for unfamiliar projects. For example, it is easier to determine the information needed for preparing a routine term paper than it would be to determine the information needed for performing brain surgery. The more that is understood about a project, the more quickly a detailed budget and schedule information can be derived for PERT.

Once completed, the PERT chart can reduce project uncertainties by determining what tasks are crucial, how time delays influence project completion, where slack exists, alternatives, ways to check progress, and a means for management reporting.

Constructing a PERT Chart

Excellent software tools are available through various CASE products to support the graphics and word processing associated with constructing both Gantt and PERT charts. Expert systems are also available to support the logic and reasoning processes. However, for educational purposes it is important to know the basic mechanics of the tools. Once these are understood, computer-based support can be properly used.

A completed PERT chart was presented in Figure 11.2. The question is how do we get from *here* to *there?* The first step is to identify the *events,* or *deliverables,* that will occur during the project. For example, in developing an information system, a *macro* list of events might include:

1. Preliminary proposal
2. Prototype constructed
3. Prototype approved
4. Detail design complete
5. Software developed
6. Quality assurance completed
7. Testing completed
8. Training completed
9. Conversion completed
10. Post implementation review

The events can be converted into a Gantt chart as illustrated in Figure 11.5*a*. Note that activities such as training can be overlapped or done in parallel with

---

Figure 11.5a   Gantt Chart for Sample System

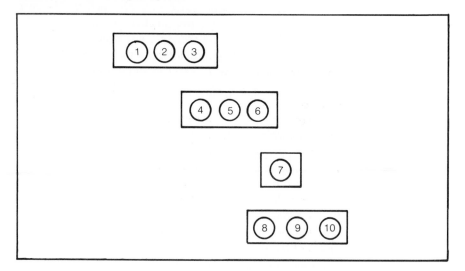

other activities. By considering the sequential dependencies and interrelationships of activities and events, a PERT chart can be constructed (Figure 11.5*b*). Remember, lines represent the activities leading up to bubbles, which are events. Therefore, event 9, "Conversion," requires that both events 7, "Testing," and 8, "Training," have occurred before the activities leading to the "Conversion" can begin.

Figures 11.5*a* and 11.5*b* are high-level project definitions. As we used general systems theory techniques to decompose or factor down large computer programs in Chapter 9, we can similarly factor down a project. For example, in Figure 11.5*b*, event 5, "Software development," can be factored into

1. Determination of computer programs to be written

2. Programs assigned to programmers

Figure 11.5b   Gantt Chart Converted into PERT Chart

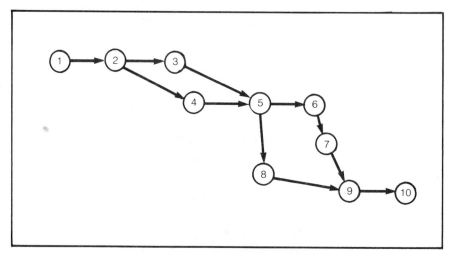

3. Design overviews completed
4. Structural charts completed
5. Program coding completed

These tasks can be further factored into the steps necessary for each single computer program. As this level of detail is reached, a PERT chart looks more like the one in Figure 11.2. Note that a management reporting system can also be constructed as depicted in the bottom of Figure 11.2.

How far should a project be factored down? This is a judgment call, because no matter how much detail is provided, much is still implied in the plan. For example, construct a list of events and a Gantt and PERT chart showing how you would move from where you are sitting as you read this text to the nearest public bathroom. Think of how many crucial details you might leave out because you would take them for granted (e.g., did you open the door before you entered the bathroom? With what? Did you consider if there was a doorknob or if the door could be pushed open? Did you check whether the bathroom was labeled men's or women's?

As a general guideline, projects should be factored to a detail level in which each task is discrete, has scheduled start and stop points, can be uniquely assigned to a single person or work unit, and can be uniquely tracked for budget and schedule purposes. Also, any implied tasks should be easily understood.

Reducing Schedule Time and Cost

Pressure usually exists to reduce the time and/or cost required to complete a project. Such reductions are desirable to the extent that they can be accomplished without unduly compromising project quality. The basic techniques for reducing project time or cost are to

1. Reduce functionality of the final system
2. Eliminate parts of the project
3. Add more resources
4. Substitute less time-consuming activities
5. Eliminate slack.

Steps 1 through 4 usually require some compromises or trade-offs in the system. Step 5, however, is a way to literally "get something for nothing." Eliminating slack time is simply tightening up the "loose ends" throughout a project by focusing on the critical path, as discussed next.

Critical Path and Slack Time

Earlier in the chapter the notion of the *critical path* was explained as the longest or most time-consuming path through the PERT chart. Any delay to the critical path translates to an equal delay to the overall project. Ways often exist to move resources from noncritical paths to the critical path, thereby reducing overall completion time. To do this requires identifying *slack time* and, when possible, making adjustments. Slack time can be defined as

$$\text{SLACK TIME} = T[L] - T[E]$$
where $T[L]$ = latest allowable date for event to occur
$T[E]$ = earliest expected date for event to occur

In other words, slack time is the differential between scheduled completion date and required completion date to meet the critical path.

Example: Reducing Project Schedule

To illustrate how to identify and reduce slack, consider Figure 11.6a, in which the critical path is identified as the darker line and slack time is documented as well.

By moving resources from task ②→④ to ②→③, task ②→③ is reduced by one week, and task ②→④ is increased by one week. Due to the three-week slack time that existed in ②→④, this adjustment causes no problem for ②→④; and due to the improvement in ②→③, the overall project is completed one week earlier because of a one-week reduction in the critical path.

The process of improving a project schedule is usually an iterative/adaptive process as depicted in Figure 11.7.

Mythical Person-Month

When making project adjustments, care must be taken to avoid the *mythical person-month* trap. A common management error is to assume that persons and person-months (or whatever time period is used) are directly interchangeable. For ex-

Figure II.6a

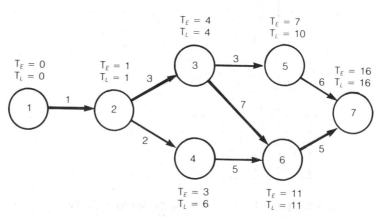

Note: Heavier line is critical path.

Figure 11.6b **Adjusted PERT Chart with Slack Time Removed and Scheduled Completion Time Improved**

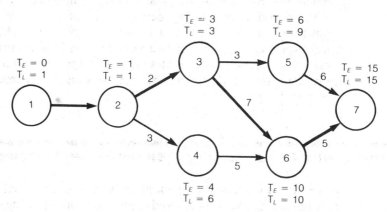

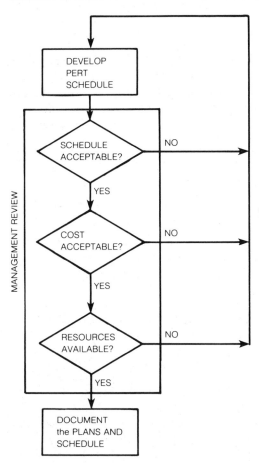

ample, a project requiring eight person-months for four people to complete (i.e., two months elapsed time) usually cannot be completed in only one month by adding four more people to the project. The complex interrelationships of most projects result in diminishing returns when more than a workable number of personnel are assigned to a project. If too many personnel are assigned to a project, they can literally get in one another's way to such an extent that the additional personnel are more a hindrance than an asset to project completion.

The relationship of adding personnel to a project to the time required to complete the project is shown in Figure 11.8. Note that additional personnel expedites project completion up to a point, but then it begins to delay the project. There's an old saying that illustrates the mythical person-month well—"A woman can have a baby in nine months, but you can't put nine women on the job and get a baby in one month."

Organizational and Behavioral Issues for Project Management

Because information systems typically transcend departmental boundaries (e.g., order-processing affects sales, credit, inventory, and shipping), the systems de-

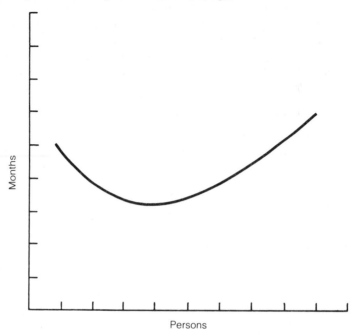

velopment effort requires input from all areas affected as well as the design expertise of information systems analysts. Information systems project management involves planning, scheduling, directing, and controlling organizational resources and managing personnel from various departments assigned to a project team. Large-scale projects can take months, and in rare cases, years to complete.

Anyone experienced with an information project team can attest that the diverse organizational perspectives and loyalties of team members and the temporary nature of a project team assignment complicate the development of a cohesive team effort. Behavioral and leadership problems have led to the demise of many well-planned information systems projects.

Most of the literature on managing an information systems project has focused on the technical and quantitative techniques of project management, offering little insight into organizational and behavioral issues. However, recently introduced team-building concepts and techniques offer some beneficial and practical tools that a project manager can use to improve the performance of a project team.

Successful Project Team Defined

A team can be broadly defined as a group of people working together in a coordinated effort. However, different teams take on quite different characteristics depending on the tasks they are to accomplish, and team members should be selected with the tasks in mind. Traditionally, participants for an information project team are selected and evaluated according to their skills, experience, and availability. Their ability to function as a part of a team has usually been undervalued or entirely overlooked. Nevertheless, this ability is crucial to the success of an information project team.

Consider a case in which one member of an information systems project team takes great pleasure in demonstrating his superior technical expertise by ridiculing nontechnical team members. He is more interested in showing off than in furthering the group effort. This behavior will likely produce antagonism in other team members and make a cohesive team effort impossible.

Project teams tend to be successful when team members

- Understand and agree on team priorities and goals
- Can contribute to the team's plan of action
- Feel free to express their feelings—negative and positive
- Trust in the support of their teammates
- Are willing to endure inconveniences for the sake of the team
- Recognize and emphasize one another's strengths instead of pointing out and criticizing weaknesses
- Are managed by a good leader who instills a sense of optimism and enthusiasm.

A Team-Building Model

Generally, a team with the desired characteristics just listed cannot be found; it must be made. A general-purpose model for team building is provided in the following sections. It is just one approach to team building, however, and there are many acceptable variations.

Ensure Team-Leader Commitment
The first step in team building is to make sure that the leader is committed to building an effective team. He or she must welcome and respond to both positive and negative feedback.

Establish Team-Building Objectives
The team should clearly define how it expects to benefit from team-building exercises. For example, team members should determine how they would like the team's behavior to differ after each team-building exercise.

Design the Interview Format
A structured interview format should be selected or designed for individually interviewing each team member. These interviews are important because they uncover the differing personalities on a team. People bring to a team different attitudes and perceptions about themselves and other team members. The more accurately team members understand themselves and one another, the greater the likelihood is that they will be able to work together effectively and efficiently. Various instruments can be used in designing the interview. Some of the more popular include

- LIFO (see Stuart Atkins, *LIFO Training Discovery Workbook,* Beverly Hills, CA: Stuart Atkins, Inc, 1978)
- Q-SORT (see Gary Dickson and James C. Wetherbe, *The Management of Information Systems,* New York: McGraw-Hill, 1985)

- Johari Window (see Jay Hall, "Communication Revisited," *California Management Review,* Spring 1973)
- Personality Profile (issued by Performax, Minneapolis)
- Myers Briggs Type Indicator (issued by Consulting Psychologists Press Inc, 577 College Ave, Palo Alto, CA)

These instruments are particularly useful for developing personality profiles. The Q-SORT method also gives perspective on an individual's motivational profile. Figure 11.9 is an instrument of *participative leadership communication skills* that can be used to evaluate the communication skills of team members.

In addition to these standard instruments, tailored survey questions should be developed that allow interviewees to articulate their perspectives on the following issues:

- Strengths and weaknesses of the team as a whole and of individual members
- Long-term critical issues
- Quality of leadership
- Other issues defined by the team.

Figure 11.9   Participative Leadership Communication Skills Survey

	Frequently	Occasionally	Seldomly
1. When listening to a subordinate, I often tune out and start thinking about something else.	☐	☐	☐
2. In a group meeting, I often think about how I will respond to what is being said before the speaker is through.	☐	☐	☐
3. I get distracted easily when people are talking to me.	☐	☐	☐
4. I let my emotions control my communications.	☐	☐	☐
5. I watch for noverbal cues that a subordinate may be giving me when he or she is speaking.	☐	☐	☐
6. When someone disagrees with me, I try to keep an open mind as they share their point of view.	☐	☐	☐
7. In a group meeting, I encourage each person to participate in the discussion.	☐	☐	☐
8. People say that I communicate clearly and openly.	☐	☐	☐
9. I am tactful and aware of people's feelings when discussing controversial issues.	☐	☐	☐
10. My subordinates would describe me as a good listener.	☐	☐	☐

Conduct the Interviews

Before the interviews begin, a disinterested interviewer who is not a member of the team should be appointed. The interviewer should have strong interpersonal skills—possibly someone from the human relations or personnel department. Team members should meet to review the objectives, process, and scheduling of the interviews; the interviewees' role; and the questions to be asked.

The individual interviews can then be conducted. Each should last about an hour. The interviewer should summarize the results of each interview to develop group and individual profiles.

Conduct the Team-Building Session

After the profiles are complete, the interviewer can conduct the team-building session. This session, which generally takes a day, focuses on the following issues:

- Styles of team leadership
- Qualities of effective teams
- Summarized survey results (without citing responses from individual team members)
- Action planning for critical team issues
- Sharing personal profiles (but not personal feedback).

Provide Individual Feedback

Feedback pertaining to individual strengths and weaknesses should be communicated confidentially. The interviewer should provide written feedback in a sealed envelope to each team member and should be available to discuss the feedback with the individuals if they so desire.

Some managers are concerned that team members' feelings may be hurt by the anonymous feedback from other team members. However, most individuals receive an inadequate perspective on how their behavior is helping or hurting their career. As Kenneth Blanchard and Spenser Johnson point out in *The One-Minute Manager*, "Feedback is the breakfast of champions."[1] For example, if a person views himself as cooperative, but team members view him as uncooperative, all can benefit by resolving those perceptions.

Of course, it is advisable in some instances to edit the feedback provided to individuals—for example, if the feedback is too personal or is not provided in the spirit of team building.

Conduct a Follow-up Session

Team-building efforts should include a follow-up session that should usually be held two to four months after the original team-building session and should be aimed at resolving any issues that have developed in the interim. During the follow-up session, progress on the action plans developed during the team-building session should also be evaluated.

Who Should Lead an Information Systems Project?

Project leadership has evolved through several stages during the past thirty years. In the early days, information systems project teams comprised technical staffs

---

1. Kenneth Blanchard and Spenser Johnson. *The One-Minute Manager* (New York: William Morrow, 1982).

and were led by a company's "technical giant." Users often perceived these early project teams to be groups of commandos who preyed on the user community. Although systems developed by the teams often worked correctly from a technical standpoint, they rarely solved the users' problems.

The late 1960s witnessed the era of user involvement. Users were included on information systems project teams; however, projects were usually led by an information systems professional. Undoubtedly, the user involvement had some value, but users generally were not influential team members for two reasons. First, they were often intimidated by the technology; second, they had other responsibilities that took priority over project membership.

The next major step in project leadership was to assign leadership to a user. Although user management agreed with the idea in principle, it balked at the time commitment required—usually at least twenty hours a week for six to nine months. User management responded by assigning project leadership responsibility to their most expendable person, not necessarily the most qualified. This generally caused unsatisfactory results.

Current thinking on project leadership is that the project should be headed by a talented, high-level end-user whose responsibility transcends those business functions affected by the new system. This person must be given adequate time away from his or her current responsibilities to head the project. After all, which is more important to the user area? The next six to nine months (the life of the project) or the next six to nine years (the life of the system resulting from the project)? Furthermore, which takes more talent? Managing the status quo of the user operation or creating the future user operations? Given the answers to these questions, it does not make sense to shortchange information systems project management.

Even though a user should be in charge of a project, a capable systems analyst will quickly find that much authority and responsibility is delegated to him or her. Smart managers will wisely defer to the systems analyst, particularly on technical issues. Consequently, the systems analyst becomes an informal leader. The wise systems analyst will give much credit to the formal leader; the wise formal leader will give credit to the informal leaders.

Leadership Styles

Figure 11.10 provides a continuum of different leadership styles. No single style is appropriate for all situations. Rather, the leadership style must be adapted to the situation and take into account the experience, skills, and needs of the workers involved. An authoritarian style is appropriate when

- Making a decision or getting the work done immediately is vital.
- Tasks or decisions are simple or routine.
- Tasks are new or unfamiliar to subordinates.
- Subordinates have not previously worked as a team.

A selling style of leadership works best when

- Getting the work done quickly is less important than building or preserving work relationships.
- Tasks and decisions are familiar.
- Problems are caused by external factors.

Figure 11.10 Continuum of Leadership Behavior

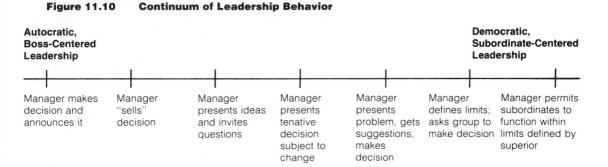

Autocratic, Boss-Centered Leadership						Democratic, Subordinate-Centered Leadership
Manager makes decision and announces it	Manager "sells" decision	Manager presents ideas and invites questions	Manager presents tenative decision subject to change	Manager presents problem, gets suggestions, makes decision	Manager defines limits; asks group to make decision	Manager permits subordinates to function within limits defined by superior

A participative style is advisable when

■ Getting work done and building strong work relationships are both important.

■ Complex problems must be solved.

■ The leader must draw from the expertise of subordinates to solve a problem or complete a task.

■ Decisions must be accepted by team members.

A delegating style is appropriate when

■ The work is relatively unimportant.

■ Tasks or decisions are simple and routine.

■ The risk of failure is low.

■ The consequence of failure is relatively inconsequential.

In some cases, a project leader must use aspects of all these leadership styles. However, experience shows that most effective leaders rely heavily on participative leadership and occasionally resort to supportive and directive techniques. An effective project leader adopts the appropriate style for a given situation.

Characteristics of Effective Team Leadership

In addition to mastering the various types of leadership styles and knowing when to implement each, effective team leaders must have certain attitudes and skills.

Perspective on People

To be effective, team leaders must recognize that people are complex and unique. The personality and motivational profiles developed during team building should help clarify differences among members. One person's reward might be another's punishment. For example, one person might enjoy making a presentation to management whereas another might view such an assignment as traumatic.

Clarity of Mission and Purpose

An effective team leader thoroughly understands the mission of the project and communicates it to the team. The mission should be reviewed frequently and revised as circumstances require. The team must understand and accept the mission; if the team and its leadership are not sure where they are going, they may take the wrong route.

Communication and Feedback Skills

In *Iacocca: An Autobiography*, Lee Iacocca says, "The only way you can motivate people is to communicate to them. The difference between mediocre companies and great companies is the great ones listen effectively to their people."[2] Communication skills are generally regarded as the most important skills for an effective manager. The following communication tips are particularly useful to team leaders:

- Avoid tuning people out by thinking excessively while they are talking.
- Don't assume that listeners understand you—assume they don't.
- Don't allow your emotions to control your communication.
- Don't be self-centered; focus on the needs of others.
- Make sure that you are attuned to nonverbal communication (e.g., body language, voice inflections, facial expressions).
- Don't interrupt people.
- Take notes while listening.

Figure 11.11 provides some good insight on communication.

Ability to Manage Conflict

Whenever more than one individual is involved in planning, problem solving, or decision making, conflict will arise. It should not be avoided; rather, it should be viewed as a natural and essential ingredient to the group process. Conflict has various sources, including differences in orientation, experience, ability, and job expectations.

Some useful steps in managing conflict are to

1. Isolate the issues causing conflict; don't confuse them with other issues.
2. Define the goals or desired results.
3. Isolate areas of agreement and disagreement and list the unresolved issues.
4. Establish priorities for the issues.
5. Brainstorm to find solutions.
6. Evaluate possible solutions.
7. Make a decision and develop an action plan.

Understanding of Power and Authority

Many leaders fear that if they share their power, they will lose it. In fact, the opposite is true. Power is not a limited resource. An effective team leader understands his or her sources of power and helps team members discover their sources of power.

There are three primary sources of power: The first source is *clout*, which comes from official or formal authority. The second source is *competence*, which comes from personal expertise or knowledge. The third is *credibility*, which comes from other people's perception of one's ability and trustworthiness.

---

2. Lee Iacocca. *Iacocca: An Autobiography* (New York: Bantam Books, 1984).

Figure 11.11   **Communication Guidelines**

A Few Words of Wisdom Are Also Useful in Improving Communication Skills:

"It is better to remain silent and be thought a fool, than to open your mouth and remove all doubt."

"God gave us two ears and one mouth for a reason."

You will never convince me that you are concerned about me or care about me if you are unwilling to listen to me."

"It is hard to start an argument when you are listening."

Objective Self-Awareness

To be effective, the team leader should perceive himself or herself in much the same way that team members do. Unfortunately, subordinates often believe that it is in their best interest to tell leaders what they want to hear rather than what they need to hear. Consequently, project managers can get the false impression that their ideas are more profound, their performances more outstanding, and their jokes funnier than they really are. The result is a perception gap. To minimize the perception gap, leaders must be willing to solicit and process feedback from team members. In particular, they must never chastise a team member for providing negative feedback even if the feedback is inaccurate. Figure 11.12 provides a list of distinctions between a good leader and a bad boss.

Figure 11.12   Distinctions between a Good and Bad Leader

BOSS OR TEAM LEADER?

The boss drives his people;
  The team leader coaches them.
The boss depends on authority;
  The team leader on goodwill.
The boss inspires fear;
  the team leader inspires enthusiasm.
The boss says "I";
  The team leader says "We".
The boss says: "Get here on time";
  The team leader gets there ahead of time.
The boss fixes blame for the breakdown;
  The team leader fixes the breakdown.
The boss knows how it is done;
  The team leader shows how.
The boss says "Go".
  The team leader says "Let's Go".
The boss uses people;
  The team leader develops them.
The boss sees today;
  The team leader also looks at tomorrow.
The boss commands;
  The team leader asks.
The boss never has enough time;
  The team leader makes time for things that count.
The boss is concerned with things;
  The team leader is concerned with people.
The boss lets his people know where he stands;
  The team leader lets his people know where they stand.
The boss works hard to produce;
  The team leader works hard to help his people produce.
The boss takes the credit;
  The team leader gives it.

Summary

Project management has become the dominant organizational structure for developing information systems in the foreseeable future. Project teams should include members from all of the functional areas that will be affected by an information system, and members should be chosen for their ability to work cooperatively with others as much as for their technical skills. Similarly, the project team leader should be evaluated according to his or her ability to work with and manage various personality types. Nothing less than the success of systems development efforts hinges on these seemingly mundane considerations. The technical skills of project management include planning and scheduling, directing and controlling skills properly using the formal tools—milestones, PERT, and Gantt charts.

Exercises

1. Discuss the difference between milestones, Gantt, and PERT planning.

2. Why is it generally a mistake to establish project schedules or budgets prior to establishing project specifications or deliverables?

3. Discuss how project structure, technology stability, project size, user proficiency, and developer proficiency affect project risks and selection of project management techniques.

4. Construct a milestone, Gantt, and PERT chart for completing a term paper, and use it next time you do a paper.

5. Discuss the concept of the mythical person-month and how it can complicate project scheduling.

6. Discuss the importance of team building for project management.

7. Complete the instrument provided in Figure 11.9 for yourself. If you are not a manager, for items 1 and 10 replace "Subordinates" with classmates, friends, or whatever is appropriate. Next, have several people, including your family, if possible, complete the form based upon how they perceive you. Compare your results with those of others.

8. Assume you are going to develop a system to serve the vice-president of personnel. Before you know what he wants from the system, he tells you that he wants it operating in three months and that he has budgeted $75,000 for its development. How would you handle the situation?

Selected References

Anett, P. L. and Wetherbe, J. C. "Addressing Behavioral and Leadership Issues to Improve Project Management." *Information Strategy*, Spring (1986), pp. 26–31.

Ansoff, H. I. "State of Practice in Planning Systems." *Sloan Management Review* 18 (1977), pp. 1–24.

Anthony, R. N. *Planning and Control Systems: A Framework for Analysis*. Division of Research, Graduate School of Business. Boston, Mass.: Harvard University, 1965.

Archibald, R. D., and Villoria, R. L. *Network-Based Management Systems*. New York: Wiley, 1967.

Davis, G. B. *Management Information Systems: Conceptual Foundations, Structure, and Development*. New York: McGraw-Hill, 1974.

Diamond, S. "Contents of a Meaningful Plan." *Proceedings of the Tenth Annual Conference of the Society for Management Information Systems*, Chicago, 1979.

Dickson, G. W., and Wetherbe, J. C. *The Management of Information Systems*. New York: McGraw-Hill, 1985.

Drucker, P. F. *Management: Tasks, Responsibilities Practices*. New York: Harper & Row, 1974.

Gitomer, Jerry, and Umbaugh, Robert E. "Controlling Projects: PERT/CPM." *The Handbook of MIS Management*. Boston: Auerbach Publishers Inc., 1986, pp. 271–281.

Gurry, E., and Bove, R. "Effective Data-Processing Planning." *CPA Journal* 47 (1977), pp. 46–47.

McFarlan, F. W. "Problems in Planning the Information Systems." *Harvard Business Review* 49 (1971), pp. 74–89.

McLean, E. R., and Soden, J. V., eds. *Strategic Planning for MIS*. New York: Wiley-Interscience, 1977.

Miller, R. W. *Schedule, Cost and Project Control with PERT*. New York: McGraw-Hill, 1963.

Moder, J. G., and Philips, C. R. *Project Management with CMP and Pert*. New York: Van Nostrand Reinhold, 1964, 1970.

Nolan, R. L. "Managing the Computer Resource: A Stage Hypothesis." *Communications of the ACM* 16 (1973), pp. 399–405.

Stanley, F. J. "Coping with Change: Project Management." In *The Handbook of MIS Management*, edited by Robert E. Umbaugh. Boston: Auerbach Publishers Inc, 1985.

Weist, J. D. and Levy, F. K. *A Management Guide to PERT/CPM*. Englewood Cliffs, N.J.: Prentice-Hall, 1977.

Strategic, Administrative, and Higher-Level Concepts and Techniques

3

I n this final section of the book, advanced techniques for information systems development are reviewed. In Chapter 12 general administrative issues for information systems development are discussed, including a historical perspective on the way information systems have evolved in organizations. Also discussed are organizational issues such as the location of the information systems department, centralization and decentralization, disappointment among top management, and the total contribution of information systems to organizational productivity.

Chapter 13 will provide some advanced techniques for information systems planning. This will start with the strategic level of information systems planning and factor all the way down to project planning.

Chapter 14 will discuss decision-support systems. This chapter can provide an understanding of the concepts of decision making and the role that decision-support systems can play in improving management decision making.

Chapter 15 will review the trends in end-user computing and provide effective means for supporting it.

Chapter 16 will cover the future of organizations and computer technology and an evaluation of the implications of the future. ∎

Chapter

Systems Administration

12

Introduction

The commitment to information systems projects and the resulting systems development cycles must be properly administered under management's direction. In this chapter, basic concepts and approaches to information systems administration are discussed.

Though systems analysts are not directly involved in the overall administration of information systems, they should be knowledgeable of systems administration processes for the following reasons:

1. Systems analysts are subject to the procedures and controls resulting from systems administration. Therefore, their effectiveness is increased if they understand systems administration.

2. Systems analysts can often make suggestions that constructively influence top management's approach to systems administration. This is particularly true if management is inexperienced in systems administration.

3. Systems analysts opten aspire to move into management. An understanding and demonstration of administrative skills can facilitate such advancement.

This chapter is organized into three sections: historical perspective, organization, and management expectations.

Historical Perspective

The results of early attempts at computer-based information systems in most organizations fell substantially below management's original expectations. Most information systems projects failed to deliver what was promised for the cost and/or in the time frame expected. For example, Weyerhauser Company, the far-flung forest products complex, has probably made as effective a use of computer-based information systems as any other U.S. corporation.[1] However, they have experienced their disappointments. When they began to develop a new on-line inventory system, they were led to believe they could do it with ten people in one year. Instead, the project took fifty people and three years. Such experiences are not uncommon.

A partial, but unfortunate, indicator of the failure of many computing efforts is the high turnover of data processing and information systems managers—as high as 40 to 50 percent annually in the early 1970's. Further contributing to management's disenchantment with the computing effort is the increasingly high cost associated with it. Prior to 1960, computer-based information systems were virtually nonexistent in most organizations. Today, large organizations frequently spend millions of dollars annually on this technology.

---

1. *Business Week* Editorial, "Business Takes a Second Look at Computers," *Business Week*, 5, June 1971, p. 63.

Difficulties of Planning and Control

Perhaps the most common management criterion for evaluating the success or failure of a project is the degree of variation between planned performance and actual performance. However, to be able to compare planned performance with actual performance, a plan that defines the objectives the project is to achieve and the criteria for measuring achievement must be established. It is unfair and even unrealistic of management to be critical of computing achievements unless management's expectations have been clearly articulated in planning documents that can be used to properly direct the computing effort. Ironically, the computing effort has historically been conspicuously void of planning and control activities in many organizations. Organizations that do plan and control information systems can expect to enjoy greater success with their computing efforts.[2]

In view of the overwhelming impact of computers on organizations, it is alarming for them undergo less management scrutiny than other functions. However, closer analysis into the uniqueness of computing provides some insight into management's reluctance and/or inability to cope with planning and controlling the computing effort. Computing as an organizational activity is complicated due to the following factors:[3]

1. The computing resource is integrated into virtually every dimension of modern organizations in both explicit and obscure ways. Indeed, it has become increasingly difficult to define exactly where computing activities begin and other activities end.

2. The computer resource has a complex set of supply/demand characteristics:
   a. The ratio of fixed to variable costs is high.
   b. Computer hardware usually offers economies of scale.
   c. Incremental computing capacity often must be acquired in large blocks.
   d. Needs for information services grow rapidly in complexity and sheer size.
   e. Processing tends to be cyclical.
   f. One computer system is unable to serve all of the diverse demands that a large organization can place on it.
   g. Processing priorities are highly variable, depending on the application, the users, and the timing.

3. Computing technology is extremely dynamic and is changing at an accelerating rate. The economic and technical feasibility of new computer applications is continually improving, resulting in a proliferation of additional computer applications.

4. Staffing creates considerable uncertainty. Personnel requirements are often as dynamic as the computing industry itself. Skills and experience that were highly valued ten years ago may be obsolete and even potentially damaging today.

---

2. McKinsey & Company, Inc., "Unlocking the Computer's Profit Potential," *The McKinsey Quarterly.* Fall 1968, pp. 17–31.

3. John Dearden and Richard Nolan, "How to Control the Computer Resource," *Harvard Business Review,* Nov–Dec 1973, pp. 68–78. And, F. Warren McFarland, "Management Audit of the EDP Department," *Harvard Business Review,* May–June 1973, p. 131–42.

Figure 12.1 **Nolan's Six Stages of Data-Processing (DP) Growth**

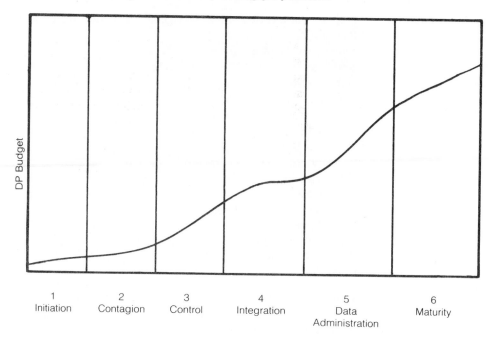

1	2	3	4	5	6
Initiation	Contagion	Control	Integration	Data Administration	Maturity

Six Stages of EDP Growth

Considerable enlightenment into both the impact of computer-based information systems and the need for improved organizing, planning, and control has resulted from research efforts headed by Richard L. Nolan.[4] These research findings indicate that organizations go through six stages of EDP growth as portrayed in Figure 12.1 and discussed here.

Stage 1: Initiation

In this stage, the computer is introduced within the organization. Users are encouraged to use the system but, due to unfamiliarity, do not yet flock to request applications. The applications that are developed are simple, and typically of an accounting orientation. During this stage, the MIS organization is often centralized because they, like the users, must also learn the new technology.

Stage 2: Contagion

Soon the users become superficially enthusiastic about the computer and request the development of all sorts of applications. Computer services are often "free" to them since computing expenses are often carried as an overhead expense during this stage, and new developments are encouraged. Pressure is exerted by MIS staff to expand both computer hardware and computer staff during this stage to keep up with the demand for services. The budget in the MIS department rises

---

4. Richard Nolan, "Managing the Crisis in Data Processing," *Harvard Business Review* 57 (1979): 115–26.

rapidly. The management of the computer department can be characterized as lax, since little planning is done and much control is lacking.

Stage 3: Control

The organization has entered the control stage when senior management becomes very concerned about the level of benefits being received from computer applications versus the cost of the MIS function. When this occurs, a halt is called to budget expansion. The total MIS budget is either held constant, or the growth rate is sharply reduced. The focus is on giving the department the type of professional management found in other parts of the organization. Planning and control systems are initiated. Emphasis is placed upon documenting existing applications and moving them toward middle management and away from a focus on strictly operational functions. It is also during this stage that an attempt is made to make the users accountable for their computer use by introducing charge-out systems.

Stage 4: Integration

The integration stage is characterized by an attempt to take advantage of new technology, typically database, by integrating existing systems. The MIS function is set up to service users much as a utility. There is, according to Nolan, a significant transition point in an organization's computer use once this stage is reached.

Stage 5: Data Administration

In the data-administration stage, the database technology is in place and a data-administration function is created to plan and control the use of an organization's data. By this time, users are effectively accountable for computer resource use, and the emphasis is upon common, integrated systems in which data is shared among various functions in the organization.

Stage 6: Maturity

When an organization reaches maturity (and few have), they have truly integrated the computer into their managerial processes. At this time the data resource is meshed with the strategic planning process of the organization. Applications mirror the information flows of the organization. Finally, joint-user and data-processing accountability exist regarding the allocation of computing resources within the organization.

Overview of the Six Stages

Nolan's stage hypothesis has a number of uses as a conceptual framework. One way of using the framework is to classify organizations in the aggregate into stages. By doing this, one can see how organizations have passed through the stages. The initiation of business computing, for example, took place in the late 1950s and the early 1960s. Contagion occurred from the early 1960s up to the late 1960s, say 1968–69. The control stage was entered by about the late 1960s and lasted until at least the mid 1970s for many firms. Some organizations are still in this stage. Many firms, however, have integrated applications and have become oriented toward database technology, which marks entry into the next stage.

Firms entering the data-administration stage most often did so in the late 1970s. There are some aspects of a few organizations that currently represent maturity, but it would be difficult to point to any organization and say that it was totally mature in all aspects of MIS. These time periods for the stages of U.S. organizations are added to the schematic of Nolan's framework as shown in Figure 12.1. The reader must recognize that the time periods are generalizations and that there are still stage 2 firms, stage 3 firms, and so on.

Another use of the Nolan framework is to go into an organization, gather data about its computer use, and identify how the computer use in the organization fits into the framework. This type of exercise is very useful as a precursor to an information systems planning activity. It is important to understand the current status of the MIS organization before embarking on ambitious new plans.

Organizations may be identified as to what stage they are in by several subcategories, including: (1) their applications portfolio (how they are using the computing); (2) their type of MIS organization; (3) how they do their MIS planning and control; and (4) the way users fit into the applications development process and their responsibility regarding the allocation of computer resources. It is thus possible to say that a firm is in stage 3 on one aspect and stage 4 on another. Few organizations would be at the same level of sophistication on all subparts of the framework.

The concepts and approaches discussed in the next two sections of this chapter provide additional insight into the issues raised in Nolan's stage model. In particular, the final section of this chapter, entitled "Management Expectations," explains some of the difficulties involved in achieving the more mature stages of the stage model. Chapters 13, 14, and 15 provide concepts and techniques to help achieve the more advanced stages.

Organization

The two major organizational issues regarding information systems function are

1. The centralization versus decentralization of technology, personnel, and authority.
2. The location and reporting relationship of information systems function.

Centralization versus Decentralization

Centralization or decentralization of organizational functions was a widely discussed and hotly debated topic long before computer technology entered the scene. However, the sudden permeating effect that computer technology has had on organizational operations has highlighted the issue.

Historical Perspective
During the 1960s, computer technology tended to favor centralized operations. That is, all computer processing and file storage was conducted on one or more computers (perhaps connected) in a single location under the direct supervision of the information systems department. The most common argument was economy of scale. However, the advent of distributed data processing and lower-cost, end-user technology in the 1970s changed that. Today the issue of centralization versus decentralization is based on the way the organization desires to operate.

Viable computer technologies are available for either centralized or decentralized computing operations. Therefore, arguments now focus more on organizational issues and less on technology.

Arguments for Centralization

Arguments in favor of centralization include:

1. *Organization-wide consolidation of operating results*. Financial and operating data are readily consolidated for reporting and evaluation purposes. Without centralization, consolidation is usually obstructed by incompatibilities of different systems designs, coding schemes, and data formats.

2. *Shortage of quality information systems personnel*. Computer professionals are scarce and often have more allegience to their profession than to their organization. Consequently, turnover tends to be high, which is potentially damaging to an organization. Centralization reduces the impact of both shortages and turnover by permitting one staff of larger size. That is, centralization of staff reduces the dependence on a few individuals. In addition, it is easier to recruit and retain computer professionals because they have a greater opportunity to interact with a variety of staff members. Consequently, they feel more closely affiliated with other information professionals.

3. *Ease of control*. Top management can control operating divisions more easily when uniform information reporting systems are used. When divisions develop their reporting systems on an individual basis, there are usually discrepancies in data used, data definitions, and reporting formats. Through centralized systems development, uniformity can be enforced.

4. *Economies of scale*. Traditionally, the computer processing capabilities of large systems have increased exponentially as their costs have increased arithmetically. Therefore, economies of scale have resulted when several small, decentralized computers are replaced by one large, centralized computer. Economies of scale also apply to centralization of staff. There is less duplication of effort, as well as more efficient allocation of systems analysis and programming activities.

Arguments for Decentralization

Arguments in favor of decentralization include the following:

1. *Familiarity with local problems*. The closer computer professionals are to problems, the more likely they are to devise good solutions. Centralization of equipment and staff tends to isolate them from the problems of the organizational functions they are attempting to support.

2. *Rapid response to local processing needs*. Computer equipment and staff can be more responsive to both development and production requirements when they are decentralized to user departments. User departments have more discretion in scheduling their resources. They do not have to compete with other departments for centralized computing services.

3. *Profit-and-loss responsibility*. When computer equipment and staff are decentralized to user departments, their costs can also be easily decentralized to users. This tends to make their departments more sensitive to cost/benefit considerations because the computing costs directly affect their department's profitability.

Technical and Organizational Issues

The preceding discussion indicates that centralization offers greater efficiency and control, while decentralization offers more flexibility to users. Unfortunately, these objectives are often in direct conflict. Resolving the conflict is often determined by organizational politics or computer vendor marketing tactics.

In actuality, many of the unique advantages of either approach have been reduced with the advent of new, computer-based information systems. For example, low-cost minicomputers and microcomputers have greatly offset the economy-of-scale argument for centralization. Better programming standards and documentation procedures have reduced the impact of turnover. In other words, technology and personnel are not quite the deciding issues they were previously. So, what should the deciding factors be?

Framework for Making the Appropriate Decision

The issue of centralization versus decentralization of information systems can be rationally evaluated using a straightforward, politically sensitive framework. First, we have to keep in mind that the centralization versus decentralization issue can be categorized by

- Hardware
- Software
- Data
- Technical staff
- Planning

Each of these can be totally centralized or decentralized, or there can be a mixture. For example, an airline reservation system is usually totally centralized, whereas a conglomerate may have totally decentralized information system efforts. As an example of a mixed structure, the phone company develops software centrally using centralized technical staff; software is then distributed to decentralized processing facilities where hardware and data are located.

But what is the proper way to make these decisions? There are three factors to consider

- Interdependency
- Homogeneity versus heterogeneity
- Corporate culture

The first factor to consider is the degree of interdependence of business operations. In other words, how important is it for the "left hand to know what the right hand is doing?" If that need is high—such as in an airline reservation system, or different functions within a manufacturing facility—centralization is necessary.

If interdependency is low, the next question is how homogeneous are the different business activities? For example, each McDonald's Hamburgers franchise is doing the same information processing, albeit in a decentralized location. Due to the homogeneity, it would make little sense to have each franchise acquire its own hardware, develop its own software, and so on. That would be reinventing the wheel. Rather, it makes sense to centralize selection of hardware, development of the software and management of technical staff and then *cookie-cutter* the software to be run on decentralized, uniform hardware.

Homogeneity versus heterogeneity has to be evaluated carefully. For example, one large hospital "chain" assumed all of its some eighty hospitals were alike and that software should be developed centrally. However, some significant heterogeneity existed in different hospitals—such as elective surgery, trauma centers, outpatient services—that meant some systems just didn't work properly in all hospitals.

Dayton Hudson recognized the heterogeneity between their prestigious department stores (Dayton's) and their low-cost high-volume stores (Target) and did not try to impose the same systems on them. Dayton's and Target have separate information systems departments. The Dayton Hudson Corporate information systems function is small and focuses on consolidation of financial information and planning support for decentralized information systems efforts. This centralized planning pays off. For example, when Target wanted to introduce credit cards, there was sufficient homogeneity to cookie-cutter what Dayton's already had available.

The ability to seize such opportunities leads to one important conclusion. You can decentralize hardware, software, data, and technical staff—but you should never completely decentralize planning.

Corporate culture is the final issue in considering the centralization versus decentralization issue. Regardless of what makes sense logically, corporate culture may, and often should, prevail. 3M is an excellent example. Recognized for outstanding management by Peters and Waterman in their book, *In Search of Excellence*, 3M has a corporate culture which nurtures the entrepreneurial flair of a small firm in a large corporate setting. This is accomplished by having over sixty separate divisions operating like small companies independently in a large corporation. 3M wants the advantage of a large corporation's corporate-wide financial and human resource management while allowing the quick, competitive, opportunistic behavior advantage of a small firm for each of its sixty divisions. For example, the company wants career-path planning possibilities to be corporate-wide, and it wants to allow economies of scale and technology transfer, but it does not want centralized bureaucracy to inhibit decentralized progress.

3M resolved these multiple objectives by having a centralized system and staff for financial and human resource system and a decentralized system and staff for logistical/operational systems. However, all decentralized technology is, to the greatest extent possible, compatible to facilitate technology transfer and movement of staff without requiring retraining.

For the logistical/operational system, the corporate information systems staff plays a key planning and coordinating role to opportunistically facilitate syndication of development efforts when homogeneity is found. They also facilitate technology transfer when an application developed by one division can be used by one or more other divisions.

This centralized planning of compatible hardware and software allows both syndication and technology transfer to occur easily, but does not enforce it. In other words, corporate culture takes priority over the homogeneity criteria.

In an organization where there is no interdependency, homogeneity or centralized corporate culture—that is, a diversified conglomerate—it is best to let each decentralized business unity "do its own thing" with information systems. However, the central organization should review all plans and be sensitive to interdependent, homogeneous opportunities that are consistent with corporate culture.

The framework of interdependency, homogeneity versus heterogeneity, and consistent corporate culture provides a straightforward method of sorting through the centralization issue. It has been used successfully in a number of delicate, confused corporate settings.

Figure 12.2   Alternative Locations for the Information Systems Function

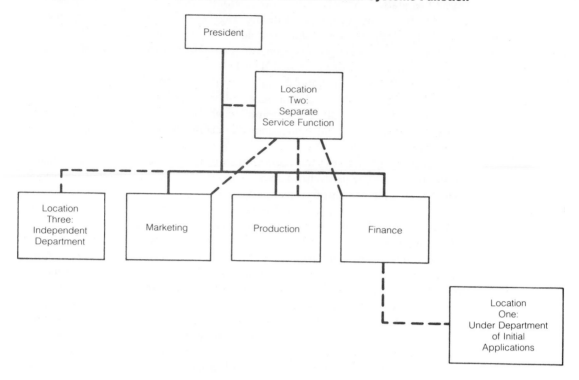

Organizational Location

The location of the information systems function is also an issue of debate and discussion. Most computing efforts are initially located in the department where the computer-based information systems originate. This is usually finance and/or accounting. However, the proliferation of computer applications in other functional areas of the organization generates interest in both a higher status and a broader orientation for the information systems function. Naturally, political as well as practical issues play a factor as the computing effort "percolates" up the organization.

Three basic approaches are used in positioning the computing effort.[5] Its possible locations are shown in Figure 12.2 and discussed in the following pages.

Location One: Under Department of Initial Applications
The information systems function is located in the finance and/or accounting department in many organizations. This positioning is due to the early use of computers for financial applications. In organizations involved in large-scale scientific or engineering projects, the computer may be originally sponsored by the research or engineering department.

As long as information systems is isolated to the department in which the computer is situated, this location is satisfactory. However, this isolation seldom exists for long; the computer's potential for other applications tends to proliferate quickly.

---

5. Donald Sanders, *Computers and Management*, New York: McGraw-Hill, (1974), pp. 318–21.

As computer applications develop outside the department in which the computer is originally located, the following problems may emerge:

1. *Lack of objectivity in setting priorities:* Information systems personnel may tend to cater to the requirements of the department in which they work rather than to those of other departments.

2. *Limited vantage point:* Information systems personnel may tend to view the organization in terms of the department in which they are located. If so, they may be unaware of problems or opportunities elsewhere in the organization.

3. *Lack of organizational status:* Organizational status and authority are lacking when the chief information executive reports to a location several echelons down in one functional area of the organization. It is therefore difficult for him or her to influence top management in such areas as systems integration and interdepartmental information requirements.

Location Two: Separate Service Function

Problems of objectivity in priority setting can be overcome by setting up the information systems function as a separate service center to support the overall organizational information requirements. The information systems function may report to a top-level executive or to an executive committee.

In this location, the information systems function occupies a position that is peripheral to the main organizational structure. The chief information executive generally has little status or authority outside of his or her direct responsibilities. This can result in an unwieldy situation as he or she attempts to allocate what are usually scarce computing resources among executives who have high status and authority.

Location Three: Independent Department

A majority of information-processing authorities contend that to fully realize the potential of computer technology, the information systems function should be established as a separate, independent department. Their major claims are

1. This location reflects the organization-wide scope of information because it does not isolate the information systems function in one functional area.

2. This location provides organizational status because it establishes the information systems function as a major organizational function.

3. This location encourages innovation and integration because it allows the information systems function to freely interact with other organizational functions.

Factors such as organizational size, complexity, extent of computer usage, and managerial politics and personalities make it impossible to single out any one of the three locations as preferable for all organizations. However, location three is generally regarded as appropriate for medium to large organizations desiring to achieve the advanced stages of EDP growth.

Management Expectations

One of the greatest expectations of computer-based information systems by management was that computers would place massive amounts of information at management's immediate disposal. Many managers envisioned their offices as equipped with computer terminals that would allow them to monitor the operations of their organizations by the mere entering of a few commands into the terminals. All organizational data would be integrated into a single, integrated database.

These expectations and their lack of fulfillment have produced one of the longest "expectation gaps" associated with computer technology. Shortly after the recognition that this initial perception of a management information system (MIS) would not be forthcoming, John Dearden wrote an article entitled "MIS Is a Mirage"[6] published in the early 1970's. This article created a great deal of discussion and debate about whether MIS was a realistic concept.

As with many new concepts or ideas, the initial enthusiasm for MIS fostered inflated expectations and often gross misconceptions. More experience and research with MIS has provided a more realistic perspective of what role computer-based information systems are to play in organizations. Some of the original expectations for MIS have been or are being realized; others have not been and may never be realized. However, it is now accepted that MIS is a meaningful and valid concept.

Two factors have inhibited or prevented the achievement of the original concept of MIS. These factors are based upon the characteristics of management's information requirements and the practicalities of integrating all organization information, as will be discussed here.

Characteristics of Management's Information Requirements

Prior to attempts to design comprehensive MISs for management, there was somewhat of a dearth of information about what executives did or how they did it. However, as discussed in Chapter 3, it is now recognized that most of the decisions made by middle management and, particularly, by top management are (1) judgmental and subjective in nature and (2) based upon information that is largely external to the organization and therefore usually incomplete.

The computer's ability to perform extremely fast retrieval of, and/or calculations on, highly structured data is of less applicability in the realm of top management. For example, though computers have been highly productive in processing sales orders and posting accounting transactions, they have been less successful at determining what actions competitors, customers, or the government may take.

The computer is, however, demonstrating increasing potential in the area of decision-support systems. This increased potential is a result of both a better understanding of management's decision-making process and the enhanced sophistication of computer technology. For example, the computer can be used to simulate the results of different decisions, given different sets of events. Such a system allows management to play "what if" in considering different decisions. That is, the computer can simulate, with varying degrees of accuracy, what would happen in each case. Consequently, management can more accurately consider the likely results of several alternative decisions with respect to a given problem or opportunity, before actually selecting the course of action to be taken.

---

6. John Dearden, "MIS Is a Mirage," *Harvard Business Review*, January–February 1972, pp. 90–99.

Decision-support systems are upgrading the use of computer technology from the operational levels of transaction processing and reporting to activities that directly support management. Decision-support systems offer great potential for achieving many of the unfulfilled expectations of MIS. A more extensive treatment of decision-support systems is presented in Chapter 14.

Integration of Organizational Information

Though the difficulty of structuring and processing information for management has been one factor constraining the development of MIS, a second, and often more perplexing, problem has been the inability to integrate organizational information. The need for integrated information for higher levels of management is best illustrated by the pyramidal structure of most organizations: there exists an ever-widening succession of levels from the top to the bottom of such organizations.

If an organization is to remain viable and operational, the activities of lower-level functional units must be coordinated and conflicts resolved by the integrative activities of higher levels of management. Ultimately, the chief executive officer must ensure the integration of the activities of the overall organization. For example, production activities must be coordinated with marketing activities in order to maintain a balance between products produced and products sold. Coordination of these different organizational activities requires the integration of information about the activities such that they can be properly managed and controlled.

The speed and capabilities of computer systems appeared to offer great promise in overcoming the information integration constraint inherent with manual information systems. But managers of many organizations that have invested in computer technology have been quite disappointed; much of the information stored in their computer systems cannot be easily integrated.

A Banking Example

The reasons for the lack of integration of organizational information are illustrated by the following example. Consider a bank that installs its first computer system. The largest and most cost-effective application to computerize is demand-deposit accounting (DDA), or checking accounts. A systems analyst is assigned to work with users to automate this system. DDA is a highly structured process that lends itself to computer processing. Account numbers can be used as keys, and the data to be stored in computer files is primarily the data used in the manual system, with some enhancements. Figure 12.3 depicts the data contained in a DDA file. Once stored in the computer system, the DDA file can be used to generate a variety of reports and/or terminal displays associated with the management and control of DDA.

During the development and implementation of the DDA system, the bank decides to begin implementation of a computer-based savings accounting system. Another systems analyst is assigned to work with users to automate this system. Input transactions, reports, and files are designed, based upon the manual savings accounting system, with several enhancements. The contents of a file designed for the savings accounting system are shown in Figure 12.4.

The bank continues its development of computer-based information systems by automating its installment loan system and its mortgage loan system. Existing

Figure 12.3   Contents of a DDA File

1. Account Number	17. Officer Code
2. Name	18. Date DDA Opened
3. Spouse's Name	19. Current Balance
4. Social Security Number	20. Last Statement Date
5. Telephone Number	21. Balance Forward—Last Statement
6. Date of Birth	22. Amount of Credits—Current Period
7. Sex	23. Number of Credits—Current Period
8. Marital Status	24. Amount of Debits—Current Period
9. Number of Children	25. Number of Debits—Current Period
10. Number of Dependents	26. Number of Returned Checks Year-to-Date
11. Rent or Own Home	27. Date of Last Deposit
12. Occupation Code	28. Service Charge
13. Years Employed	29. Date of Last Overdraft
14. Income Range	30. Amount of Last Overdraft
15. Credit Rating	31. Number of Overdrafts
16. Line of Credit	

Figure 12.4   Contents of a Savings Accounting File

1. Account Number	17. Officer Code
2. Name	18. Date Savings Account Opened
3. Spouse's Name	19. Current Balance
4. Social Security Number	20. Last Statement Date
5. Telephone Number	21. Balance Forward—Last Statement
6. Date of Birth	22. Date of Last Deposit
7. Sex	23. Amount of Last Deposit
8. Marital Status	24. Low Balance—Current Period
9. Number of Children	25. Number of Withdrawals—Current Period
10. Number of Dependents	26. Amount of Credits—Last Statement
11. Rent or Own Home	27. Number of Credits—Last Statement
12. Occupation Code	28. Amount of Debits—Last Statement
13. Years imployed	29. Number of Debits—Last Statement
14. Income Range	30. Excess Activity Charges
15. Credit Rating	31. Anticipated Interest—Current Period
16. Line of Credit	32. Interest Earned Year-to-Date

account numbers are used for each system, and existing data-capture methods are expanded to obtain all of the data shown in Figures 12.5 and 12.6.

The bank also automates its payroll system. The file for this system includes the data shown in Figure 12.7.

The development of these five information systems has spanned several years and cost several million dollars. Bank employees and managers responsible for DDA have their own information system; bank employees and managers responsible for savings accounts have their own information system; and so forth. However, there are some serious problems. Each system was developed as an independent subsystem; no thought was given to interfacing the subsystems to provide integrated processing or reporting.

Redundancy

The first problem with the bank's information systems is that there is a great deal of redundancy in the data collected, stored, and processed. Consider a bank

Figure 12.5   Contents of an Installment Loan File

1. Account Number
2. Name
3. Spouse's Name
4. Social Security Number
5. Telephone Number
6. Date of Birth
7. Sex
8. Marital Status
9. Number of Children
10. Number of Dependents
11. Rent or Own Home
12. Occupation Code
13. Years Employed
14. Income Range
15. Credit Rating
16. Line of Credit
17. Officer Code
18. Collateral
19. Amount of Payment
20. Mode of Payment
21. Number of Times Refinanced
22. Late Charge Indicator
23. Maturity Date
24. Number of Payments
25. Interest Rate
26. Original Amount of Loan
27. Life Insurance Premium
28. Balloon Payment
29. Date of Last Activity
30. Current Balance
31. Interest Earned Year-to-Date
32. Number of Payments to Date
33. Number of Late Notices Sent
34. Number of Returned Checks
35. Last Statement Date

Figure 12.6   Contents of a Mortgage Loan File

1. Account Number
2. Name
3. Spouse's Name
4. Social Security Number
5. Telephone Number
6. Date of Birth
7. Sex
8. Marital Status
9. Number of Children
10. Number of Dependents
11. Rent or Own Home
12. Occupation Code
13. Years Employed
14. Income Range
15. Credit Rating
16. Line of Credit
17. Officer Code
18. Original Date of Loan
19. Property Identification
20. FHA or VA Number
21. Late Charge Code
22. Amount of Payment
23. Mode of Payment
24. Escrow Payment
25. Original Amount of Loan
26. Maturity Date
27. Interest Rate
28. Terms in Months
29. Date of Last Appraisal
30. Amount of Appraisal
31. Purchase Price
32. Percent of Loan Guaranteed
33. Late Charge Rate
34. Date of Last Activity
35. Escrow Balance
36. Number of Late Notices
37. Number of Months to Go
38. Interest Paid Year-to-Date
39. Principal Remaining

Figure 12.7   Contents of the Payroll File

1. Social Security Number
2. Name
3. Address
4. Department Code
5. Job Title
6. Date of Hire
7. Marital Status
8. Number of Dependents
9. Pay Rate
10. Sick Leave
11. Vacation Leave
12. Federal Withholding
13. FICA
14. State Taxes
15. Insurance
16. Regular Earnings Year-to-Date
17. Federal Withholding Year-to-Date
18. FICA Year-to-Date
19. State Taxes Year-to-Date
20. Net Earnings Year-to-Date

customer who has a checking account, a savings account, an installment loan, and a mortgage loan. Though the bank is dealing with one customer, its information systems treat this customer as four customers. The customer has four account numbers (one in each information system).

Certain data are redundantly collected, stored, and processed in each system. Specifically, data elements 1 through 16 are identical in each of four files (see Figures 12.3 through 12.6.)[7] If a customer has more than one checking account, savings account or loan, the redundancy becomes even greater. This redundancy creates the following problems:

1. Customers are required to supply much duplicate data for each account they open, even when some or all of the data needed has been collected previously.

2. Storage space is wasted because the same data are stored in different places in the same file and/or in different files.

3. Processing time is wasted. For example, a customer's address may, due to redundancy, be stored in eight different places in the bank's computer files. If the customer's address changes, all eight addresses must be updated to keep the data used in all information systems current.

4. Inconsistencies and/or other errors develop in data files. The majority of information systems fail to update all redundant data. For example, a customer's address may be updated in the DDA file but not in the savings accounting or loan files. Consequently, there are inconsistencies as to the address of that customer.

Integration

Besides creating inefficient redundancies, the independent subsystem structure causes difficulties in integrating information. The systems and files have been developed along departmental or functional boundaries. The account numbers are not logically related and cannot be used for cross-referencing customer's acounts. This seriously limits reporting capabilities. For example, a loan officer may want to check information pertaining to a loan applicant's checking and savings accounts. However, there is no linkage to these data from the loan system. Indeed, the loan officer may have to ask the loan applicant if he or she has checking and/or savings accounts with the bank and what his or her account numbers are.

Consider a case where the management of the bank wants to increase mortgage loans to offset several large savings deposits. Management decides to send letters encouraging specific customers to consider buying homes, using convenient financing available through the bank. Management also decides that the best customers to send such letters to are customers meeting the following criteria:

1. Customers who do not have mortgage loans

2. Customers who have good checking account records (i.e., few or no overdrafts)

3. Customers with sufficient funds in their savings accounts to make down payments on homes

---

7. These data elements were made identical for the purposes of this example. In actuality, the systems would share many common data elements, but files would tend to vary some in their names, sequence, and inclusion of elements.

4. Customers who have good payment records on any installment loans with the bank

Though the data necessary to identify such customers are available in the different files of the different information systems (see Figures 12.3 through 12.6), there is no convenient way to integrate them. Extensive programming and clerical work are required to satisfy such an information request. Management is understandably disappointed.

The problem is that the bank's information systems are not designed to integrate information to serve management's needs. Integration of information is more readily achievable when considered prior to developing and implementing systems. After the fact, integration can be unwieldy; it usually requires new systems development.

It is important to point out that not all organizational information needs to be integrated. For example, the contents of the payroll file in Figure 12.7 are independent of the contents of other files. Though bank employees may have checking accounts, savings accounts, and/or loans with the bank, there is little incentive to endure the expense of structurally relating these data in the computer system (i.e., because a bank usually has several thousand accounts but only a few hundred employees).

Thus organizational information may not be integrated due to oversight or because the data cannot be integrated practically. The situation is caused by lack of planning. Systems were developed in a fragmented, piecemeal manner with no thought given to eventual integration requirements. Figure 12.8 shows the relationships of the data contained in the bank's five files.

The scenario of the bank can be translated into other organizational settings. For example, in a university, student data, course data, faculty data, and classroom facilities data should all be integrated. In a manufacturing plant, sales data, inventory data, production resources data, and purchasing data should be integrated.

Failure to Integrate

In retrospect, it is not difficult to see that, when appropriate, information systems should be integrated. However, few organizations attempted such integration in

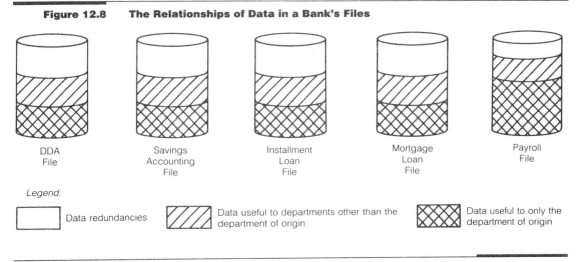

Figure 12.8   The Relationships of Data in a Bank's Files

| DDA File | Savings Accounting File | Installment Loan File | Mortgage Loan File | Payroll File |

Legend:

☐ Data redundancies   ▨ Data useful to departments other than the department of origin   ▧ Data useful to only the department of origin

their initial attempts to develop computer-based information systems. The failure to approach information systems in an integrated fashion illustrates a key concept discussed in Chapter 5: *the difference between reacting to problems and recognizing opportunities.* Many information systems are developed as high-speed automated versions of existing manual systems. Since manual systems do not lend themselves to integration with other manual systems, the computer versions of these systems retain this non-integration orientation. The opportunity to integrate the manual systems as they are computerized is not exploited. Rather, the personnel developing the systems react to management's demands for integrated information after it becomes apparent that there is a structural deficiency within their initial development.

Difficulties in Integrating Data

Most organizations have attempted to integrate their information systems either initially or during revisions of their initial systems. Though their insight and efforts have enabled them to more readily integrate data and therefore reduce redundancies, they have still encountered problems. The programming of data relationships that transcend departmental or functional boundaries is extremely complex. Integration requires consolidation of files and/or linking of data stored in separate files.

In our example of a bank's system, four systems of the bank (DDA, savings accounting, installment loans, and mortgage loans) can be integrated. One way this can be accomplished is by assigning to each customer a single account number and then using suffixes to designate the accounts the customer has with the bank. Figure 12.9 illustrates such a customer account structure. Note that a customer can have several checking accounts, savings accounts, and loans. The unique identity provided for each customer account allows the accounts to be consolidated as shown in Figure 12.10.

Figure 12.9   **Use of Suffixes to Achieve Centralized Account Numbers at a Bank**

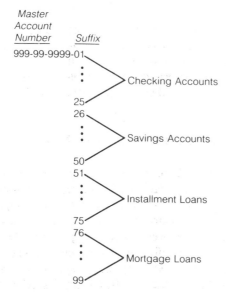

Figure 12.10   A Customer Record from a Consolidated File of a Bank

```
┌─────────────────────────────┐
│        326-47-9867-00       │
│                             │
│          Customer           │
│            Data             │
├─────────────────────────────┤
│        326-47-9867-01       │
│                             │
│      Checking Account 1     │
│            Data             │
├─────────────────────────────┤
│        326-47-9867-02       │
│                             │
│      Checking Account 2     │
│            Data             │
├─────────────────────────────┤
│        326-47-9867-26       │
│      Savings Account 1      │
│            Data             │
├─────────────────────────────┤
│        326-47-9867-52       │
│                             │
│      Installment Loan 2     │
│            Data             │
├─────────────────────────────┤
│        326-47-9867-76       │
│                             │
│       Mortgage Loan 1       │
│            Data             │
└─────────────────────────────┘
```

The bank is unlike most organizations in that, generally, an organization's separate files cannot practically be consolidated into a single file. Where consolidation is not possible, integration of data is sometimes accomplished by linking the data from separate files as was done in Chapter 7 when relationships were established between customers, orders, and items for the order-processing system.

Technology and Techniques to Achieve MIS

Prior to the availability of database management systems, the programming necessary to link files was extremely difficult. However, with the implementation of database management systems (and CASE products, including code generators), the programming task of integrated systems was greatly reduced. A database management system in itself, however, does not provide integrated systems. A great deal of effort must go into information systems planning, to ensure that the various systems that get developed within an organization over time can be integrated as they need to be. In other words, a great deal of conceptual work has to go into the various files that make up the database of an organization in order to achieve integration. To simply purchase a database management system and install it in an organization only gives programmers a more powerful access technique for their fragmented, piecemeal information systems. Database is a concept; database management systems is a technology. To fully exploit the technology, database planning is required as well as database technology. And, of course, integrated systems are required to provide the capabilities really necessary to support top management, including decision-support capabilities.

It is through well integrated information systems that have on-line decision-support capabilities that management expectations can be met. Planning techniques and decision techniques for achieving integrated systems and decision-support systems are provided in the next two chapters.

Summary

Information systems have historically been a difficult functional activity to develop in organizations. This is true for a variety of reasons, including the fact that the computing resource is integrated into virtually every dimension of modern organizations in ways that make it very difficult to isolate and plan for. Also, the computing resource has a complex set of supply/demand characteristics. The extremely dynamic nature of the technology and the shortage of qualified people further complicate good information systems planning.

A great deal of insight into the way information systems emerge and develop in organizations comes from the six-stage model of EDP growth. This model shows how information systems step through initiation, contagion, control, integration, data administration, and maturity. A great deal of organizational learning occurs during these stages as management learns how to properly plan and control information systems.

Organization of the information systems function is an issue that must be carefully evaluated within an organization. Issues that must be resolved include centralization vs. decentralization, location and reporting relationships of the information systems function.

A major administrative issue of information systems is the general failure of the systems to meet management expectations. The two major causes for this stem from the inability to achieve integrated information systems and decision-support systems to support management decision making. Failure to achieve management expectations in these areas can be attributed both to conceptual failures of those designing systems to understand these requirements and to the inadequacies of the early technologies to support integrated systems and decision-support systems. Fortunately, today's technologies make such capabilities possible. Careful conceptual and planning work, however, is required to take advantage of this technology.

Exercises

1. Discuss some of the difficulties associated with planning and controlling information systems.

2. Identify and discuss the six stages of EDP growth.

3. Discuss the arguments for and against centralizing and decentralizing the information systems function, and discuss general guidelines to be used in resolving this issue.

4. Discuss the different reporting locations for the information systems function and discuss the pros and cons of each.

5. Discuss why management has been often disappointed in the capabilities provided to them by information systems.

6. Discuss both the conceptual and technical reasons for failing to achieve integrated information systems.

Selected References

Brooks, Frederick P., Jr. *The Mythical Man Month—Essays on Software Engineering.* Reading, Mass.: Addison-Wesley, 1974.

Business Week Editorial, "Business Takes a Second Look at Computers." *Business Week.* 5 June, 1971, pp. 59–136.

Dearden, John, and Nolan, Richard. "How to Control the Computer Resource." *Harvard Business Review,* November–December 1973, pp. 68–78.

Dickson, G. W., and Simmon, John K. "The Behavioral Side of MIS, Some Aspects of the People Problem." Bloomington, Ind.: Indiana University Press, 1970.

Drucker, Peter F. *Managing for Results.* New York: Harper & Row, 1964.

Glaser, George. "The Centralization vs. Decentralization Issue: Arguments, Alternatives, and Guidelines." *Data Base* 2, (1970): 1–7.

Golub, Harvey, "Organizing Information System Resources: Centralization vs. Decentralization." From *The Information Systems Handbook,* edited by Warren McFarlan and Richard Nolan. Homewood, Ill.: Dow-Jones-Irwin, 1975.

Guthrie, A. *Attitudes of Middle Managers Toward Management Information Systems.* Ph.D. Diss. Seattle: University of Washington, 1971, pp. 61–67.

Head, Robert V. "The Information Systems Manager as the Administrator of a Major Corporate Function." From *The Information Systems Handbook.* Edited by Warren McFarlan and Richard Nolan. Homewood, Ill.: Dow-Jones-Irwin, 1975, pp. 61–64.

Knutsen, K. Eric, and Nolan, Richard. "On Cost/Benefit of Computer-Based Systems." *Managing the Data Resource.* Edited by Richard Nolan. St. Paul, Minn.: West Publishing, 1974, pp. 277–92.

Lawrence, Paul and Lorsch, Jay. *Organization and Environment—Managing Integration and Differentiation.* Boston: Harvard Business School Division of Research, 1969.

Leavitt, H. "Applied Organizational Change in Industry: Structural, Technological, and Humanistic Approaches." In *Handbook of Organizations.* Chicago: Rand McNally, March 1965, pp. 1144–70.

McFarland, F. Warren. "Effective EDP Project Management." *Managing the Data Resource,* edited by Richard Nolan. St. Paul, Minn.: West Publishing 1974, pp. 273–307.

"Management Audit of the EDP Department." *Harvard Business Review.* May–June 1973, pp. 131–42.

McKinsey & Company, Inc. "Unlocking the Computer's Profit Potential." *The McKinsey Quarterly.* Fall 1968, pp. 17–31.

Miller, James C. "Corporate Organization and Information Systems." From *The Information Systems Handbook* edited by Warren McFarlan and Richard Nolan. Homewood, Ill.: Dow-Jones-Irwin, 1975, pp. 32–60.

Moder, Joseph J. and Phillips, Cecil R. "Project Management with CPM and PERT." New York: Litton, 1970 pp. 20–36.

Mulvihill, Dennis E., and Cohen, Burton J. "Strategy Formulation and Information Systems: Setting Objectives." From *The Information Systems Handbook*, edited by Warren McFarlan and Richard Nolan. Homewood, Ill.: Dow-Jones-Irwin, 1975, pp. 19–31.

Nolan, Richard L. "Controlling the Cost of Data Services." *Harvard Business Review*, July–August 1977, pp. 114–124.

_____. "Effects of Chargeout on User/Manager Attitudes." *Communications of the ACM* 20 (1977) 177–85.

_____. "Managing the Computer Resource: A Stage Hypothesis." *Communications of the ACM* 16 (1973): 399–405.

_____. "Plight of the EDP Manager." *Harvard Business Review*, May–June 1973, pp. 143–52.

Nolan, Richard L., and Gibson, Cyrus F. "Managing the Four Stages of EDP Growth." *Harvard Business Review*, January–February 1974, pp. 76–88.

Porter, L. W. "A Study of Perceived Need Satisfaction in Bottom and Middle Management Jobs." *Journal of Applied Psychology* 45 (1961): 1–10.

Reichenbach, Robert, and Tasso, Charles. *Organizing for Data Processing*. AMA Research Study 92. 1968, pp. 71–86.

Sanders, Donald. *Computers and Management*. New York: McGraw-Hill, 1974, pp. 318–21.

Wetherbe, James C. *Executive's Guide to Computer-Based Information Systems*, Englewood Cliffs, N.J.: Prentice-Hall, 1983.

_____. *Systems Analysis for Computer-Based Information Systems*, 1st ed. St. Paul, Minn.: West Publishing, 1979.

_____. "A Zero-Based Approach to Allocating MIS Resources." In *Proceedings of the Ninth Annual Conference of the Society for Management Information Systems*, September 1977.

Wetherbe, James C. and Dickson, Gary W. *Management of Information Systems*, New York: McGraw-Hill, 1984.

Wetherbe, James C. and Alavi, Maryam, "Reducing Complexity in Information Systems Planning." *Systems, Objectives, Solutions*, August 1982, pp. 143–157.

Wetherbe, James C. and Whitehead, Carlton J. "A Contingency View of Managing the Data Processing Organization." *Management Information Systems Quarterly* 1 (1977): 19–25.

Wetherbe, James C.; Bowman, Brent; and Davis, Gordon B. "Three Stage Model of MIS Planning." *Information and Management* Vol. 6, No. 1 (1983).

_____. "Modeling for MIS." *Datamation*, July 1981, pp. 155–164.

Portions of this chapter are adapted from *Management of Information Systems*, McGraw Hill, 1984, by Gary W. Dickson and James C. Wetherbe.

Introduction

Chapters 5 through 11 provide the concepts and techniques for developing a single information system. However, as illustrated in the last section of Chapter 12, it is naive to develop a single information system without considering the long-run strategic information systems requirements of an organization. Failure to consider these requirements results in systems that cannot readily be integrated and combined with DSS (decision-support systems) technology to support top management decision making.

This chapter reviews MIS planning and its problems. A four-stage model of MIS planning, consisting of strategic planning, organization information requirements analysis, resource allocation, and project planning, is discussed. Methodologies that are used in MIS planning are classified according to their use in one of the four stages of the MIS planning model.

The material in this chapter is more abstract, conceptual, and therefore more difficult to grasp than previous material. Keep in mind, however, that the purpose of the chapter is to make systems analysts aware of what is involved in good MIS planning, not to make them experts at doing it. Also, the techniques of MIS planning are not yet as well understood or defined as are the techniques for systems analysis and design.

Evolution of MIS Planning

The forces necessitating effective MIS planning are well documented and accepted. The "need" to plan seems clear. The "how" to plan is less obvious. The stages of the EDP growth model developed by Nolan and discussed in Chapter 12 provide valuable insights into the evolution of MIS planning. During stage 1 (initiation) the organization is oriented to its new computer resource through accounting transaction applications. As the potential benefits of the computer become apparent, increased demand for services results in entry into stage 2 (contagion). During stage 2 there is a proliferation of applications throughout the organization. In these two first stages, there is little or no formal planning or control of information systems activities. This deficiency, complicated by ever increasing demands for information service systems, results in a skyrocketing MIS budget. In response, upper management directs the MIS manager with designing and implementing adequate planning and control systems. This marks emergence into stage 3 (control). Most organizations that have been involved in MIS for several years are in stages 4 (integration) and 5 (data administration). According to Nolan, few organizations have achieved stage 6 (maturity).

It is instructive to consider the nature of these initial efforts to establish planning and control systems. MIS resources are expended on new application development projects and existing operational application systems. These become the focal objects of initial planning and control systems. System development methodologies are adopted, and project management systems are installed to assist with the planning of new applications. These include the use of well-defined project phases, specified deliverables, formal user reviews, and sign-off points. Techniques such as structured design, HIPO, structured programming, and walk-through are used to better manage the systems development process. Additionally, attention is focused on processing completed systems efficiently. High availability and reliability is emphasized, and computer operations planning and scheduling is initiated.

These initial mechanisms address *operational* MIS planning. As the organization becomes more sophisticated in its use of MIS, emphasis shifts to *management* or resource allocation control. A manifestation of this shift is the organization of the MIS function into a corporate computing utility. Some form of chargeout (i.e., users pay for computing and information services) is implemented in an attempt to shift accountability for MIS expenditures to the users. There is some question concerning the effectiveness of chargeout as a cost control tool, but in theory, chargeout fosters greater user attention to benefits versus costs and results in more effective planning.

Collectively, these measures have an effect on planning, and a process for identifying demands for MIS services is developed. Typically, annual planning cycles are established to identify potentially beneficial MIS services, some form of cost/benefit analysis is performed, and the portfolio of potential projects is subjected to some resource allocation process. This is often in the form of an MIS steering committee composed of key managers representing major functional units within the organization. The steering committee is created to oversee the MIS function, ensure that adequate planning and control processes are present, and direct MIS activities in support of long-range organizational objectives and goals. The steering committee reviews the project portfolio, approves those projects thought to be beneficial, and assigns relative priorities. The approved projects can then be mapped into a development schedule, usually encompassing a one- to five-year time frame. This schedule becomes the basis for determining MIS support requirements: long-range hardware/software, personnel, and facilities financial requirements.

The planning process described above is typical of the traditional approach to MIS planning currently practiced by many organizations. The specifics of MIS planning processes will, of course, vary among organizations. For example, not all organizations have a high-level steering committee. Project priorities may be determined by the MIS manager, his or her superior, company politics, or even on a first-come-first-served basis. Organizations with decentralized MIS functions often employ integrative mechanisms such as formal review and consolidation meetings to determine their overall MIS plan. In cases of strong divisional autonomy, no centralized planning may be attempted; rather, a process similar to that just described may be utilized by each divisional MIS group. Acknowledging variations, the model reasonably represents traditional MIS planning.

Problems of MIS Planning

The most common difficulties experienced in MIS Planning are the following:

1. Difficulty in aligning the MIS plan with the overall strategies and objectives of the organization
2. Designing an information system architecture for the organization in such a way that databases can be integrated
3. Problems in allocating information system development and operations resources among competing applications

4. Completing information system projects on time and on budget
5. Selection and use of methodologies for performing the first four processes

Alignment of the MIS Plan with Organizational Plan

The first problem of the MIS planning process is to make sure it identifies and selects information systems applications that fit the priorities established by the organization. However, organizational strategies and plans may not be written, or they may be formulated in terms that are not useful for information system planning. Therefore, it is often difficult to ascertain the strategies and goals to which the information system plan should be aligned. But, without this alignment, the information system plan will not obtain long-term organizational support. If the selection and scheduling of information system projects is based only on proposals submitted by users, the projects will reflect existing computer-use biases in the organization, aggressiveness of some managers in submitting proposals, and various aspects of organizational power struggles, rather than reflecting the overall needs and priorities of the organization.

Design of an Information System Architecture

The term *information system architecture* refers to the overall structure of all information systems combined. This structure consists of the applications for the various managerial levels of the organization (operations, management control, and strategic planning) and applications oriented to various management activities such as planning, control, and decision making. The system architecture also includes databases and supporting software. An information system architecture for an organization should guide long-range development, but it should also allow response to diverse short-range information system demands.

Allocation of Development Resources

The rational, optimum organizational allocation of development resources among competing organizational units likewise is difficult, especially if the portfolio of potential applications does not fit into an overall organizational plan and if the functional/organizational unit requirements do not fit into some orderly framework that establishes completeness and priority. Organizational dynamics such as relative power, aggressiveness, and so forth, may be used in place of some rational allocation. This can result in a precarious political situation for MIS management.

Completing Projects on Time and on Schedule

Few information system projects are completed on time or on schedule. Consequently, MIS managements' credibility suffers. Project plans are seldom accurate, as time and resource requirements are generally underestimated. This results in credibility problems for systems analysts as well as MIS management. Techniques for improving project planning and management are discussed in Chapter 11.

Selection of Methodologies

The last major problem is the selection of one or more planning methodologies from the set of competing methodologies (especially methodologies for developing the application portfolio and allocating resources). In their promotional literature, each of the methodologies tends to be presented as "the solution." Enthusiastic developers (and even some users) provide testimonials of the power of the methodologies in MIS planning processes. But, even though the techniques are competing, they are not directly equivalent; presumably, each methodology has a set of circumstances under which it is superior. However, there is very little guidance in their literature to make such a selection, taking into account the contingencies an organization is facing. In fact, no overall framework exists for classifying methodologies.

As these discussions of problems in MIS planning suggest, a comprehensive model of MIS planning is needed so that the process can be researched, explained, and applied.

Four-Stage MIS Planning Model

A basic, generic MIS planning model has been formulated based on observation of planning efforts, promotional literature, and an analysis of methodologies being used in the planning process. The basic MIS planning model depicted in Figure 13.1 and described in Figure 13.2 consists of four major, generic activities: strategic MIS planning, information requirements analysis, resource allocation, and project planning.

Most organizations engage in each of four stages, but their involvement tends to be evolutionary and influenced by problems as they occur, rather than by a plan for engaging in each stage as appropriate. During these stages planning methodologies often are chosen based on the persuasive power of methodology developers, rather than on a reasoned choice of a methodology for a given stage of MIS planning. The basic MIS planning model presented here provides a framework for study and evaluation of the MIS planning process and for mapping methodologies to the basic activities.

The four-stage basic MIS planning model can be illustrated by a case study in which an organization followed the steps described in the model. A Fortune 100 company, based on recommendations from its external auditors, was upgrading its computing capabilities from predominately batch, second-generation systems. Major problems were being encountered in the accounting area in terms of processing speed and ability to integrate data. For example, processing was so slow in accounts payable that the company's credit rating was being affected. Therefore, the company made a *strategic* decision to upgrade its computing ca-

Figure 13.1   Basic Four-Stage Model of MIS Planning

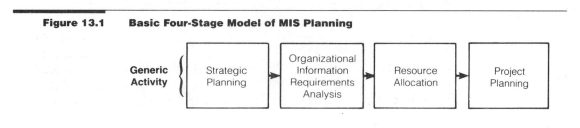

Figure 13.2   **Description of MIS Planning Stages**

Major MIS Planning Activity	*Description*
Strategic MIS Planning	Establishing the relationship between the overall organizational plan and the MIS plan
Organizational Information Requirements Analysis	Identifying broad, organizational information requirements to establish a strategic information architecture that can be used to direct specific-application system development projects
Resource Allocation	Allocation of both MIS application development resources and operational resources
Project Planning	Developing a plan that expresses schedules and resource requirements for specific information system projects

pabilities to an on-line, database environment with initial emphasis on improving accounting processing. Other applications were also to be reviewed. Although they did not use a formal approach, the organization had, at this point, gone through the strategic stage. The top management of the organization was strategically determining MIS objectives.

During the next six months the organization hired new MIS management to conduct analysis on the new system and look at overall information requirements. The MIS group used IBM's BSP (business system planning) methodology to conduct a comprehensive study of information requirements for the entire organization. This period of analysis can be characterized as an organizational information requirements analysis planning stage.

During the following eighteen months, several systems were implemented both in the accounting and in the operational areas. User management began to complain about two issues: (1) they wanted more systems, and (2) they wanted faster responses for new systems. But MIS costs had proliferated during the past two years. Top management, and, consequently, MIS management, had become concerned about allocating limited resources to increasing demand. This put the organization in the resource allocation stage. They decided to install a chargeout system to allocate resources.

This case illustrates how, based on organizational requirements, MIS planning moves from one stage to another. It also illustrates how specific methodologies may be selected for use in each of the three stages. In the case situation, the methodologies were BSP in the organizational information requirements analysis stage, and chargeout in the resource allocation stage. There was no use of a methodology in the strategic planning stage, but strategic MIS decisions were a function of overall company strategy. They used Gantt charts (discussed in Chapter 11) for project planning.

Detailed Four Stage Model

The very general four-stage model presented in Figure 13.2 can be expanded to include major activities and outputs of the three stages as shown in Figure 13.3. By adding this detail, the model moves from a high level of abstraction to a more

Figure 13.3   Major Activities and Outputs in Four Stages of MIS Planning

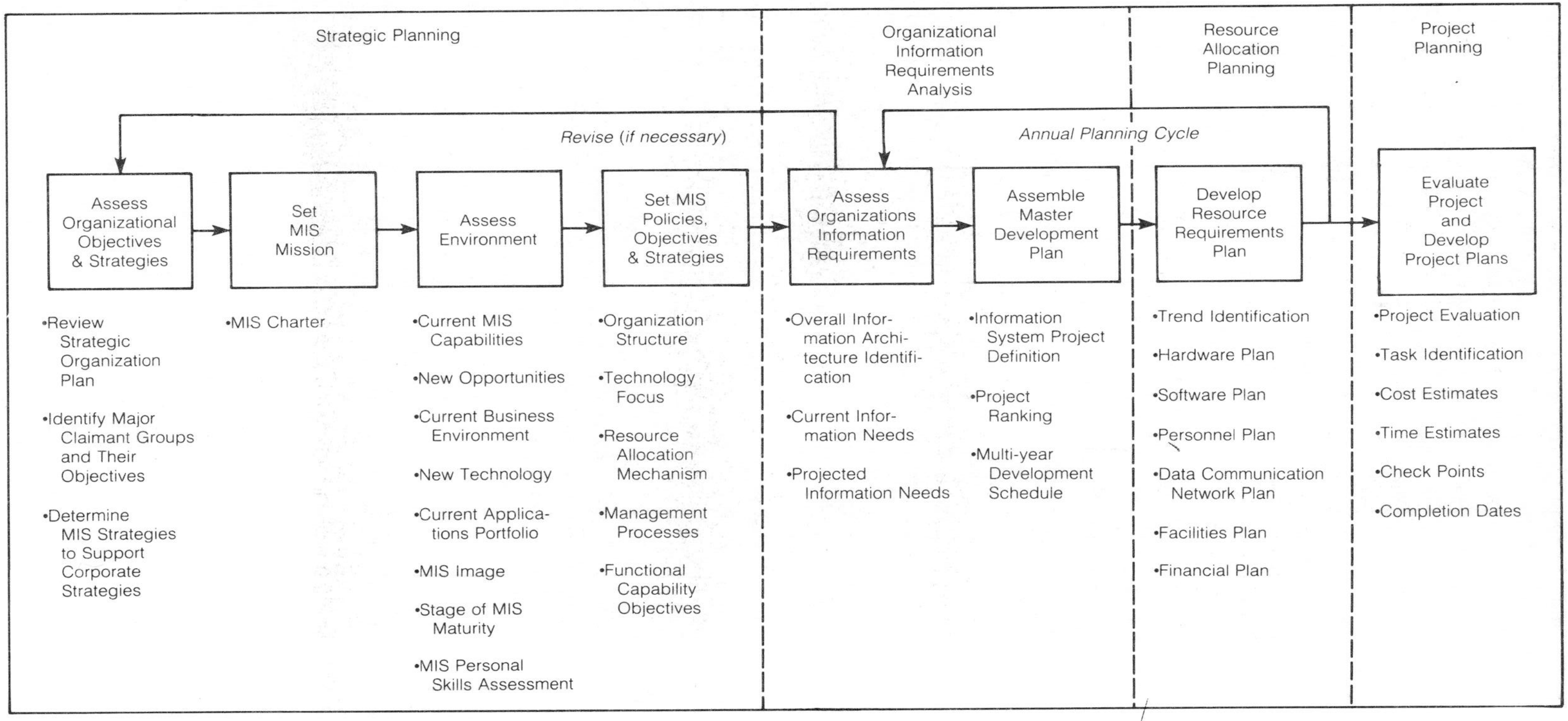

concrete formulation of MIS planning activities. The expanded stages are discussed in the following pages. The four-stage planning model sets the stage for different information systems to be developed using the system development life cycle discussed in Chapters 5 through 10. Since several systems will usually be under development at one time, the relationship between the four-stage and the system development life cycle is portrayed in Figure 13.4.

Strategic MIS Planning

During the strategic planning stage, it is critical to align MIS strategic planning with overall organizational planning. To accomplish this, the organization must

- Assess organizational objectives and strategies
- Set MIS mission
- Assess environment
- Set MIS policies, objectives, and strategies

The output from this process should include the following: an accurate perception of the strategic aspirations and directions of the organization; a new or revised MIS charter; an assessment of the state of the MIS function; and a statement of policies, objectives, and strategies for the MIS effort.

Organizational Information Requirements Analysis (OIRA)

The first phase of the *organizational information requirements analysis* (OIRA) stage consists of assessing current and projected information needs to support decision making and operations of the organization. This effort is not to be confused with or to replace the detail information requirements analysis associated with application system specifications (e.g., report and terminal display layouts). Rather, this is a higher level of information requirements analysis aimed at developing an overall information architecture for the organization or a major sector of the organization.

Figure 13.4   **Relationship of Planning Model to System Life Cycles**

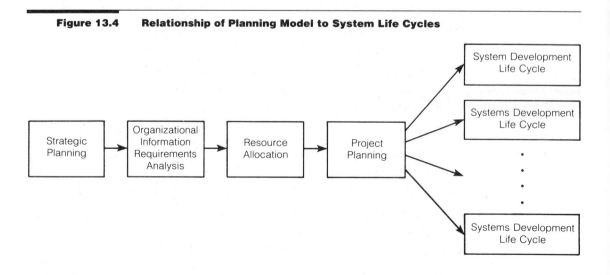

The second phase of the OIRA stage consists of assembling a master development plan. This plan is derived from the information architecture and defines specific information system projects, ranking of projects, and a development schedule.

Resource Allocation

Resource allocation consists of developing hardware, software, data communications, facilities, personnel, and financial plans needed to execute the master development plan defined in the OIRA stage. This stage provides the framework for technology procurement, personnel planning, and budgeting to provide appropriate service levels to users.

Project Planning

The first step of *project planning* consists of evaluating the project in terms of requirements and difficulty. Providing definitions of tasks that need to be performed is the next step. Finally, time, cost, and completion estimates must be developed, and check points to be used for evaluating project progress must be defined. Milestone, Gantt, or PERT plans must be developed as discussed in Chapter 11.

Staging of the Model

As depicted in Figure 13.3, the activities within stages and the stages themselves have a sequential flow starting with "assess organizational objectives and strategies" and ending with "evaluate project and develop project plan." A complete execution of the model is not necessary at each planning effort. As Figure 13.3 portrays, the annual planning cycle may consist only of assessing organizational information requirements, assembling a master plan, and developing a resource allocation and project plan. The time involved in executing the entire model is a function of how rapidly the organization's overall objectives and strategies are changing in ways that impact MIS requirements. The interval between comprehensive planning cycles may be from one to five years.

Methodologies for Use in MIS Planning

The next issue is the relationship of the various planning methodologies to the four-stage planning model. Several of the most publicized planning methodologies are the following:

- Competitive strategy
- Customer resource life cycle
- Strategy set transformation
- Business systems planning (BSP)

- Critical success factors (CSFs)
- Ends/means analysis (E/M)
- Return on investment (ROI)
- Charge-out
- Zero-based budgeting (ZBB)
- Milestones
- PERT
- Gantt charts

The last three techniques—Milestones, PERT, and Gantt charts, used for project planning—are covered in detail in Chapter 11. A brief description of the other methodologies together with a discussion of their relationship to the four major planning activities follows.

Competitive Strategy

Much of the current work, particularly at the Harvard Business School, on strategic use of information systems has evolved from Michael Porter's work on *competitive strategy*.[1] Competitive strategy does not focus specifically on MIS strategies; rather, it focuses on corporate strategy in general. However, many organizations have found the competitive strategy framework particularly useful as a means for determining how MIS can contribute to corporate strategy.

Competitive strategy identifies five major competitive forces faced by all organizations:

1. Threat of new competitors
2. Intensity of rivalry from existing competitors
3. Pressure from substitute products
4. Bargaining power of buyers
5. Bargaining powers of suppliers

Porter proposes that organizations wishing to gain a strategic advantage should consider building defenses against them by formulating specific courses of competitive action that can directly influence these forces.

Three generic strategies that an organization may choose to determine competitive strategy are

1. *Be a low-cost supplier:* Information systems technology can be very helpful in this area by reducing clerical, scheduling, inventory costs, and so forth.
2. *Differentiate product or service:* Information systems technology can help by adding features to products or services. For example, a pharmacy checks prescriptions for customers to make sure that no unhealthy combination of drugs has been prescribed. They also keep records for customers on all tax-deductible purchases and send them a report prior to income tax preparation.
3. *Focus on a specialized niche:* Information systems technology can help by identifying specific customers with specific needs. For example, frequent-flyer

---

1. Michael Porter, *Competitive Strategy.* New York: The Free Press, 1980.

programs allow airlines to identify their most important customers and offer them special packages for travel, hotels, rental cars, and so on.

Customer Resource Life Cycle

The *customer resource life cycle* (CRLC) is an innovative framework proposed by Ives and Learmouth that focuses directly on the relationship to the customer.[2] The idea of CRLC is that an organization differentiates itself from its competition in the eyes of the customer. Therefore, focusing on the relationship to the customer is the key to achieving strategic advantage. CRLC postulates that the customer goes through thirteen fundamental stages in its relationship to a supplier, and each of those stages should be examined to determine if information systems can be used to achieve a strategic advantage.

The thirteen stages and examples are as follows:

1. *Establish customer requirements:* Owens Corning Fiberglass uses data on energy efficiency to help builders evaluate insulation requirements for new building designs. Evaluations are provided free of charge to builders meeting minimum standards of energy efficiency, provided they agree to purchase insulation from Owens Corning.

2. *Specify customer requirements:* A greeting card distributor developed a system for automatic reordering that frees retailers from involvement in specifying particular cards. When a particular card is sold out, the retailer returns the reorder ticket at the back of the stack; the system determines the type of card (e.g., father's birthday) and resupplies, not with the same card, but with one specified by the system, usually the best-selling card in the particular category.

3. *Select a source (match customer with supplier):* The customer must locate an appropriate source for the required resource. The Phone-In Drive Thru Market of Los Angeles permits customers to order groceries by phone and pick them up at store loading docks. The system suggests alternative products if a requested product is currently out of stock.

4. *Place order:* Distributors such as American Hospital Supply, McKesson Pharmaceuticals, General Electric Supply Company, and Arrow Electronics have all established sophisticated round-the-clock order-entry systems that accept a customer-entered order without human intervention.

5. *Authorize and pay for good or service:* Exxon is pilot testing a debit-card network that immediately debits customer bank accounts for purchases made with Exxon's AutoCard. The retail customer is provided with most of the conveniences of a credit card but at the same per-gallon discount they receive for a cash purchase. The retailer, meanwhile, is spared most of the additional work and inconvenience of credit-card sales.

6. *Acquire good or service:* Automated teller machines normally deliver money to customers but also dispense airline and ski-lift tickets and store coupons.

7. *Test and accept good or service:* Western Union provides a service for matching freight shippers with motor-freight carriers and checks to ensure that responding carriers have appropriate authority and insurance to qualify for the prospective load.

---

2. Blake Ives and Gerald Learmouth, "The Information System as a Competitive Weapon," *Communications of the ACM*, December 1984, pp. 1193–1201.

8. *Integrate into and manage inventory:* General Electric Supply Company assists customers in inventory management by committing to stock, for a specified time, a prearranged quantity of items that are purchased repetitively. The distributor, in effect, keeps inventories for customers.

9. *Monitor use and behavior.* ARA Services distributes magazines to a variety of retail establishments. Unsold magazines are returned to ARA for a refund. Returns are machine-processed, which permits ARA to monitor sales information so as to make appropriate decisions about which magazines the customer should stock.

10. *Upgrade if needed.* Bergen Brunswig monitors product sales data for druggists and then uses those data to support its "Space Management" product. For a monthly fee, customers are advised on shelf arrangement and potential upgrades related to product choice.

11. *Maintenance.* Sears uses information systems to support its maintenance service by sending out annual postcard reminders to people whose maintenance contracts are about to lapse. Sears also offers special package deals to customers who have made multiple appliance purchases but have yet to purchase maintenance contracts. The program has boosted service revenues and benefited marketing by improving customer goodwill.

12. *Transfer or dispose.* The Washington Hotel in Tokyo is using an automated registration "robot" to check guests in with no human assistance and, on the way out, to "eat the key and inform you of how much you owe for cold drinks and telephone calls."

13. *Accounting of purchases.* AMEXCO has combined traditional credit services with a travel-agency function. Large corporations can obtain travel services from American Express Travel Related Services, pay with a corporate credit card, and receive reports from AMEXCO on how travel funds are being spent. Also, many retail drugstores, as a service to their customers, provide a detailed accounting of their customers' drug purchases for income tax purposes. Even a small local druggist, with the aid of a personal computer, can offer this competitively attractive service.

Strategy Set Transformation

One expert, W. R. King proposes an approach to the strategic phase of MIS planning that he terms *strategy set transformation.*[3] The overall organizational strategy is viewed as an *information set* consisting of the mission, objectives, strategies, and other strategic variables (e.g., managerial sophistication, proclivity to accept change, important environmental constraints, etc.). Strategic MIS planning is the process of transforming the organizational strategy set into an *MIS strategy set* consisting of MIS system objectives, constraints, and design strategies.

Business Systems Planning (BSP)

Business systems planning (BSP) is a comprehensive planning methodology developed by IBM. BSP was initially developed for IBM's internal use, but as IBM's

---

3. W. R. King, "Strategic Planning for Management Information Systems," *MIS Quarterly* 3, (1979): 16–21.

customers expressed interest in learning how they might better manage their MIS resources, BSP was released as a generalized methodology to assist in this task. It is supported by IBM manuals and training courses.

BSP basically involves a two-phase approach. It is conducted by a BSP planning team composed of both user and MIS personnel. Phase I focuses on developing a broad overall understanding of the organization, identifying how MIS currently supports the business, specifying the gross network of information systems required to support the business, and identifying the highest priority subsystems to be implemented within the network. Data is primarily gathered through interviews with numerous managers to determine their environment, objectives, key decisions, problems, and perceived information needs. The analysis concentrates on business processes without regard for organizational structure.

The objective of phase II is to develop a long-range plan for the design, development, and implementation of a network of information systems to support the business process identified in phase I. The current information systems are assessed, and weaknesses and deficiencies are noted. Processes and users that share data are identified, and the potential for common information systems across organizational boundaries is determined. The output of phase II is the *information systems plan*, which this plan describes the overall information systems architecture and defines the scheduling implementation of individual systems within the overall network. It serves as the "blueprint" for development of an integrated MIS.

Critical Success Factors (CSFs)

A framework advocated by J. F. Rockart argues that the information needs for top managers can be derived from *critical success factors*,[4] that is, the key areas for any organization in which performance must be satisfactory if the business is to survive and flourish. Critical success factors (CSFs) differ among industries and for individual firms within a particular industry.

As an example, Rockart cites the four industry-based CSFs of supermarkets: (1) have the right product mix available at each store; (2) keep it on the shelves; (3) provide effective advertising to attract shoppers to the store; and (4) develop correct pricing. As these areas of activity are major determinants of a supermarket chain's success, the status of performance in these areas should be continually measured and reported. The Rockart research team at MIT has identified the four primary sources of CSFs as

1. Industry-based factors
2. Competitive strategy, industry position, and geographic location
3. Environmental factors
4. Temporal factors

The CSF approach involves a series of interviews conducted in two or three sessions. In the first session, the manager is queried as to his or her goals and the CSFs that underlie those goals. Considerable discussion may be required to ensure that the analyst thoroughly understands the interrelationships between the goals and CSFs. Every effort is made to combine or eliminate similar CSFs,

---

4. J. F. Rockart, "Chief Executives Define Their Own Data Needs," *Harvard Business Review,* March–April 1979, pp. 81–93.

and an initial set of performance measures is developed. The second session is a review of the first, and it primarily focuses on identification of specific performance measures and possible reports. Additional sessions are held as necessary to obtain agreement on the CSF measures and reports for tracking them. The reports and related information systems required to provide them are designed by the MIS group.

Ends/Means (E/M) Analysis

Ends/means analysis is a planning technique developed by Wetherbe and Davis at the MIS Research Center at the University of Minnesota.[5] The technique can be used to determine information requirements at the organizational, departmental, or individual manager level.

Based upon general systems theory, this technique focuses first on the *ends*, or outputs (goods, services, and information), generated by an organizational process. Next, the technique is used to define the *means* (inputs and processes) used to accomplish the ends.

The ends, or output from one process, whether the process be viewed as an organizational, departmental, or individual process, is the input to some other process. For example, the inventory process provides a part to the production process, the accounting process provides budget information for other organizational processes, and the marketing process provides products to customer processes.

Ends/means analysis is concerned with both the effectiveness and the efficiency of generating outputs from processes. *Effectiveness* refers to how well the outputs from a process fill the input requirements of the other processes. *Efficiency* refers to resources required and the use of those resources to transform an input into an output.

A model of ends/means analysis is provided in Figure 13.5. As shown in the model, *effectiveness* information is based upon (1) what constitutes output effectiveness and (2) what information, or feedback, is needed to evaluate this effectiveness. *Efficiency* information is based upon (1) what constitutes input and

Figure 13.5   **Model of Ends/Means Analysis**

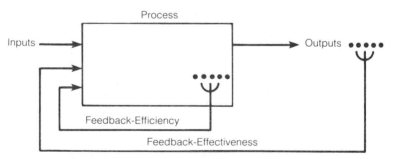

5. Gordon B. Davis, and James C. Wetherbe, "Developing a Long-Range Information Architecture," *National Computer Conference*, Los Angeles, May 1983.

transformation efficiency and (2) what information, or feedback, is needed to evaluate this efficiency.

As an example of information requirements in ends/means analysis, an inventory manager might specify the following:

1. *Ends specification:* The outputs, or end result, of the inventory management function is an inventory kept as low as possible but at an acceptable level of availability.

2. *Means specification:* The inputs and processes to accomplish the ends are the following:
   - Forecasts of future needs
   - Amounts on hand and on order
   - Items that are obsolete or in unuseable condition
   - Safety stock policy
   - Demand variations
   - Cost of ordering and holding inventory
   - Cost of items
   - Stockouts

3. Efficiency measures needed for inventory management are the following: Number and cost of orders placed, cost of holding inventory, and loss from disposal of obsolete or unuseable inventory.

4. Effectiveness measures needed for inventory management are the following: Number and seriousness of stockouts.

Ends/means analysis has been used in diverse industrial settings with positive results. Information requirements determined by this means are usually more extensive than those generated using other techniques. The problem with most information planning tools is that they usually result in information systems that provide only efficiency-oriented information. However, managers agree it is more important to be effective than to be efficient. Ends/means analysis brings out effectiveness information requirements. Such requirements typically transcend departmental boundaries, and, therefore, ends/means analysis is especially useful for a database planning effort.

Return on Investment (ROI)

Return on investment (ROI) is a cost/benefit analysis technique widely used in a variety of planning applications. Typically, projects are ranked in descending order by ROI, and the highest ranked projects that provide an acceptable rate of return are selected. For example, a project offering a ROI of 15 percent would be ranked over a project offering 10 percent. Considerations other than ROI such as resource constraints, organizational priorities, or politics may alter the selection process.

Many organizations apply ROI analysis to MIS projects in an attempt to make them pass the same criteria as other organizational undertakings. To the extent that cost and benefits are quantifiable, ROI is a useful planning tool. Unfortunately, MIS projects often do not lend themselves to easy quantification and estimation of costs and benefits. The costs and benefits of MIS projects are variable, complex, interrelated, and difficult to estimate. This often precludes a meaningful ROI analysis.

Chargeout

Some form of chargeout system is frequently used as a basis for planning and controlling MIS. In large organizations, the MIS function is often organized as a service bureau charged with providing MIS services to all organizational sub-units. Fee schedules are developed for each unit of service (e.g., CPU seconds, DASD I/Os [direct-access storage device input/outputs], lines printed, programming time, etc.) with the objective of recovering (or partially recovering) MIS expenditures. Users are charged for those MIS services rendered for their particular subunit. In theory, holding users responsible for the cost of their information systems fosters greater planning and control of those systems.

Chargeout-based planning systems are typical of the traditional approach to MIS planning discussed earlier. In addition to the chargeout system, this approach usually includes guidelines, procedures, and schedules to specifically direct planning efforts, but the focus is frequently toward justifying the costs relative to benefits of proposed information systems. Planning decision making is decentralized to user departments. This can tend to limit the search for potentially beneficial new information systems, especially with respect to integrated systems affecting multiple departments and to applications areas with intangible benefits. There are a number of problems associated with chargeout-based planning, including high expense (in terms of both administrative and computer-processing overhead), complexity, market imperfections, and difficulties with the development of integrated, multi-departmental systems.

The nature of chargeout-based planning systems varies among organizations. However, without specific procedures to the contrary, there are no systematic mechanisms linking information system planning based on chargeout to broader organizational strategy and objectives. This may result in strictly bottom-up development of information systems with short-range time horizons (e.g., consider the bank case study presented in Chapter 12).

Zero-Based Budgeting (ZBB)

Zero-based budgeting (ZBB) is a highly structured planning technique developed by Peter Pyhrr as an alternative to incremental budgeting.[6] Its use has been fairly widespread by various government and private organizations.

Wetherbe and Dickson suggest the use of ZBB as an MIS planning and control tool and as an alternative to chargeout-based systems.[7] The first step in this process involves conceptually reducing all MIS activities to zero-base, that is, no development or maintenance of information systems. Next, all potential information systems applications are identified and structured into sequentially dependent incremental service levels. Expected benefits and MIS resource-support requirements are listed for each service level. The projects are combined into an applications portfolio and submitted to a steering committee (or some other resource-allocation mechanism) for priorities to be established. Through discus-

---

6. P. A. Pyhrr, "Zero-Base Budgeting," *Harvard Business Review*, November-December, 1970, pp. 111–12.

7. J. C. Wetherbe and G. W. Dickson "Zero-Based Budgeting: An Alternative to Chargeout Systems," *Information and Management*, November 1979, pp. 203–13.

sion and debate, the projects are ranked in order of priority, and cumulative resource requirements are calculated.

This technique is particularly useful in identifying applications that have outlived their usefulness. It has a strong bottom-up orientation, and the service level concept could conceivably result in a logical evolutionary design of the MIS. This methodology has a strong focus on resource allocation, but again, there is no explicit strategic-planning cycle or direct link to the host organization's overall planning process. Compared with ROI, ZBB allows a more subjective analysis that does not require quantification of all costs and benefits. Compared with chargeout, using an MIS Steering Committee to establish priorities adds a centralized, high-level perspective to planning decision making. However, the lack of explicit consideration of strategic MIS planning may yield a planning process with a short-range time horizon.

The amount of personnel time required to utilize the ZBB approach can be significant. Information analysts must devote a considerable amount of their time to interacting with users in identifying information system projects and structuring proposed systems into incremental service levels. Also, preoccupation with service-level definition may narrow the search for alternatives.

Relationship of Methodologies to MIS Planning Model

The methodologies that have just been reviewed fit into the framework of the basic MIS planning model. Each may be classified as applying primarily to one of these four generic activities (see Figure 13.6)

1. Strategic planning
2. Organizational information requirements analysis
3. Resource allocation
4. Project planning

As depicted in Figure 13.6, competitive strategy, customer resource life cycle, and strategy set transformation fall into the strategic planning category. The first

Figure 13.6 **Alternative MIS Planning Methodologies Classified by Stage of MIS Planning of Most Significant Impact**

two are more for generating strategic ideas. Strategy set transformation is more for ensuring proper linking to different business areas.

BSP, CSF, and E/M analysis fall into the organizational information requirements analysis (OIRA) category. The three approaches differ in their method and comprehensiveness. BSP is the more comprehensive and labor-intensive approach and generates a more extensive definition of total information requirements. CSF and E/M analysis are less labor-intensive and direct information requirements analysis toward higher-level management requirements.

Charge-out, ROI, and ZBB fall into the resource allocation category. All three approaches are concerned with allocating resources; however, their orientations are quite different. Charge-out advocates decentralized "marketplace" decision making with cost recovery. ZBB advocates centralized planning committee decision making with MIS expenses carried as overhead. ROI can be used in either centralized or decentralized decision making.

PERT, Gantt, and milestone planning techniques (discussed in Chapter 11) fall into the realm of project planning and management. PERT is the more formal and structured of the three techniques, and Gantt is more structured and formal than the milestone technique.

Classifying the various planning methodologies within the four-stage model adds clarity to their use and purpose. Since each of the planning methodologies has been implemented in a number of organizations, the indication is that under certain circumstances each of the planning methodologies performs a useful and needed function.

Guidelines for MIS Planning

At the beginning of the chapter, the major problems of MIS planning were identified as residing in the following endeavors:

1. Aligning MIS strategy with organizational strategy
2. Developing an information architecture that allows integration of databases
3. Allocating resources
4. Completing information systems projects on time and on budget
5. Selecting a methodology for the preceding steps

These problems are addressed directly by the four-stage MIS planning model. The first four problems correspond to the four stages of the model. Given the framework of the model, the set of appropriate methodologies is specified for each stage. This aids in selecting one methodology for each stage.

Practical guidance for MIS planning can be gained from the model. It can aid in recognizing the nature of MIS planning problems as well as in selecting the appropriate stage of planning. Too often, this is not done. For example, some organizations may view their MIS function as making minimal contributions to organizational objectives. In seeking to resolve this problem, some organizations have installed a chargeout system (resource allocation planning) to make MIS pay its own way. Other organizations have conducted a BSP (OIRA planning) exercise to resolve the same problem. While these activities may result in improved MIS services, the MIS planning model suggests they are probably not the appropriate methodologies for the situation. If the MIS effort is not responsive to the organization, the four-stage MIS planning model indicates that a strategic-oriented planning effort should precede OIRA and resource allocation planning exercises.

To establish MIS planning, an organization should conduct a stage assessment to determine the extent to which each stage of MIS planning has been accomplished. This can be performed by analyzing the major activities and outputs of the four-stage planning model depicted in Figure 13.5. After the MIS planning needs at each stage have been established, appropriate methodologies can be selected.

Stage Assessments

For each of the four stages of the MIS planning model a strategic stage assessment should be performed in which the following questions are asked:

1. Is there a clear definition of organizational objectives and strategies?
   a. Has the strategic organizational plan been reviewed?
   b. Are the major claimant groups and their objectives identified?
   c. Have strategic applications been identified to improve strategic advantage?
2. Is there an MIS mission expressed in an MIS charter?
3. Is there an assessment of the MIS environment?
   a. Are MIS capabilities adequately assessed?
   b. Are new opportunities identified?
   c. Is the current business environment understood?
   d. Is the current applications portfolio defined and documented?
   e. Is the MIS image healthy?
   f. Is the stage of EDP growth understood?
   g. Are MIS personnel skills accurately inventoried?
4. Are MIS policies, objectives, and strategies established?
   a. Is the MIS organization appropriate to the overall organization?
   b. Is the MIS technology focus appropriate to the technology focus of the organization?
   c. Are the objectives for allocating MIS resources appropriate?
   d. Are the MIS management processes appropriate?
   e. Are the functional capability objectives appropriate?

If answers to these questions indicate a strategic stage weakness, a strategic planning exercise is in order. Competitive strategy, customer resource life cycle, and strategy set transformation offer formal methodologies for conducting such an exercise.

Before conducting an OIRA stage assessment an organization should ask the following questions:

1. Is there an adequate assessment of organizational information requirements?
   a. Is the overall organizational information architecture identified?
   b. Is there a good understanding of current information needs of the organizations?
   c. Is there a good understanding of projected information needs of the organization?
   d. Are the major databases and their relationships defined?

2. Is there a master MIS development plan?
   a. Are MIS projects defined?
   b. Are projects ranked by priority?
   c. Is there a multi-year development schedule?

If an organization does not have acceptable answers to the OIRA stage questions, an OIRA planning exercise is in order. Examples of formal planning methodologies available to conduct such an exercise are BSP, CSF, and ends/means (E/A) analysis.

To evaluate the current status prior to conducting a resource allocation stage assessment an organization should ask the following questions:

1. Does the organization have a resource requirements plan?
   a. Are trends identified?
   b. Is there a hardware plan?
   c. Is there a software plan?
   d. Is there a personnel plan?
   e. Is there a data communications plan?
   f. Is there a facilities plan?
   g. Is there a financial plan?
2. Does the organization have an adequate procedure for resource allocation?

If an organization does not have acceptable answers to the resource allocation stage questions, a resource allocation planning exercise is in order. Formal planning methodologies available to conduct such an exercise are charge-out, ROI, and ZBB.

To evaluate the status of project planning an organization should ask the following questions:

1. Is there a procedure for evaluating projects in terms of difficulty or risk?
2. Are projects tasks usually identified adequately?
3. Are project cost estimates generally accurate?
4. Are project time estimates generally accurate?
5. Are checkpoints defined to monitor progress of projects?
6. Are projects generally completed on schedule?

If an organization does not get satisfactory answers to the project planning questions, a review of project planning techniques is in order. Techniques available to improve project planning include PERT, Gantt, and milestones.

Selecting a Methodology

The four-stage planning model provides considerable insight into the MIS planning issues. This should reduce confusion among competing planning methodologies. For example, use of the planning model can prevent an organization from using a resource allocation methodology when an OIRA or strategic methodology is more appropriate. However, the planning model does not indicate which of several methodologies categorized within a planning stage should be used for any particular planning stage. Most organizations find using the methodologies in a combination provides different, useful perspectives that are missed when only one methodology is used.

There has been almost no research to evaluate the comparative advantages of one technique or combination of techniques over another for the first three stages of the model. Organizations must evaluate the methodologies available in the context of the specific issues they are facing. Most organizations use more than one approach. For example a company might use *competitive strategy* and *CRLC* to generate strategic ideas for information systems.

By far the most time-consuming, extensive part of the information planning is the OIRA stage. Valuable experience and insight for OIRA planning has evolved over recent years. To facilitate a deeper appreciation of the "what" and "how" of OIRA planning, the techniques and an example are provided in the following pages.

Developing a Long-Range Information Architecture

Some interesting work in the OIRA planning stage has resulted in the development of a hybrid technique for conducting an OIRA.[8] This methodology, which follows, is based upon comparative research involving three methods of organizational requirements analysis: BSP, CSF, and E/A analysis.

Figure 13.7 portrays the model for conducting an OIRA. To provide concreteness to the methodology, the results of a case study are used to illustrate documents generated during the study. The company agreeing to share the results of an OIRA study is EPIC Realty Services Inc., leasers of single-family dwellings. Headquartered in Washington, D.C., with offices in major cities throughout the United States, the company manages over 14,000 homes.

Define Underlying Organizational Subsystems

The first phase of the OIRA is to define underlying organizational subsystems. An organizational subsystem is a fundamental organizational activity that is necessary for the operation of the organization. For EPIC Realty Services Inc., the major subsystems are as follows:

1. Credit
2. Leasing
3. Maintenance
4. Evictions/delinquency
5. Marketing
6. Advertising
7. Accounts receivable/collections
8. Corporate accounting
9. Market and product analysis
10. Client reporting
11. Appraisal
12. Insurance
13. Sales
14. Personnel/administration
15. Inspections
16. Audit
17. Inventory
18. Legal

These subsystems were obtained by the iterative process of discussing all organizational activities and defining them as belonging to broad categories of sub-

---

8. Wetherbe and Davis, "Developing a Long-Range Information Architecture." National Computer Conference, Anaheim, May 1983.

Figure 13.7   **Organizational Information Requirements Planning Model**

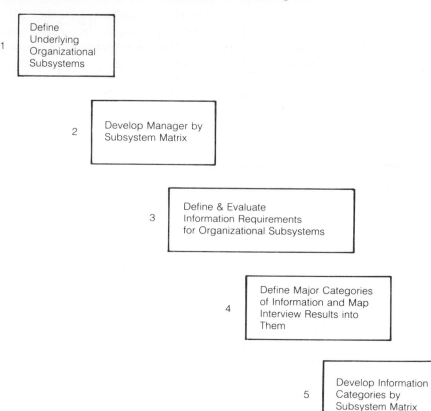

systems. As new activities are considered, they should be placed either in previously defined categories or in newly created categories.

Develop Subsystem Matrix

Once the underlying organizational subsystems are defined, the next phase of the OIRA planning exercise is to relate specific managers to organizational subsystems. The resulting document, called a *manager subsystem matrix*, is illustrated in Figure 13.8. Note that the subsystems on the left column of the matrix are the same as those identified in phase 1.

The matrix is developed by reviewing the major decision responsibilities of each middle to top manager and relating that decision making to specific subsystems. The matrix denotes the managers having major decision-making responsibility for each specific subsystem. Note that personnel changes or organizational changes can easily be reflected in an adjusted matrix.

Define and Evaluate Information Requirements for Organizational Subsystems

This phase of the planning model obtains the information requirements of each organizational subsystem by group interviews of those managers having major

Figure 13.8   **Manager Subsystem Matrix**

Organizational Subsystems	Managers				
	Manager 1	Manager 2	Manager 3	Manager 4	Manager n
Leasing					
Maintenance					
A/R	X				
Credit	X	X			
Evictions/ Delinquent		X		X	
Inspection				X	
Inventory					
Marketing			X		
Advertising					
Insurance					
Sales					
Audit					X
Appraisal			X		
Personnel/ Administration					X
Legal					X
Market & Product Analysis			X		
Corporate Accounting					X
Client Reporting			X		

decision-making responsibility for each subsystem. The interview method is identical to the structured interview described in Chapter 6. BSP, CSF, and E/M analysis questions are used for each subsystem interview. (See Chapter 6 for a sample interview.)

Define Major Information Categories and Map Interviews into Them

The process of categorizing information categories is done in the same manner that the data dictionary for an information system was factored into entities and attributes in Chapter 7. The difference is that the task is much larger when doing an organization-wide entity/attribute analysis. By placing the information cate-

Figure 13.9   **Interviews Mapped to Information Categories**

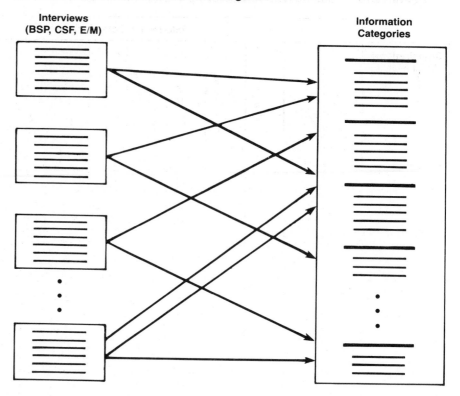

Interviews
(BSP, CSF, E/M)

Information
Categories

gories defined from the organizational subsystem interviews into broad, generic categories of information, an overall profile of information categories needed by the organization can be developed. Figure 13.9 illustrates this process. Broad categories of information that can be identified as entities (i.e., customers, contracts, vendors), and attributes (i.e., names, addresses, phone numbers) are categorized within their respective entities.

Develop Information/Subsystem Matrix

By mapping information categories against the organizational subsystems, an information-categories-by-organizational-subsystem matrix can be developed. Figure 13.10 illustrates such a matrix for EPIC.

Note that at the intersections of information categories and subsystems there are coded values defined in the following manner:

During the interview, managers are asked both the *importance* and the current *availability* of different types of information. Responses for both importance and availability are recorded as high, medium, or low, and these responses are quantified as follows where

	Low	Medium	High
Importance	1	2	3
Availability	3	2	1

Figure 13.10   Information Categories by Organizational Subsystems Matrix

Organizational Subsystems	Contract	Policy Training	Customer Financial	Customer Demographics	Complaint	Leasing/ Transactions	Vendor	A/P	A/R	Maintenance	Warranty	Inventory
Leasing	9	9	6	6	4	4				9		9
Maintenance	9	9			9		9	2	1	9	2	4
A/R	9	6	3	4	4	9	4	1	3	9		2
Credit	9	9	9	4		6			3			
Evictions/ Delinquencies	9	9	9	6	6		9	2	3			9
Inspection	4	9			9		6			9	2	6
Inventory	6		9	9	6		6		4	9	3	6
Marketing	9	9										9
Advertising	2		9	9			9	6				4
Insurance	9	9			4		2		3	9	3	
Sales	9	9	9	9	9		3	1	4	9	2	3
Audit	6	9	4		9	4	2	6	9	9	2	
Appraisal	9	9			2		6	6		4		
Personnel/ Administration		9			2			9				
Legal	9	4			6	2	6	2	6	1		1
Market & Product Analysis	9	4	6	9	9					4		6
Corporate Accounting	9	4			6	3	6	3	6	4	2	9
Client Reporting	9	4			9	2		3	4	9		
Total Score	135	121	64	56	94	30	68	41	46	94	16	68

Note: Scores in boxes are computed as follows:

$$\text{Score} = \begin{matrix}\text{Importance}\\ \text{of}\\ \text{Information}\\ \text{Category}\end{matrix} \quad \times \quad \begin{matrix}\text{Current}\\ \text{Availability}\\ \text{of}\\ \text{Information Category}\end{matrix}$$

where responses are quantified as follows.

	Low	Medium	High
Importance	1	2	3
Availability	3	2	1

A score is computed for each category of information using the following formula:

$$\text{Score} = \text{Importance} \times \text{Availability}$$

Note that the reversed scaling of responses results in those information categories with the highest importance and lowest availability getting the highest score (e.g., $S = 3 \times 3 = 9$). Conversely, information that is not that important and is readily available gets the lowest score (e.g., $S = 1 \times 1 = 1$).

This scoring procedure gives a good indication of the value of a category of information to a single business subsystem. By totaling all scores for a category of information, a composite score can be totaled and used as a rough indicator of the value provided to all organization subsystems that intersect with the information category.

Use of the OIRA Planning Results

The results of the OIRA exercise are two-fold:

1. It identifies high-payoff information categories.
2. It provides an architecture for information projects.

Identify High Payoffs

Evaluating composite scores for information categories allows selection of the categories with the highest scores for first consideration for feasibility studies. Note that the information category by subsystem matrix does not tell you whether it is technically, economically, or operationally feasible to improve an information category. The matrix merely indicates relative value of information. Feasibility studies and project definitions must still be done as usual.

Provide Architecture

By clearly defining the intersection of information and subsystems, an organization can avoid the problem of building separate, redundant information systems for different organizational subsystems. When an organization decides to improve information for one organizational subsystem, other subsystems that need such information can be taken into consideration. This avoids building separate information systems for each subsystem, which often requires reworking or duplicating what has already been done. By doing the conceptual work first, an organization can identify information system projects that will do the most good and lead to cohesive, integrated systems. This is far better than randomly selecting projects that result in fragmented, piecemeal systems that are continually being reworked or abandoned because they do not mesh with the organization's overall requirements. This means planning from the top down rather than from the bottom up.

Using the data modeling analysis techniques discussed in Chapter 7, the information architecture can be used to construct a conceptual data model for the entire organization. The information categories defined in Figure 13.10 can either be entities or subdivided into entities for constructing the model. This is a complex task that usually requires assistance from database administrators.

Notice how an architecture such as that portrayed in Figure 13.10 could have kept the bank discussed in Chapter 12 from developing nonintegrated systems. For example, when the bank developed a savings account system, it would have known that checking, installment loans, and mortgage loan decision makers needed access to savings account information.

Executive's Perspective

Perhaps the best way to illustrate the value of an organization's having an organizational information architecture for MIS is by quoting the president of EPIC a year after he personally led the development of their architecture:

> I had worked in top management in one of our other subsidiaries and experienced the disappointment that comes from developing systems in the traditional FIFO [First In First Out processing], piecemeal way with the consequences of redundant, nonintegrated and inaccessible information.
>
> When I took over a new subsidiary, I decided there must be a better way. There was. By developing an information architecture before developing systems we have been able to pull all our systems together. Our short-run system decisions are dovetailing into our long-range systems. We know where we are going and [we are] getting there.
>
> Beyond that, just the process of going through an organizational information requirements analysis gave me and my management invaluable insight into our business.

Resource Allocation

A good information architecture greatly reduces the difficulty of making resource allocation decisions by identifying high-payoff information categories. Whether ROI, ZBB, or charge-out are used, having an architecture from which to make resource allocation decisions is most helpful. Once resource allocation decisions are made, project planning can begin as discussed in Chapter 11.

Summary

The four-stage model of MIS planning provides a framework for addressing critical issues and problem areas of MIS planning. The first stage of the model, strategic planning, addresses the problem of alignment of the MIS effort with the overall strategic objectives of the organization. The second stage, organizational information requirements analysis, addresses the problem of development of a long-range information architecture for the organization. The third stage, resource allocation, addresses the allocation of information system development and operational resources among competing applications. The fourth stage, project planning, addresses project management techniques and strategies.

This four-stage model provides a framework in which competing and diverse planning methodologies can be categorized. The model can thus lead to better MIS planning and aid MIS planning research.

Exercises

1. Discuss the five major problems identified in this chapter pertaining to MIS planning.

2. Describe the four-stage model of the MIS planning and discuss the basic issues as they relate to each stage of the model.

3. Briefly discuss each of the methodologies used for MIS planning and categorize them within the four-stage model of MIS planning.

4. Discuss what a person uses stage assessment for, when evaluating MIS planning within an organization.

5. Describe the basic steps involved in organizational information requirements analysis and the function of each of the documents generated from such a study.

Selected References

Ansoff, H. I. "State of Practice in Planning Systems." *Sloan Management Review* 18 (1977): 1–24.

Anthony, R. N. *Planning and Control Systems: A Framework for Analysis.* Division of Research, Graduate School of Business. Boston, Mass. Harvard University, 1965.

Beard, L. "Planning a MIS: Some Caveats and Contemplations." *Financial Executive* 45 (1977): 34–39.

Benjamin, R. I. "When Companies Share, It's Virtually a New Game." *Information Systems News*, December 26, 1983, p. 24.

Benjamin, R. I., Rockart, J. F., Scott Morton, M. S., and Wyman, J. *Information Technology: A Strategic Opportunity.* Working paper 108, MIT Center for Information Systems Research, Cambridge, Mass., December 1983.

Bowman, B. Davis, G. B., and Wetherbe, J. C. "Modeling for MIS." *Datamation*, July 1980, pp. 155–62.

Burnstine, D. C. *BIAIT An Emerging Management Discipline.* New York: BIAIT: International, 1980.

Bush, R. L. and Knutsen, I. E. "Integration of Corporate and MIS Planning: Its Impact of Productivity." *Proceedings of Ninth Annual Conference of the Society for Management Information Systems*, Chicago, 1977.

————."Business is Turning Data into a Potent Strategy Weapon." *Business Week*, August 22, 1983, pp. 92–98.

————."A Coupon Machine at the Supermarket." *Business Week*, March 5, 1984, p. 68.

Canning, R. G. "Developing Strategic Information Systems." *EDP Analyzer*, May 1984, pp. 1–8.

Carlson, W. M. "Business Information Analysis and Integration Technique (BIAIT): The New Horizon." *Data Base* 10 (1979): 309.

Davis, G. B. *Management Information Systems: Conceptual Foundations Structure and Development.* New York: McGraw-Hill, 1974.

_____. "Strategies for Information Requirements Determination." *IBM Systems Journal* 22 (1982): 4–30.

Diamond, S. "Contents of a Meaningful Plan." *Proceedings of the Tenth Annual Conference of the Society for Management Information Systems*, Chicago, 1979.

Drucker, P. F. *Management: Tasks, Responsibilities Practices.* New York: Harper & Row, 1974.

Ein-Dor, P. and Segev, E. "Strategic Planning for Management Information Systems." *Management Science* 15 (1978): 1631–41.

Forster, A. J. "Effective Strategies and Techniques for the Development of MIS Master-Plans for Top Management Approval." *Proceedings of the Tenth Annual Conference of the Society for Management Information Systems*, Chicago, 1978.

Gibson, D. G. and Nolan, R. L. "Managing the Four Stages of EDP Growth." *Harvard Business Review*, January–February 1974, pp. 76–88.

Gurry E. and Bove, R. "Effective Data Processing Planning." *CPA Journal* 47 (1977): 46–47.

Head, R. V. "Strategic Planning for Information Systems." *Information Systems* 25 (1978): 46–47.

Holloway, C. and King, W. R. "Evaluating Alternative Approaches to Strategic Planning." *Long-Range Planning* 12 (1979): 74–78.

Hootman, J. T. "Basic Considerations in Developing Computer Charging Mechanisms." *Data Base* 8 (1977): 1–13.

IBM Corporation. *Business Systems Planning: Information Systems Planning Guide.* Publication no. GE20-0527.

Kerner, D. V. "Business Information Characterization Study." *Data Base* 10 (1979): 10–17.

King, W. R. "Strategic Planning for Management Information Systems." *MIS Quarterly* 2 (1978): 27–37.

Lyles, M. A. "Making Operational Long-Range Planning for Information Systems." *MIS Quarterly* 3 (1979): 16–21.

McFarlan, F. W. "Problems in Planning the Information Systems." *Harvard Business Review* 49 (1971): 74–89.

McFarlan, F. W. "IS and Competitive Strategy." Note 0-184-055, Harvard Business School, Cambridge, Mass., 1983.

McFarlan, F. W., and McKenney, J. L. *Corporate Information Systems Management.* Homewood, Ill.: Richard D. Irwin, 1983.

McFarlan, F. W., McKenney, J. L., and Pyburn, P. "The Information Archipelago—Plotting the Course." *Harvard Business Review* 61 (January–February) 1982, pp. 145–46.

McLean, E. R. and Soden, J. V., eds. *Strategic Planning for MIS.* New York: Wiley-Interscience, 1977.

Mulvihill, D. E. and Cohen, B. J. "Strategy Formulation and Information Systems: Setting Objectives." In *The Information Systems Handbook*, Edited by F. W. McFarlan and R. L. Nolan, Chicago: Dow Jones-Irwin, 1975, pp. 19–31.

Munro, M. C. and Wheeler, B. R. "Planning, Critical Success Factors, and Management Information Requirements." *MIS Quarterly* 4, (1980): 27–38.

Nolan, R. L. "Managing the Computer Resource: A Stage Hypothesis." *Communications of the ACM* 16 (1973): 399–405.

_____. "Managing the Crises in Data Processing." *Harvard Business Review* 57 (1979): 115–26.

Parsons, G. L. "Information Technology: A New Competitive Weapon." Note 0-183-121, Harvard Business School, Cambridge, Mass., 1983.

Parsons, G. L. "ARA Services Inc.—Periodicals Distribution Group (A)." HBS Case Serv. 9-182-157, Harvard Business School, Cambridge, Mass., 1982.

Porter, M. *Competitive Strategy.* New York: The Free Press, 1980.

Pyhrr, P. A. "Zero-Base Budgeting." *Harvard Business Review* 48 (1970): 111–121.

Rockart, J. F. "Chief Executives Define Their Own Data Needs." *Harvard Business Review*, March–April 1979, pp. 81–93.

Rush, R. L. "MIS Planning in Distributed Data Processing Systems." *Journal of Systems Management* 30 (1979): 17–26.

Schwartz, M. H. "MIS Planning." *Datamation* 16 (1970): pp. 28–31.

Shidal, J. G. "Long-Range DP Planning." *Journal of Systems Management* 29 (1978): 40–45.

Soden, J. "Pragmatic Guidelines for EDP Long Range Planning." *Data Management* 13 (1975): 8–13.

Soden, J. and Tucker, C. "Long-Range MIS Planning." *Journal of Systems Management* 27 (1976): 28–33.

Steiner, G. A. and Miner, J. B. *Management Policy and Strategy.* New York: Macmillan, 1977.

Wedley, W. "New Uses of Delphi in Strategy Formulation." *Long-Range Planning* 10 (1977): 70–78.

Wetherbe, J. C. and Dickson, G. W. "Zero-Based Budgeting: An Alternative to Chargeout." *Information and Management* 2 (1979): 203–13.

Wetherbe, J. C. and Davis, G. B. "Strategic MIS Planning Through Ends/Means Analysis." Working paper, MIS Research Center, 1982.

———. "Developing a Long-Range Information Architecture," *Proceedings of the National Computer Conference*, Los Angeles, May 1983.

Zachman, J. A. "Control and Planning of Information Systems." *Journal of Systems Management* 28 (1977): 34–41.

———. "The Information Systems Management System: A Framework for Information Systems Planning." *Proceedings for the Ninth Annual Conference of the Society for Management Information Systems, Chicago, 1977.*

Zani, W. M. "Blueprint for MIS." *Harvard Business Review*, November–December 1970, pp. 95–100.

Introduction

In Chapter 12 one of the most common reasons cited for management's disappointment in the concept of MIS was the limited support that computer-based information systems have provided for top-management decision making. The fact that decision-support systems (DSS) are increasingly providing capabilities that are significantly offsetting this disappointment was also pointed out. These capabilities represent an advanced dimension of computer-based information systems that are more useful and better accepted by operating, middle, and top management. A DSS goes beyond transaction processing and reporting. It actually assists management in making decisions.

A well-done requirements determination as described in Chapter 6 will often reveal the need for DSS capabilities in a new information system. For example, managers may express the need to

1. Assess the profitability impact of price increase on products
2. Determine how adding temporary employees will increase productivity
3. Evaluate whether adding another work shift or subcontracting work is the best way to handle a backlog problem.

Since DSS requirements are often needed in new systems, it is important for a systems analyst to be knowledgeable of them. Accordingly, this chapter is included as an advanced topic in this book.

Herbert A. Simon's model for describing the decision-making process was introduced in Chapter 6.[1] This model is shown in Figure 14.1. Decision making is a three-phase, continuous process, flowing from intelligence to design to choice but capable of returning to a previous phase when necessary. The need for a decision-making process is invoked by the recognition of a problem or an opportunity. The resultant decision-making activity is directed at resolving the problem or taking advantage of the opportunity.

A DSS serves to directly support the intelligence and design phases of decision making. In so doing, it indirectly supports the choice phase. However, a DSS does not actually make the choice. When a system makes choices for management, it moves out of the realm of a DSS and into the realm of programmed decision making.

The topic of DSS is discussed in four sections of this chapter. The first section reviews management decision making and its relationship to DSS. The second and third sections discuss DSS as they pertain to the intelligence and design decision-making phases, respectively. The fourth section discusses the impact of DSS on choice. And a final section of the chapter examines the recent emergence of expert systems to support management.

It is not the objective of this chapter, nor would it be realistic, to provide comprehensive training for designing and developing a DSS or an expert system. Rather, the objective is to provide exposure to the nature and potential of a DSS and possibly stimulate interest in further study of techniques required to develop a DSS or an expert system. A systems analyst who contributes in the area of

---

1. Herbert A. Simon. *The New Science of Management Decision*, (Englewood Cliffs, N.J.: Prentice-Hall, 1960), p. 54.

Figure 14.1   The Decision-Making Process

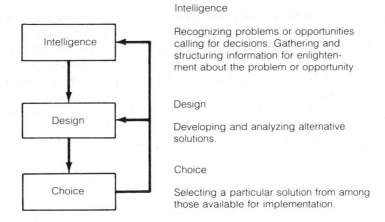

Intelligence

Recognizing problems or opportunities calling for decisions. Gathering and structuring information for enlightenment about the problem or opportunity

Design

Developing and analyzing alternative solutions.

Choice

Selecting a particular solution from among those available for implementation.

DSS or expert systems should be skilled in such areas as management science, modeling, simulation, advanced statistics, and knowledge engineering. To comfortably read the material in this chapter, one needs only a basic background in statistics and mathematics. The selected references at the end of the chapter are sources of more advanced discussions of the topics presented.

Decision Making

The nature and characteristics of decision making are concepts discussed frequently in this book. In this section of the chapter, these concepts are reviewed and expanded to provide a framework for discussing DSS.

Characteristics of Organizations and Organizational Processes

In Chapter 3 it was pointed out that organizations and organizational processes can be characterized as existing on a continuum between closed/stable/mechanistic and open/adaptive/organic. The more closed/stable/mechanistic an organizational process is, the more structured, computational, and routine is the decision making associated with planning and controlling it. Consequently, the decision making associated with closed/stable/mechanistic processes can be more readily supported with computer processing.

Generally, higher levels of management are dealing with more open/adaptive/organic decisions, increasing the difficulty of making a good (let alone, a perfect) decision. The lack of structure and routine, as well as the external and incomplete nature of information, at higher organizational levels of decision making significantly constrains the ease of providing computer support. Table 14.1 (duplicated from Chapter 3) reviews both the characteristics of decision making and the information required at different organizational levels.

Table 14.1   **Characteristics of Decision Making and Information at Different Management Levels**

	Operating	Management Level — Middle	Top
Characteristics of Decision Making	Computational/Objective ——————————————→		Judgmental/Subjective
Examples	Inventory Reordering Production Scheduling Credit Approval	Short-Term Forecasting Budget Preparation Capacity Planning	New Product Planning Location of New Factory Mergers and Acquisitions
Characteristics of Information	Internal/Complete ——————————————→		External/Incomplete
Examples	Sales Order Production Requirements Customer Credit Status	Sales Analysis Budget Analysis Production Summaries	Market Conditions Industry Forecasts Government Regulations

Knowledge of Outcomes

An important dimension of decision making is the degree of certainty with which decision makers can predict the outcomes, or results, of their decisions. The knowledge of the outcomes of decision making can be categorized as follows:

1. *Decision making under certainty:* There is complete, accurate knowledge of the outcome of each choice available. Each decision has only one unique outcome (or set of events) associated with it.

2. *Decision making under risk:* Alternative decisions can result in more than one outcome. The possible outcomes and their probabilities of occurrence can be identified.

3. *Decision making under uncertainty:* Alternative decisions can result in more than one outcome. Some of these outcomes can be identified, but their probabilities of occurrence are unknown.

There is a relationship among the degree of certainty in decision making, the characteristics of organizational processes, and management levels. This relationship is depicted in Figure 14.2. The relatively open/adaptive/organic processes that higher-level management must cope with increase the uncertainty associated with decision making. Conversely, the relatively closed/stable/mechanistic processes that lower-level management must cope with reduce the uncertainty associated with decision making. The examples in Table 14.1 convey the varying degrees of uncertainty associated with decision making at different management levels.

In **decision making under certainty** the objective is to determine which is the best solution to a problem or the best opportunity. Since the decision variables and their relationships are known, the best solution can usually be computed. Accordingly, tools based upon mathematics are useful. In **decision making under risk** the problem is to determine, within acceptable levels of probability, the decision that provides an optimal (or, more likely, a satisfactory) outcome. Accordingly, tools based on probability or statistics are useful for decision making under risk. In **decision making under uncertainty** the probabilities of outcomes

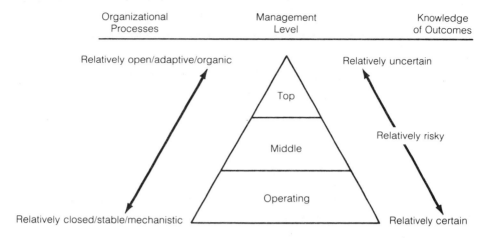

are not known. Therefore, the decision maker must assign arbitrary probabilities or assume no probabilities.

A more extensive discussion of the use of mathematical, statistical, and other decision-support tools is presented later in the chapter.

Behavioral Considerations

The cognitive style of the decision maker, as discussed in Chapter 3, is an often overlooked but significant dimension of the decision-making process. The trial-and-error tendency that is a key part of most managers' problem-solving approach needs to be recognized and accommodated for in developing a DSS.

A DSS should allow managers to explore a wide set of alternatives that depart from predefined models. Managers often prefer, however, to rely on their hunches and/or information based on discussions with peers or subordinates. In such situations, managers can use the DSS to predict various outcomes and thereby play "what if"—trying out their hunches prior to implementing them.

Sometimes, systems analysts are critical of the intuitive/heuristic approach used in higher-level decision making. However, there are many cases where an intuitive/heuristic approach is appropriate, or even necessary, irrespective of management's cognitive style. For extremely open/adaptive/organic problems or opportunities, when there is limited time, or when a systematic/analytic approach is too expensive and too time-consuming relative to the importance of the decision, an intuitive/heuristic approach is usually preferable.

For example, consider a company that has an option to buy the patent rights for a revolutionary solar energy device. The company is not certain of the marketability of the device or if other superior devices are forthcoming. The patent option costs $100,000. Potentially, the company could make several million dollars. To adequately research and prepare for this decision requires three months of work at a cost of $120,000, but the company must make a decision within two days or lose its option to a competitor. An intuitive approach, therefore, must be used.

To summarize, systems analysts should strive to consider cognitive styles of managers and design DSS's that are flexible enough to accommodate different managerial approaches. They should also accept that, due to the limitations of DSS technology and/or the limitations of managers, many good and poor decisions must be made without the assistance of computer technology.

DSS and Decision Intelligence

The intelligence phase of decision making involves becoming aware of problems or opportunities. Initially, this awareness is usually at a symptomatic level. Therefore, the intelligence phase usually requires delving deeper into the problem or opportunity to fully identify and examine the underlying variables and their relationships. For example, a symptom of a problem may be a decrease in sales, but the actual cause of the problem may be one of the following: a change in consumer behavior, a new product introduced by competitors, a morale problem in the sales force, or a combination of these factors or additional ones.

The more *decision intelligence* that management has of the cause of a problem or the effort required to capitalize on an opportunity, the more success it is likely to have in arriving at an optimum, or at least a good, decision. Indeed, it is the degree of intelligence that to a great extent determines whether a particular decision will be made under certainty, risk, or uncertainty.

Exception and Summary Reporting

Since the first step of decision making is identifying problems or opportunities, it logically follows that a DSS should assist in this identification process. The routine reporting provided by an information system plays a key role in providing management with intelligence about problems or opportunities. A common and successful means of identifying a great many organizational problems and opportunities is the use of *exception reporting* derived from routine transaction processing of the organization's internal database. For example, inventory items that experience an unanticipated decrease or increase in activity may be identified as problems or opportunities.

Another means for identifying problems and opportunities is through the use of summarized information, or *summary reporting*. In particular, graphical presentations of summarized data have gained great popularity. Some board-of-directors' conference rooms are now equipped with video equipment that graphically displays various cost and profit trends from data generated by computer systems. Unusual shifts in summarized data (either favorable or unfavorable) alert management to further investigate detail data to assess the causes of these shifts.

Both exception and summary reporting have the advantage of presenting information in a decision-impelling format. That is, they highlight important information that encourages management action, rather than overload management with details.

DBMS

A constraint of routine exception, summary, and detail reports is that their content and format must be determined in advance (i.e., so they can be programmed).

However, the open/adaptive/organic characteristic of many organizational processes makes it extremely difficult to anticipate all information requirements. Consequently, upper-level managers may request "crash" projects to produce special reports for decision making. Unfortunately, the complexity and frequency of such requests often make it impossible for information systems personnel to respond to the requests within the time constraints placed upon them.

With the advent of DBMSs (database management systems) the ability to query a database in ad hoc (rather than predetermined) ways has been greatly enhanced. Most DBMSs include a general-purpose query language or fourth-generation language that can be used for information-retrieval purposes by persons who are not technicians. These languages allow users to specify the values or ranges of attributes to be used for including and/or excluding a group of data for a particular query. For example, a credit manager may easily specify that he or she needs a list or display of all credit customers as follows:

Include if:

BALANCE-DUE > 500
DAYS-PAST-DUE = 30 to 90
SALES-DISTRICT = 4

Exclude if:

VIP-CODE = 1

The query language facility is an additional layer of software between the user and the DBMS. This relationship is shown in Figure 14.3.

Statistics

During the intelligence phase of decision making, management attempts to reduce, to the extent possible, the uncertainty associated with decision making.

Figure 14.3   **Relationship of Query Language to Database Management System (DBMs)**

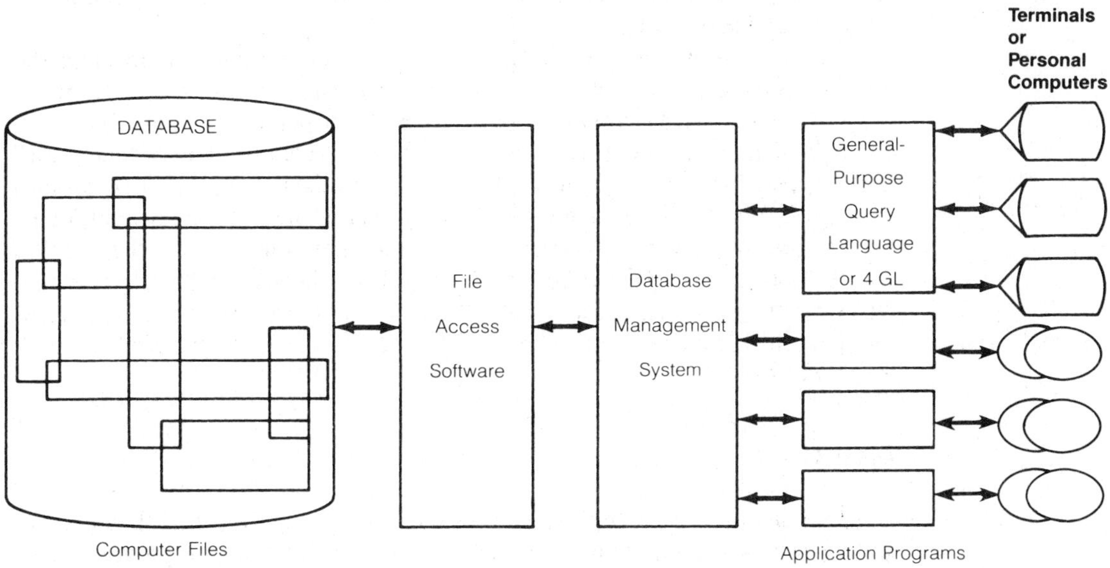

Statistical analysis is a useful tool for developing a better understanding of variables surrounding a decision-making process. Statistics can be used to structure disorganized detail data into descriptive measures such as range, mean, mode, median, and frequency distributions. Analyses can be conducted on such data as employee salaries, total sales, production yields, and the like. They can be useful for analyzing a group of data or for comparing one or more groups of data (e.g., comparing sales in different stores).

Management, particularly top management, does not usually have access to all data pertinent to a decision. Consequently, management must frequently deal with incomplete information. This is especially true when the information needed is external to the organization. For example, it is not practical to survey all potential consumers of a new economy car to determine their attitudes toward its styling, engineering, and performance. However, management can take a statistical sample from the population of potential consumers and use inferential statistics to generalize about overall consumer behavior.

In other cases, management may want to determine the similarities and differences between two or more populations for which data on only a sample of each population can be collected. Again, statistical tests can be used to compare the samples and generalizations can be inferred about the similarities and differences. Such tests include t-tests, chi-square, and analysis of variance.

A number of statistical packages have been developed. They provide an excellent repertoire of commonly used statistical tools. The packages can be used to conduct complex statistical analyses with minimal effort. The addition of one or more statistical packages to an organization's computer-based information systems can greatly enhance the ability to make prompt statistical analyses in support of decision making. Four of the commonly used statistical packages and their suppliers are listed as follows:

- BMD (biomedical computer programs)
- Omnitab
- SAS (statistical analysis system)
- SPSS (statistical package for the social sciences)
- SYSTAT

In practice, few managers use statistical packages directly. Rather, their staff and/or members of the information systems staff perform the necessary technical tasks.

DSS and Decision Design

The design phase of decision making involves developing and analyzing decision alternatives. In recent years, quantitative models have gained in popularity as aids for decision making, particularly among managers with systematic/analytic orientations. Numerical models are quantitative representations of objects or processes. They can be visualized as systems capable of transforming input data about an object or process into informative output. The transformation process is conducted in accordance with one or more processing statements (e.g., formulas) defined within the model.

Consider as a simple numerical model the mathematical equation used to compute the area of a rectangle.

Area = Length × Width

The inputs to the model are values for the variables of length and width. The transformation process consists of multiplying length times width and assigning the resulting value to the variable area. The output from the model is the computed area for a given length and width. This model can be used to compute the area of any rectangle for which the width and length are known. This general-purpose characteristic of models makes them extremely useful. Once a model has been developed to represent an object or process, it can be used repeatedly for similar applications, no matter how complex the model is.

The use of mathematical equations provides considerable precision and convenience. For example, mathematical equations can be used to compute the areas of triangles, circles, squares, and so on. However, many objects or processes are difficult or impossible to represent using mathematical equations. Consider the irregularly shaped object in Figure 14.4*a*. There is no convenient or accurate mathematical equation available to compute the area of this or similarly structured objects. However, the area can be reasonably estimated by using numerical techniques of probability as follows:

1. Enclose the irregularly shaped object in a square.
2. Place the numbers 1 through 10 at equally spaced intervals on the *Y*- and *X*-axes of the square.
3. Place ten pieces of paper numbered 1 through 10 in a hat.

Figure 14.4　**Using Probabilities to Estimate the Area of (a) an Irregularly Shaped Object with the technique of (b) Randomly Selected Coordinates**

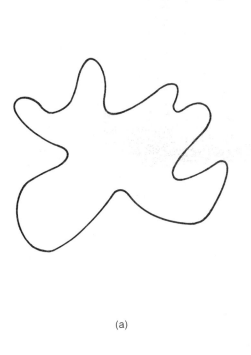

(a)

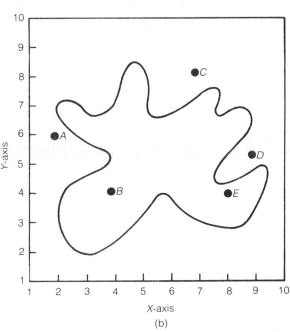

(b)

4. Randomly generate pairs of X and Y values by repeatedly drawing numbers from the hat, being careful to return each number to the hat and to shuffle all numbers before the next drawing. This ensures that each number has an equal chance of being drawn each time.

5. Plot each pair of X and Y values using the scales of their respective X- and Y-axes. For example, the following values for X and Y are plotted in Figure 14.4.

Point	A	B	C	D	E
X	2	4	7	9	8
Y	6	4	8	5	4

6. After several (X, Y) coordinates have been plotted, determine the percentage of points located within the irregularly shaped object.

7. Compute the area of the square, and multiply it by the percentage of points within the irregularly shaped object.

8. The result is an estimation of the area of the irregularly shaped object.

If this entire process is repeated several times, the computed area estimates are likely to differ slightly. This variability in outputs is due to the use of probabilities in the modeling process. Note that the more coordinates that are plotted, the greater the probability that the estimate of the area will be more precise.

Numerical models based strictly on equations always generate the same output, given the same input. They are called *deterministic models*. Numerical models that contain dimensions of probability produce different, though usually predictable, outputs. These models are called *stochastic models*.

Both deterministic and stochastic models are used to support management decision making. Certainly, these models are of greater complexity and size than the preceding two examples. Fortunately, the computational capabilities of computer technology allow complex, large models to be manipulated conveniently.

The execution of a model to represent an object or process is called *simulation*. Because decision making is concerned with outcomes, the ability to simulate outcomes of possible situations is particularly helpful in designing decision alternatives.

The ability to simulate the results of decisions using a model is a powerful tool. Models are usually much less expensive to experiment with than are the objects or processes they represent. Therefore, decisions are tested and outcomes analyzed using the model rather than the real objects or processes.

For example, the management of a hotel is planning to re-carpet the entire hotel. Management needs to know how much carpet is required and how much it will cost. It would be foolish to buy the carpet and then see how much carpet is required. Rather, hotel employees can measure the lengths and widths of all rooms and then use the model for computing area to determine the total number of square feet of carpet required. By including the cost per square foot of carpet as an input variable to the model, they can simulate what it will cost to carpet the entire hotel using various grades of carpet. The equations within this deterministic model are the following:

Area = Length × Width
Total Cost = Area × Cost per Square Foot

The type of model constructed and the simulation conducted for a particular decision are a function of whether the decision is made under certainty, risk, or uncertainty. A discussion of modeling and simulation for these decision-making categories follows.

Decision Design under Certainty

Decision making under certainty refers to situations where there is only one outcome for each decision alternative. Therefore, models used to simulate outcomes for decisions made under certainty must produce only one outcome for each decision alternative. Accordingly, deterministic models based on mathematical equations are used.

Usually, the objective of decision making under certainty is to optimize some objective function (e.g., to maximize productivity or minimize costs). Several general-purpose mathematical models for computing optimum solutions for a variety of organizational functions are available. To illustrate their type and variety, some of the models are listed here; however, it is beyond the scope of this chapter to explain these techniques.

1. Systems of equations
2. Linear programming
3. Integer programming
4. Dynamic programming
5. Break-even analysis
6. Return on investment
7. Queuing models
8. Inventory models

In many cases, models must be tailored to particular organizational processes. Consider the simplified model for computing net profit in Figure 14.5. The model requires two inputs: sales and operating expenses. A manager can use a terminal to interact with a computer-program representation of the model. He or she enters either actual or predicted values for the two variables and receives simulated profit statements. Figure 14.6 shows how this person/machine dialogue progresses. By simply re-executing the model, the manager can assess the effects of different levels of sales and operating expenses on net profit.

Figure 14.5   Model for Computing Net Profit

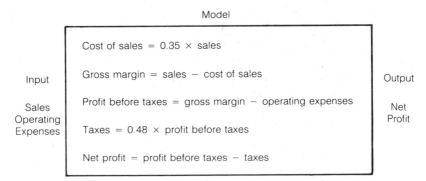

Model

Input

Sales
Operating
Expenses

Cost of sales = 0.35 × sales

Gross margin = sales − cost of sales

Profit before taxes = gross margin − operating expenses

Taxes = 0.48 × profit before taxes

Net profit = profit before taxes − taxes

Output

Net
Profit

Figure 14.6   Use of a Computer-Supported Model for Computing Net Profit

Manager:	EXECUTIVE PROFIT-MODEL
Terminal:	ENTER SALES
Manager:	125,750
Terminal:	ENTER OPERATING EXPENSES
Manager:	58,100

Terminal: PROFIT STATEMENT

SALES	$125,750
LESS: COST OF GOODS SOLD	44,013
GROSS PROFIT	$ 81,737
LESS: OPERATING EXPENSES	58,100
PROFIT BEFORE TAXES	$ 23,637
LESS: TAXES	11,346
NET PROFIT	$ 12,291

The design and development of DSS models usually begins with simple models calling for high-level inputs. As a better understanding of a process emerges, basic models are expanded and/or modules are added to provide greater detail and capability. The model can then generate high-level items from more basic inputs.

For example, operating expenses have to be inputted to the model in Figure 14.6. However, an additional model could be developed to compute operating expenses. Figure 14.7 shows such a model. The additional inputs (long-term debt, average short-term debt, and accounts receivable balance) could be generated by another model, and so on. By interfacing such models, a sophisticated profit model can be established.

Other models can be developed for other organizational processes-financing, staffing requirements, facility requirements, and so on. This is accomplished by developing mathematical equations that are representative of the processes and then translating them into computer programs.

Decision Design under Risk

Decision making under risk pertains to situations where alternative decisions can result in more than one outcome and the probabilities of these outcomes are known. This means that at least one dimension of the object or process under

Figure 14.7   A Model for Computing Operating Expenses

Model

Input	Model	Output
	Administrative expense = $0.08 \times$ sales	
Sales	Advertising expense = $0.04 \times$ sales	
	Interest expense = $0.08 \times$ long-term debt + $0.09 \times$ average short-term debt	Output
Long-term Debt Average Short-term Debt	Bad-debt expense = $0.01 \times$ accounts receivable balance at beginning of period	
Accounts Receivable Balance at Beginning of Period	Operating expenses = Administrative + advertising + interest + bad debt	Operating Expenses

consideration has an element of chance associated with it. Therefore, models used to simulate outcomes for decision making under risk must be able to represent elements of probability within the model to reflect the probabilities associated with outcomes. Accordingly, stochastic models using numerical techniques based on probability are used for decision making under risk.

Stochastic models generally contain mathematical equations along with probability functions. To the extent that a process can be represented mathematically, it is advantageous to do so because of the greater precision afforded. Probability functions should be used only when a process cannot be conveniently and accurately represented by one or more mathematical equations.

Considerable perception is required to develop a mathematical equation or set of equations to model an object or process. In many cases, it is not possible, given existing mathematical techniques, to model objects or processes. In other cases, objects and processes are modeled using probability functions until they can be described mathematically. In such cases, the probability functions serve as a surrogate until the mathematical functions can be defined.

As a fictitious illustration of this sequence, assume that the mathematical equation for computing the area of a triangle is unknown. The probability technique illustrated in Figure 14.4b can be used to estimate the area of a triangle. Once it is discovered that a mathematical equation (area = ½ × base × height) can be used to compute the exact area of a triangle, it is no longer necessary, and certainly not advantageous, to continue using the probability technique.

Many organizational processes are difficult or impossible to describe by mathematical equations. Therefore, techniques of probability are the norm rather than the exception. However, the systems analyst should watch for opportunities to replace probability functions with more accurate mathematical equations whenever possible.

Airline Ticket Counter

The application of stochastic modeling is explained best by example:

The market research department of an airline has determined that a major area of dissatisfaction among airline passengers is having to wait in line ten to twenty minutes for ticket purchases and luggage check-in. The management of the airline feels that the airline could gain a competitive advantage by significantly reducing this wait time and developing an advertising program to inform consumers of such a reduction.

To determine what measures are necessary to reduce the wait time, it is decided to develop a general-purpose model of the line-waiting process at the numerous ticket counters at various airports. Such a model can be used to simulate the wait time being encountered by passengers and then to experiment with ways to improve it. The key processes to be simulated are the arrival of customers, their time spent waiting in line, the time spent servicing customers, and the departure time of customers (see Figure 14.8).

Management is concerned with keeping line-waiting time to a minimum. This can be accomplished by having an adequate number of attendants and computer terminals available at the ticket counters. However, management also wants to avoid overstaffing ticket counters, resulting in idle employees and equipment, an unnecessary cost.

Probability Distributions and Random Numbers

The first step of stochastic modeling is to develop probability distributions for a process. These distributions can be based upon a frequency distribution from

Figure 14.8   **Processes Associated with an Airline Ticket Counter**

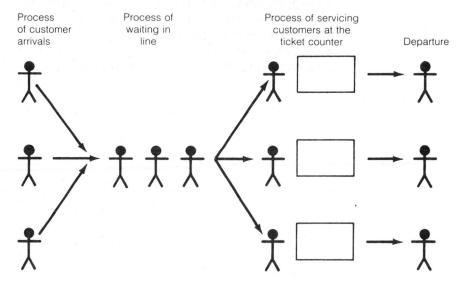

Process of customer arrivals   Process of waiting in line   Process of servicing customers at the ticket counter   Departure

actual observations of the process, or they can be arrived at subjectively. For example, a frequency distribution of interarrival times (i.e., times since last arrival) of customers at an airline ticket counter is given in the first two columns of Table 14.2. The third and fourth columns of Table 14.2 illustrate the relative and cumulative frequencies of the observations. It is now possible to simulate arrivals of customers by the use of random numbers that are merely a quantitative reflection of chance. To be considered random, any number from a range of numbers must have an equal chance of occurring (e.g., as in rolling dice, spinning a roulette wheel, or drawing a number from a hat).

Rather than draw numbers from a hat (as was done earlier in the chapter), the computer is used to generate random numbers. A computerized random number generator is simply a computer program that produces a sequence of numbers that appear to be random. Computer programs capable of generating a series of numbers having the statistical properties of random numbers are relatively easy to create and are available on most computers.

A random number generator can quickly generate a stream of random numbers that can be translated into values for interarrival times by comparing the

Table 14.2   **Observed Frequency Distribution of Interarrivals of Customers at an Airline Reservation Counter**

Interarrival Time (in minutes)	Observed Frequency	Relative Frequency	Cumulative Frequency	Categories of Random Numbers
1	15	15%	15%	1–15
2	20	20%	35%	16–35
3	30	30%	65%	36–65
4	20	20%	85%	66–85
5 and more	15	15%	100%	86–100
	100 *total*	100%		

values to the cumulative probability distribution. To accomplish this, the possible range of the random numbers is set (from 1 to 100 in this example). The values of the random numbers actually generated are categorized into the ranges defined in the last column of Table 14.2. For example, a random number value of 6 falls between 1 and 15; therefore, it causes an interarrival time of one minute to be generated. Other random numbers produce interarrival times as follows:

Random Number	Resulting Interarrival Time
98	5
38	3
16	2
1	1
84	4

If the simulation model is functioning properly, a stream of 100 random numbers will result in a simulated frequency distribution that is very close to the one originally observed.

Once a model has been developed to simulate the arrivals of customers, the next step involves developing a model of the times required for ticket counter attendants to process customer tickets and luggage. This is accomplished using the same techniques as used for developing a model for arrivals of customers. Once both models are operational, the arrival model can be used to feed simulated passenger arrivals to the ticket counter model. To the extent that the ticket counter model has adequate capacity to service all simulated passengers, no lines will develop. To the extent that the passenger arrival model gets ahead of the ticket counter model (i.e., generates simulated passengers faster than they can be serviced), lines will develop.

A computer program of this model can quickly simulate hours, days, weeks, or even years of passenger processing. Statistics can be readily gathered on the amount of time that customers spend in line and the amount of idle time that ticket counter attendants have.

Though the computer is just pretending to process passengers, the statistics about the simulation are representative of the statistics that would result from the real process. Management can use the model to conveniently experiment using different numbers of attendants and different arrival rates at the different airports. The results of their experimentation can be used to estimate the number of attendants and terminals needed at the airports to keep line-waiting times at an acceptable level (e.g., below three minutes).

Simulation Applications and Languages

The airline ticket counter simulation is concerned with the balancing between line waiting and idle capacity. A number of organizational processes fall into this type of simulation. Examples include processing of customer orders, checking out customers at supermarkets or retail stores, assembly line processes, unloading docks for freight trucks, and checking patients into hospitals.

Of course, stochastic simulation is applicable to any organizational process that can be described in terms of probability. For example, simulation can be used to model weather conditions, equipment failures, customer attitudes, and actions by competitors. Stochastic models can also be used in conjunction with deterministic models. For example, additional analyses may make it possible to develop mathematical equations to represent the process of servicing customers at an airline ticket counter. This deterministic model may be based upon how each customer pays (e.g., cash, credit card, or check), how many connecting flights must be scheduled, and how much luggage must be checked. The stochastic model of generating passenger arrivals could feed this deterministic model with simulated passenger arrivals, the same as it fed the original stochastic model of the ticket counter.

Initially, most simulation experiments were programmed in FORTRAN, using random number subroutines to generate random numbers when needed. Though FORTRAN is still used, a number of high-level simulation languages that minimize the difficulty of programming simulation models are more common. These simulation languages are usually designed to simulate particular types of processes; they are of little value out of the areas for which they are designed. However, when a simulation language fits a situation, it is usually preferable to a language such as FORTRAN. For example, GPSS (General-Purpose Systems Simulation) is a simulation language developed by IBM specifically for line-waiting problems. Other popular general-purpose simulation languages are SIMSCRIPT, DYNAMO, and GASP. It is considerably easier to simulate line-waiting problems and the like using simulation languages rather than FORTRAN. There are also languages designed specifically for financial modelling such as Interactive Financial Planning System (IFPS), LOTUS, and MBA.

Decision Design under Uncertainty

Decision making under uncertainty refers to situations where alternative decisions can result in several identifiable outcomes, but their probabilities of occurrence are unknown. The potential contribution of a DSS is lower in this area of decision making.

In some cases, management may assign arbitrary probabilities to outcomes. For example, management may simply assign equal probabilities. When this is done, the problem can be treated as decision making under risk, and stochastic modeling can be used. However, caution is warranted in analyzing the results, because of possible inaccuracies in the probabilities assigned.

Without assigning arbitrary probabilities and using stochastic models, decision making under uncertainty is of necessity intuitive/heuristic. This type of decision making tends to be trial-and-error oriented. Therefore, the major contribution a DSS can make in this environment is to provide rapid feedback about outcomes after a decision is made, thus facilitating quick reactions to outcomes. This allows corrective decisions to be made if outcomes are undesirable. For example, there may be uncertainty about the demand for a new product. Therefore, its inventory level is determined on an intuitive basis. By monitoring the demand for the new product on a daily basis (i.e., with short-interval feedback), management can adjust inventory levels heuristically. This short-interval feedback allows a bad decision to be recognized and a corrective decision made.

By combining DSS tools with DBMS, distributed processing, and an integrated database, an organization can achieve the concept illustrated in Figure 14.9. This is what management has wanted since the late 1960s.

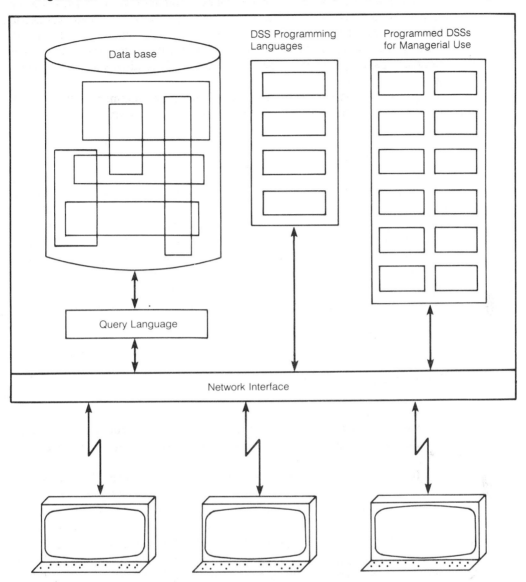

Personal computers with personal data and DSSs on small disks and "downloaded" data and DSSs from central computer.

DSS and Decision Choice

The choice phase of decision making involves selecting a particular course of action to resolve a problem or to capitalize on an opportunity. DSS support of decision choice results from the output of the models during decision design. The difficulty of the choice phase is, as might be expected, a function of the degree of certainty associated with outcomes.

Decision Choice under Certainty

If a decision is made under certainty, the process of choosing a course of action is, to a great extent, computational. Once the optimum solution has been determined, it is the rational choice.

Decision making under certainty is very similar to programmed decision making. In fact, once managers develop confidence in a model for a particular decision-making process, they often allow future decisions to be programmed, based on the solutions generated by the model.

Decision Choice under Risk

The output from a stochastic model varies each time the model is executed. Therefore, the usual approach is to execute the model repeatedly and develop probability distributions describing the outputs from the model. These statistics are used to develop probabilities of outcomes as they pertain to the real process. For example, management can determine that if there are four ticket counter attendants at a particular airport, there is a 90 percent probability that customers will never have to wait more than three minutes for service.

Decision making when the probabilities of outcomes are known is similar to decision making under certainty. However, instead of optimizing an outcome, the usual approach is to optimize the expected outcome. For example, assume a manager is faced with three choices, one offering a 10 percent probability of a gain of $100,000, the second offering a 20 percent probability of a gain of $45,000, and the third offering a 70 percent probability of a gain of $20,000. The expected values are computed as follows:

Alternative	Probability	$\times$	Outcome	$=$	Expected Value
1	.10		100,000		$10,000
2	.20		45,000		$ 9,000
3	.70		20,000		$14,000

Given these expected values, the third choice (offering an expected value of $14,000) would be selected.

Due to statistical properties, expected value as a selection criterion is most valid when the decision to be made a number of times. Since this is frequently not the case, there is evidence to suggest that behavioral factors influence decision making when a decision must be made only once. Often, depending on the optimism or pessimism a manager has, he or she will gamble that a lower-probability, but higher-valued outcome will occur. For example, an optimistic manager may select the first alternative in the preceding expected-value analysis.

Decision Choice under Uncertainty

Situations where management does not at least have subjective thoughts about the probabilities of different outcomes are, fortunately, rare. However, when they occur, decision choices must be made in an intuitive/heuristic manner.

To the extent that management is willing to assign arbitrary probabilities, stochastic models can be used to generate expected values. However, considering the arbitrary nature of these values, the process of choice should be combined with the intuitive feelings of management. When no probabilities are assigned, DSS technology is of no particular value for decision choice. Decision making must then be made on a nonquantitative basis.

However, in much the same way that election outcomes are predicted with only a fraction of the vote in, DSSs can be developed to provide preliminary feedback on how well a decision is working out. Such information can allow management early identification of a wrong decision and thereby minimize negative consequences.

Expert Systems

In Chapters 3 and 4 the concepts of artificial intelligence and expert systems were introduced. In this chapter, the recent emergence and application of expert systems to business problems is examined. Since the late 1940s, computer scientists have been conducting research on how to get computers to act more like humans. The domain of this research includes decision making, computer speech, and robotics, and is generally referred to as *artificial intelligence* (AI).

A subset of the AI research has developed techniques that can support decision makers in analyzing problems and making decisions. This subset of techniques is referred to as *knowledge-based expert systems*, and has proven very useful for business applications.

A systems analyst conducting requirements determination (as described in Chapter 6) is likely to come across situations where expert-systems technology may be needed to solve a business problem or provide a new business opportunity. Accordingly, a systems analyst should know enough about expert systems to be familiar with what they can do, what problems they are suited for solving, some of the tools available, and where to get assistance when an expert system is needed to augment an information system.

Expert Systems versus Decision Support Systems

DSS is differentiated from traditional transaction processing and reporting systems by its ability to simulate, via models, the outcomes of different courses of action being considered by a decision maker. So how is an expert system different from a DSS? Besides supporting a decision maker, an expert system can do one or more of the following:

- Make the decision (or perform the task)
- Explain the results and reasons for the actions taken
- Learn from its mistakes

Tools for Expert Systems
Powerful tools have been developed to support this higher level of sophistication. Some of the more popular include

INTELLECT

LISP

PROLOG

ROSIE

OPS-5

RLL

HEARSAY-III

These languages require a great deal of computational support. Continued improvements in cost/performance in computational speed has only recently made such tools more widely used or considered for use. With their increased availability, expert system development will undoubtedly expand.

Developing Expert Systems

If ever there was a need to use heuristic design or prototyping (discussed in Chapter 7), it is for the development of an expert system. The process of developing an expert system involves a lot of trial and error that can span years.

An expert system is constructed through a process called *knowledge engineering*, which is similar to traditional system and DSS development in that the technical person interviews the expert (or user) to determine requirements. However, there is a major difference: the traditional system or DSS designer tries to *support* the expert, whereas the knowledge engineer is attempting to *simulate* the expert's behavior.

Two cycles of knowledge engineering are required to simulate an expert's behavior: The first cycle consists of developing an inference engine and presentation mechanism. The second cycle involves developing a knowledge base. For example, to develop an expert system to help people prepare income taxes would first require capturing the logic ("mental" decision trees, etc.) used by expert tax preparers (inference engine). Also, the knowledge of the current tax code would have to be captured. Both the logic and the knowledge must be within the system.

Digital Equipment's XCON System

As with any new technology, expert systems are surrounded by a certain amount of hype. But there is enough evidence at this early date to indicate that expert systems are likely to make a major difference in business by the 1990s.

Unfortunately, many companies are keeping their expert-systems projects confidential, but some well-documented examples exist. One of the successful applications in the computing industry is Digital Equipment Corporation's (DEC) XCON System. Founded in 1957, DEC played a major role in the emergence of the minicomputer and over the years has developed an excellent reputation for configuring custom computer systems for special applications. The firm's strategy has been to provide leading technology, quality, and flexible custom systems.

As its systems became more powerful and complex, the task of configuring became overwhelming. In the late 1970s it was clear that configuring and mistakes from incorrect configuring (e.g., equipment that cannot be matched or used together) for their VAX and PDP computers were becoming a major cost of doing business. Consequently, in 1978 they began development of the eXpert CONfiguration (XCON). By 1980 the XCON could address 6 to 14 percent of configuration problems. By 1986 it could handle 86 to 88 percent (Note the evolutionary nature of the development process—the system got better over time.).

What were the business effects of XCON?

- A new configuration testing facility did not have to be built.

- In 1980 twenty-seven technical editors were needed to do 4,000 orders; by 1986, thirteen technical editors were needed to do over 60,000 orders.

- In 1980 twelve to twenty months were needed to complete an order; by 1986, less than three months were needed for most orders.

- Overall estimated economic value to DEC = \$40,000,000.

Keep in mind that XCON was an early, pioneering effort in expert systems. Consequently, it took a long time to develop and required a great deal of maintenance. However, due to improved tools and an increasing experience base, expert systems will take less time and money to develop in the future.

Summary

Decision-support systems (DSS) offer significant potential for assisting and improving management decision making. DSS support for the intelligence and design phases of decision making can lead to better choices with more predictable outcomes.

Decisions can be categorized as made under certainty, risk, or uncertainty. The category is determined by the intelligence phase of decision making. A DSS can provide intelligence through routine reporting, information-retrieval capabilities, and statistical packages.

During decision design, computer-based numerical models can be used to develop and define decision alternatives. Deterministic models are used for decision design under certainty. Stochastic models are used for decision design under risk. DSS are limited in the contribution they can make to decision design under uncertainty. However, rapid feedback mechanisms can be used to inform the decision maker of the emerging outcome of his or her decision, thus allowing adjustments if a bad decision has been made.

The choice phase of decision making involves selecting a particular course of action. DSS support of decision choice results from the output of the models during decision design. When decision making is under certainty, the process of choosing a course of action is to a great extent based on the optimum solution as computed by a deterministic model. When decision making is under risk, the course of action selected is usually a function of the expected values of the possible outcomes. However, management may select a course of action with a lower or higher expected value, due to optimistic or pessimistic considerations. When decision making is under uncertainty, a DSS can be useful by providing rapid feedback on outcomes.

Expert systems is an emerging technology that shows promise as a step beyond DSS to assist in the productivity of organizations and management. Progressive systems analysts should pay attention to and become competent in expert systems technology.

Exercises

1. Discuss the characteristics of decision making and information used for decision making at operating-, middle-, and top-management levels.

2. Define decision making under certainty, risk, and uncertainty. Give examples of each.

3. Discuss behavioral characteristics of managers and their implications for the systems analyst.

4. Discuss how a DSS can assist management during the intelligence phase of decision making.

5. Differentiate between deterministic and stochastic models.

6. What is simulation? How can it assist in decision making?

7. Why are DSS limited in their support of decision making under conditions of uncertainty?

8. Discuss the use of rapid feedback to assist in decision making.

9. What are expected values? For what type of decision making (i.e., certainty, risk, or uncertainty) are expected values used?

10. Discuss the difference between a DSS and an expert system. Why are expert systems more difficult to develop than DSS?

Selected References

Alavi, M., and Henderson, J. C. "Evolutionary Strategy for Implementing a Decision Support System." *Management Science*, November 1981, pp. 1309–23.

Alter, S. L. *Decision Support Systems: Current Practices and Continuing Challenges.* Reading, Mass.: Addison-Wesley, 1980.

Aiso, M. "Forecasting Techniques." *IBM Systems Journal* 12 (1973): 187–209.

Barrett, Michael. *Information Processing Types and Simulated Production Decision Making.* Working paper 73–2. Minneapolis: The MIS Research Center, University of Minnesota.

Barton, Richard F. *A Primer on Simulation and Gaming.* Englewood Cliffs, N.J.: Prentice-Hall, 1970.

Bennett, John L., ed. *Building Decision Support Systems.* Reading, Mass.: Addison-Wesley, 1982.

Bomball, Mark R.; Hallam, James A.; Hallam, Stephen F., and Scriven, Donald D. "Understanding Simulation." *Data Management*, February 1975, pp. 30–32.

Boulden, J. B. and Buffa, E. S. "Corporate Models: On-Line Real-Time Systems." *Harvard Business Review*, July–August 1970, pp. 65–83.

Buchanan, B. G., Barstow, D., Bechtel, R., Bennett, J., Clancey, W., Kulikowski, C., Mitchell, T., and Waterman, D. A. "Constructing an Expert System." In *Building Expert Systems*, edited by F. Hayes Roth, D. A. Waterman and D. B. Lenat. Reading, Mass.: Addison, Wesley, 1983, pp. 127–67.

Carlson, Eric D. "Proceedings of a Conference on Decision Support Systems." *Data Base* Vol. 8, No. 3 (Winter 1977).

Clark, Charles T., and Schkade, Lawrence L. *Statistical Analysis for Administrative Decisions.* 2d ed. Cincinnati: South-Western, 1974.

Davis, Gordon B. *Management Information Systems: Conceptual Foundations, Structure, and Development.* New York: McGraw-Hill, 1974.

Davis, R. "Amplifying Expertise with Expert Systems." In *The AI Business*, edited by P. H. Winston and K. A. Prendergast. Cambridge, Mass: MIT Press, 1984.

Dickson, Gary W. ed. *DSS-82 Transactions.* Austin, Texas: Execucom Systems Corporation, 1982.

Edelman, Franz. "Managers, Computer Systems, and Productivity." *MIS Quarterly*. 5 (1981): 1–19.

Emshoff, J. R., and Sisson, R. L. *Design and Use of Computer Simulation Models*. New York: Macmillan, 1970.

Fick, G. and Sprague, R. H., Jr. eds. *Decision Support Systems: Issues and Challenges*. London: Pergamon Press, 1980.

Gordon, Geoffrey. *System Simulation*. Englewood Cliffs, N.J.: Prentice-Hall, 1969.

Gorry, Anthony G. and Morton, Michael Scott S. "A Framework for Management Information Systems." *Sloan Management Review* 13 (1971): 55–70.

Hallam, Stephen F.; Scriven, Donald D.; Bomball, Mark R.; and Hallen, James A. "Basic Steps in Developing Simulation Models." *Data Management*, April 1975, pp. 26–29.

Hammond, John S., III. "Do's & Don'ts of Computer Models for Planning." *Harvard Business Review*, March–April 1974, pp. 110–123.

Harmon, Paul and King, David. *Artificial Intelligence in Business: Expert Systems*. New York: Wiley, 1985.

Holsapple, Clyde W., and Whinston, Andrew B. *Business Expert Systems*. Homewood, Ill: Irwin, 1987.

Infosystems Staff. "Management Gets the Picture." *Infosystems*, April 1977, pp. 37–42.

Jones, C. H. "At Last: Real Computer Power for Decision Makers." *Harvard Business Review*, September–October 1970, pp. 75–89.

Keen, Peter G. W. and Morton, Michael Scott S. *Decision Support Systems: An Organizational Perspective*. Reading, Mass.: Addison-Wesley, 1978.

Keen, P. G. W. and Wagner, G. R. "DSS: An Executive Mind-Support System." *Datamation* 24 (1979): 117–22.

Kim, Chaiho. *Quantitative Analysis for Managerial Decisions*. Reading, Mass.: Addison-Wesley, 1976.

Kingston, P. L. "Concepts of Financial Models." *IBM Systems Journal* 12 (1973): 113–25.

Pople, H. E. "Knowledge-Based Expert Systems: The Buy or Build Decision." In *Artificial Intelligence Applications for Business*, edited by Walter Reitman. Norwood, N.J.: Ablex Publishing, 1984.

Rockart, J. F. and Treacy, M. E. "The Chief Executive Goes Online." *Harvard Business Review* 60 (1982): 82–87.

Smith, R. "On the Development of Commercial Expert Systems." *The AI Business* 3 (1984), pp. 61–73.

Sol, H. G. *Processes and Tools for Decision Support*. Proceedings of the 1982 IFIP/IIASA Working Conference on DSS. Amsterdam: North-Holland Publishing Co., 1982.

Sprague, R. H., Jr. "A Framework for the Development of Decision Support Systems." *MIS Quarterly* 4 (1980): 1–26.

Sprague, R. H., Jr. and Carlson, E. D. *Building Effective Decision Support Systems*. Englewood Cliffs: N.J.: Prentice-Hall, 1982.

Sviokla, John. "Business Implications of Knowledge Based Systems," Part 1. *DATA BASE*, Summer 1986, pp. 5–19.

Sviokla, John. "Business Implications of Knowledge Based Systems," Part 2. *DATA BASE*, Fall 1986, pp. 5–14.

Waterman, D. A. *A Guide to Expert Systems*. Reading, Mass: Addison-Wesley, 1985.

Wetherbe, James C. *Systems Analysis for Computer-Based Information Systems*. St. Paul, Minn.: West Publishing, 1979.

———. *Executive's Guide to Computer-Based Information Systems*. Englewood Cliffs, N.J.: Prentice-Hall, 1983.

Wetherbe, James C. and Dickson, Gary W. *Management of Information Systems*. New York: McGraw-Hill, 1984.

Young, D. and Keen, P. G. W. eds. *DSS-82 Transactions*. Austin, Texas: Execucom Systems Corporation, 1981.

Chapter

15 End-User Computing

Introduction

MIS departments are struggling to keep up with the demand for computing. The traditional methods for producing systems have been overwhelmed, and long lead times have resulted, with promising computer applications being deferred or dropped because of the delay. User-managers' computing needs far exceed what information systems departments can deliver—even with dramatic improvements in development efficiency.

This unmet demand has combined with a decrease in hardware costs and improvements in software design to increase the amount of computing being done by end-users. Consequently, in a typical large U.S. manufacturing firm, end-user computing has grown from virtually nothing in 1970 to 40 percent of computer-processing capacity in 1980 and is projected to grow to 80 percent in the 1990s.

The rise of end-user computing has not gone unnoticed. In a survey of leading MIS executives, consultants, and researchers,[1] the "facilitation and management" of end-user computing was ranked as the second most important MIS management issue. Only MIS planning was ranked higher. The need to effectively manage end-user computing is being widely felt.

This chapter reviews the opportunities and risks presented by end-user computing, then describes an approach to managing end-user computing that uses the concept of service support levels to offer organizations the benefits of end-user computing while minimizing the risks. With this approach, end-user computing can be coordinated and done in conjunction with traditional information system development efforts.

End-User Computing Opportunities and Risks

Despite much discussion of end-user computing in the academic and popular press, no definition is generally agreed upon. In this book, *end-user computing* is defined as "the use and/or development of information systems by the principal users of the systems' outputs or by their staffs." The inability of the MIS department to meet computing needs gave rise to end-user computing, specifically because

1. Many applications, especially those that are retrieval- and analysis-oriented, can readily be done by the user.

2. Lead times on development requests are shorter.

3. End-users have more control over system development and use.

4. Services are not available from the MIS department.

5. MIS department procedures are not appropriate for small applications.

6. The MIS department is not perceived as being concerned about users' needs.

7. End-users want to learn about computing.

---

1. G. W. Dickson, R. L. Leitheiser, and J. C. Wetherbe, "Key Information Systems Issues for the 1980's," *MIS Quarterly* 8 (September 1984), pp. 135–54.

8. End-users gain more flexibility.

9. The information systems developed this way better meet users' needs.

10. Development costs are lower.

The rapid and continued growth in end-user computing is evidence that users believe the benefits from doing their own computing are substantial.

Besides the advantages for users, there are also benefits for the MIS department. First, the shortage of systems development personnel can be relieved. This allows MIS executives to use their expensive human resources on larger, more technical development projects. Second, if users know their requirements, they can implement them directly into a system and thus avoid the time-consuming and error-prone process of communicating requirements to an outside developer. Finally, system implementation becomes the responsibility of the users. Their ownership of the system is thereby ensured, and a major stumbling block to successful system implementation is removed. Taken together, these benefits to both users and the MIS department are substantial.

Downside

Unfortunately, end-user computing has a downside. The organization's hardware, software, and data are valuable resources that can be lost or diminished if not properly developed and protected. End-users, acting independently, cannot be expected to always use these resources in ways that are optimum for the whole organization. And since end-user computing bypasses the monitoring and control mechanisms built into the MIS department, there is no formal check on user behavior. To illustrate these risks, consider three scenarios where end-user computing leads to problems for three unwary firms.

1. A faulty corporate model leads to a disastrous acquisition that later forces the firm into bankruptcy.

2. Errors in retrieving data for an analysis cause a substantial underestimation of the cost of introducing a new employee benefit. The actual costs are so high that the firm is forced to cancel the program, with disastrous results for employee morale.

3. An employee who developed an important end-user application leaves a company to establish his own business. Since the system was not documented, no one in the organization knows how to use or maintain it.

Problems with inadequate documentation, poor data, faulty backup procedures, and a lack of data security are common among end-user applications.

MIS executives might argue that they should not be held accountable for end-user mistakes and failures. And for certain kinds of computing and certain knowledgeable end-users, this argument may be defensible. In the majority of cases, however, the expertise and experience of the MIS department make it the obvious organizational unit for protecting the computing and information resources of the firm. Top management reasonably looks to the MIS department for leadership in computing and information systems, no matter who develops or uses the system.

Risks

The range of risks that result from end-user development is suggested by G. B. Davis:[2]

1. The elimination of the separate analyst functions of technical expert, independent reviewer of information requirements, and enforcer of standards and policies.
2. The inability of end-users to completely and accurately specify system requirements without the tools, techniques, and experience of the analyst.
3. A lack of end-user knowledge about, and commitment to, quality assurance in systems development.
4. The development of continuously changing (unstable) systems where stable systems are needed.
5. The encouragement of private information systems at the expense of shared corporate computing resources.
6. The expensive accumulation of unneeded information.

Most firms will find it too dangerous to ignore these risks. Clearly then, success in managing end-user computing requires the facilitation of beneficial computing while minimizing risks to the firm.

Rockart and Flannery[3] present recommendations for both of these management goals. To facilitate end-user computing, they suggest (1) establishing a distributed organizational support structure, (2) providing a range of software tools, (3) developing a broad educational program, and (4) establishing efficient procedures for locating and transferring data. Risk-coordination recommendations include (1) identifying critical applications, (2) controlling through user (not MIS) management, (3) using MIS experts to assist end-users in making computing decisions, and (4) developing standards for hardware and software and creating incentives for following them. These recommendations provide valuable guidelines for MIS executives, but they do not describe a systematic approach to end-user computing. Yet such an approach is important for achieving effective facilitation and coordination.

Responses to End-User Computing

Before proposing an organized approach, it is useful to lay out four general responses that an MIS department can take toward end-user computing. These general responses represent the range of specific actions available to executives.

1. *Sink or swim:* Don't do anything—let the end-user beware.
2. *Stick:* Establish policies and procedures to control end-user computing so that corporate risks are minimized.

---

2. G. B. Davis, "Caution: User Developed Systems Can Be Dangerous to Your Organization," MISRC-WP-82-04, MIS Research Center, University of Minnesota, Minneapolis, Minnesota, 1984.

3. J. F. Rockart and L. S. Flannery, "The Management of End User Computing," *Communications of the ACM*, 26 (October 1983), pp. 776–84.

---

3. *Carrot:* Create incentives to encourage certain end-user practices that reduce organizational risks.

4. *Support:* Develop services to aid end-users in their computing activities.

Each of these responses presents the MIS executive with different opportunities for facilitation and coordination. The *sink-or-swim* response is the default action. It does nothing to increase the level of end-user computing or improve its quality. All of the risks mentioned above apply, while the benefits to end-users and the MIS department are solely a function of end-user efforts.

The restrictive *stick* response forces coordination at the cost of some computing opportunities. Many of the risks associated with end-user computing could be reduced by effective policies, standards, and procedures. These same policies, standards, and procedures are likely to reduce the incentives for end-users to do their own computing (lead times are increased, user control and flexibility are reduced, and development costs are higher).

The *carrot* response motivates end-users to reduce the risks to the firm associated with their computing. By allowing end-users to retain their computing freedom, it encourages end-user computing to be done in proper, consistent ways, but it does not provide a complete program of support.

The *support* response is designed to increase end-user computing through the provision of services. These services are carefully designed to increase beneficial computing and to foster coordination. As long as end-users are able to choose to use or not use services, they retain control over their computing activities and thus view services positively. Their having a choice also provides a check on the MIS department, because services that focus too much on coordination and not enough on facilitation will simply not be used.

Each type of response is appropriate in some situations, but a combination of responses is necessary for management on a broad scale. The next section describes an organized approach to defining these combinations.

Service Support Levels

An effective approach to managing end-user computing achieves both facilitation and coordination. Service support levels do this by (1) defining computing responsibilities, (2) providing a framework for designing support services, and (3) allowing end-users to retain as much control as possible over their own computing.

Service support levels are formal divisions of computing responsibility between end-users and the MIS department. These divisions are based on a small set of critical computing decisions that are made by end-user management. The way managers make these decisions commits them to accept certain responsibilities and allows them to turn over others to the MIS department. Since end-user management make the decisions, they are free to choose the amount and kind of support they feel they need. This freedom to choose—as mentioned before—provides a check on the MIS department. If the MIS department is to meet its objectives of coordinating and facilitating end-user computing, it must develop and deliver support services to meet needs.

An approach based on service support levels offers several other advantages. First, it reduces "finger pointing" by clearly specifying department responsibilities. When a microcomputer malfunctions, everyone knows who is responsible for fixing it. Second, it provides a structure for the design and delivery of end-user services by the MIS department. This structure allows a range of support options to be provided from a common pool of services. For example, the same consultant can have clearly specified duties at each support level. Third, incentives are created for end-users to improve their computing practices, thereby reducing the computing risks to the firm. By clearly stating what has to be done to qualify for better MIS service, end-user management is in a better position to make trade-offs in its computing decisions. Finally, it provides a means for the MIS department to coordinate end-user computing. The same staff people who provide support services also monitor and report on end-user activities. Some end-user computing will remain outside these coordination efforts, but those activities are recognized to be the responsibility of end-user managers.

Establishing service support levels requires four steps:

1. Defining the levels
2. Dividing computing responsibility at each level
3. Designing services to meet MIS support-level commitments at each level
4. Implementing service support levels

The process may be applied to each of the principal computing resources: hardware, software, and data.

Defining Service Support Levels

An important part of the process of instituting support levels is determining the critical decisions that will define the levels. These decisions must be few in number and generally perceived as affecting computing accountability. A small number of decisions ensures that the requirements for obtaining levels are understandable and obvious to end-users, top management, and MIS staff.

One obvious means of defining levels, and thereby dividing responsibility, is according to the relative amounts of control that departments have over the processes that affect a computing resource. For example, if the user-manager purchases a piece of hardware, then the responsibility for ensuring that the equipment meets department and corporate needs should rest with that manager. Similarly, if the MIS department formally reviews and approves an application program, then the MIS department should be accountable for any subsequent quality problems. These indicators of control offer a fair means of assigning responsibility. Organizations must consider their own unique computing environments when defining the qualifying decisions.

For example, if a company strives to integrate its application systems, then a decision emphasizing compatibility could be defined. Concerns regarding security, sharing of applications, and documentation could lead to other qualifying decisions. Many sets of qualifying decisions should be analyzed, but no set can be final until it is formally accepted by end-user, MIS, and top managements.

Control and Accountability

Control and accountability are the basis for the qualifying decisions in establishing service levels. As Table 15–1 shows, separate decisions can be used to define

Table 15.1   Qualifying Decisions for Service Support Levels

Hardware Decisions

1. What department will purchase or lease the equipment?
2. Will the hardware be formally registered with the MIS department?
3. Will the hardware be compatible with the MIS department's computing environment?

Software Decisions

1. What department will purchase or develop the software?
2. Will the software be formally reviewed by the MIS department?
3. Will the software be formally registered by the MIS department?
4. Will the software be compatible with the MIS department's computing environment?

Data Decisions

1. What department will provide the data?
2. What will be the support level of the software that processes the data? (see above)
3. Will the data be formally registered with the MIS department?

service criteria hardware, software, and data resources. Organizations may elect to establish support levels for only one or two types of resources (e.g., software only) or may combine two types of resources (e.g., software and data) and use one set of decisions. However, the qualifying decision sets given in Figure 15.1 have general applicability and offer a good starting point.

Levels are defined by assigning outcomes to qualifying decisions. Figure 15.1 illustrates how specific support levels might be decided. Five levels are defined for each of the types of computing resources and range from almost complete MIS department responsibility (level 1) to almost no MIS department responsibility (level V). Notice that qualifying decisions can be related. In Figure 15.1, item 2 under "Data Decisions" is based upon the software level that processes the data.

Figure 15.1   Support Levels for Hardware, Software, and Data

| Qualifying Decisions | HARDWARE SUPPORT LEVEL | | | | |
	I	II	III	IV	V
1. Purchaser?	MIS	USER	USER	USER	USER
2. Register with MIS department?	YES	YES	NO	YES	NO
3. Compatible with MIS systems?	YES	YES	YES	NO	NO

| Qualifying Decisions | SOFTWARE SUPPORT LEVELS | | | | |
	I	II	III	IV	V
1. Purchaser or developer?	MIS	USER	USER	USER	USER
2. Formal review by MIS?	YES	YES	NO	YES	NO
3. Register with MIS?	YES	YES	YES	YES	NO
4. Compatible with MIS systems?	YES	YES	YES	NO	YES/NO

| Qualifying Decisions | DATA SUPPORT LEVELS | | | | |
	I	II	III	IV	V
1. Source of data?	MIS	USER	USER	USER	USER
2. Software support level?	—	I-II	III-IV	V	V
3. Register with MIS?	YES	YES	YES	YES	NO

"—" indicates that the cell does not apply at this level.

In the preceding example, the qualifying decisions were resolved before the support levels were defined. In practice, the process is iterative, with tentative decisions being suggested and then tested for their ability to define useful levels. They may subsequently be retained, modified, or discarded.

To be effective, the number of levels should be small (less than ten) while the qualifying decisions should be powerful, easily recognizable, and readily understandable. Review and modification are necessary to achieve a good balance among these factors. Once the levels have been defined and agreed upon, the specific responsibilities can be determined.

Dividing Computing Responsibilities

After defining support levels, the next step is to divide responsibilities into specific groups. This is a shared task for end-users and MIS departments. One approach is to first brainstorm a list of required computing activities, making the list as long as possible before cutting eventually it to the most important activities. These activities should then be defined as clearly as possible, since the formal definition will prevent later questions about what was actually meant by the assignment of an activity. Table 15.2 lists the activities and definitions commonly identified.

Next, a two-dimensional matrix is created to link the defined levels with the required computing activities. Figures 15.2, 15.3, and 15.4 contain these matrices for hardware, software, and data. For each activity and level an assignment of responsibility is made. In Figure 15.2 an "X" indicates that the MIS department has the major responsibility for that activity. The absence of an "X" means that

Table 15.2	**Support Activities for End-User Computing**
Support Activity	*Definition*
1. Advising	Assist in analyzing needs, planning action, and selecting products or services.
2. Backup	Provide duplicate hardware, software, or data.
3. Compatibility	Insure compatibility of hardware, software, or data.
4. Development	Create or enhance a process, program, product or data set.
5. Documentation	Provide descriptions of how to use a computing resource.
6. Hotline/debugging	Isolate and correct system problems, perhaps through the use of a special phone number.
7. List resource	List the resource (hardware, software, or data) as available to other users.
8. Maintenance	Keep the system in working order; including tests, measurements, replacements, repairs, and adjustments.
9. Data transfer	Move data from one system to another.
10. Newsletter	Produce and distribute a report on new products or services for users.
11. Purchase assistance	Provide support for users who are buying their own systems.
12. Recovery	Restore of hardware, software, or data after a failure.
13. Research products	Investigate new products and services for possible use by end users.
14. Training	Provide users with the knowledge needed to develop and/or use systems.

Figure 15.2   **Support Activities for Hardware Levels**

Qualifying Decisions	HARDWARE SUPPORT LEVELS				
	I	II	III	IV	V
1. Purchaser?	MIS	USER	USER	USER	USER
2. Register with MIS department?	YES	YES	YES	NO	NO
3. Compatible with MIS systems?	YES	YES	NO	YES	NO
Support Activities	**Turnkey**	**Full Service**	**Partnership**	**Assistance**	**Advising**
1. Advising	X	X	X	X	X
2. Backup	X				
3. Compatibility	X	X		X	
4. Documentation	X				
5. Hotline/debugging	X	X		X	
6. List resource	X	X	X		
7. Maintenance	X				
8. Data transfer	X	X		X	
9. Newsletter	X	X		X	
10. Purchase assistance	X	X			
11. Recovery	X				
12. Research new products	X	X	X	X	
13. Training	X	X		X	

''X'' indicates that the MIS department has major responsibility for the support activity.

Figure 15.3   **Support Activities for Software Levels**

Qualifying Decisions	SOFTWARE SUPPORT LEVELS				
	I	II	III	IV	V
1. Purchaser or developer?	MIS	USER	USER	USER	USER
2. Formal review by MIS?	YES	YES	YES	NO	NO
3. Register with MIS?	YES	YES	YES	YES	NO
4. Compatible with MIS systems?	YES	YES	NO	YES	YES/NO
Support Activities	**Turnkey**	**Full Service**	**Partnership**	**Assistance**	**Advising**
1. Advising	X	X	X	X	X
2. Backup	X	X			
3. Compatibility	X	X		X	
4. Development	X				
5. Documentation	X	X	X		
6. Hotline/debugging	X	X	X		
7. List resource	X	X	X	X	
8. Maintenance	X	X?	X?		
9. Data transfer	X	X		X	
10. Newsletter	X	X	X	X	
11. Purchase assistance	X	X	X	X	
12. Recovery	X	X			
13. Research new products	X	X	X	X	
14. Training	X	X			

''X'' indicates that the MIS department has major responsibility for the support activity.
''?'' indicates that responsibilities are negotiable.

Figure 15.4   Support Activities for Data Levels

Qualifying Decisions	I	DATA SUPPORT LEVELS II	III	IV	V
1. Source of data?	MIS	USER	USER	USER	USER
2. Software support level?	—	I-II	III-IV	V	V
3. Register with MIS?	YES	YES	YES	YES	NO
Support Activities	Turnkey	Full Service	Partner-ship	Assis-tance	Advising
1. Advising	X	X	X	X	X
2. Backup	X	X			
3. Compatibility	X	X			
4. Development	X				
5. Documentation	X	X			
6. Hotline/debugging	X	X			
7. List resource	X	X	X	X	
8. Maintenance	X				
9. Data transfer	X	X	X		
10. Newsletter	X	X	X		
11. Purchase assistance	X				
12. Recovery	X	X			
13. Research new products	X	X	X		
14. Training	X	X			

"X" indicates that the MIS department has major responsibility for the support activity.

the major responsibility is assigned to the end-users. For example, the MIS department has major advising responsibility at all hardware levels, but major maintenance responsibility only at the "Turnkey" level (i.e., level I in Figure 15.2).

Selecting a Support Level

Crucial to the success of the support-level approach is the ability of end-user managers to make appropriate decisions about levels of support. In making these decisions, managers should consider the benefits of the support offered and the trade-offs of additional time and expense that result when MIS services are used. Beyond this cost/benefit assessment, end-user managers should ask themselves three questions:

1. Do I have the resources required to perform the set of activities not covered by MIS services? The Xs in Figures 15.2 through 15.4 indicate activities that are the major responsibility of the MIS department. It is the responsibility of the user department to perform the other activities. Different departments have different levels of computing expertise. Rockart and Flannery, for example, found knowledgeable end-users in many departments.[4] To answer this question, managers of user areas must (1) know what their internal computing resources are and (2) decide whether they want their own personnel spending substantial amounts of time doing support activities. Choosing a particular support level involves choosing a level of internal support activity.

---

4. J. F. Rockart and L. S. Flannery, "The Management of End User Computing, *Communications of the ACM* 26 10 (October 1983), pp. 776–84.

2. How much risk is involved with this application and how much risk am I willing to assume? Some computing applications are riskier than others. If the risks are high, it might be in the end-user manager's best interests to share some of these risks. Assessing the level of risk associated with a particular application is not always easy. End-user managers will want to consider

- The size and complexity of the application
- The number and sophistication of the users
- The familiarity of end-user developers with this type of application
- The stability of the application environment

The MIS department can help end-user managers by providing training and written procedures that can assist them in making the risk assessment.

3. Are other organizational units involved in this application? Should they be? Rockart and Flannery[5] found a substantial number of end-user applications that extended beyond single departments. When different managers become involved with an application, coordination problems arise. Since the MIS department is concerned with meeting the overall needs of the firm, it is an obvious choice for the task of coordinating multi-departmental computing. Selecting support levels with high MIS department involvement is recommended for interdepartmental systems.

Information Centers

The most common approach to delivering end-user services is an information center, first mentioned in Chapter 4. An information center is an organizational unit, usually part of the MIS department, whose principal function is to facilitate and coordinate end-user computing by offering support services. Such a department supplements the traditional development staff of analysts and programmers. It allows users to make better use of existing, traditionally developed systems and to develop certain applications for themselves. End-users should not develop large-scale systems that transcend departmental boundaries (e.g., airline reservations), but they should become self-sufficient in making many ad hoc inquiries into the database of such systems.

Large numbers of information centers already exist. In two different surveys, 80 percent of respondents claimed that their firms had or would have information centers. Figures 15.5a through 15.5c illustrate just how important the information center has become to end-users. The figures also indicate how important it is for the MIS department to play a leadership role in knowledge of technology if it wants to influence and lead users. Figure 15.6 documents the services most valued by end-users.

Information Architecture and End-User Computing

Like the personal automobile, the personal computer gives its users great flexibility, power, and freedom. But just as the user of the automobile needs access

---

5. Ibid.

Figure 15.5a **Overal Importance of the Information Center.**

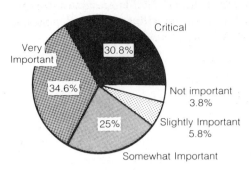

Critical
30.8%

Very Important
34.6%

25%

Somewhat Important

Not important 3.8%

Slightly Important 5.8%

Figure 15.5b **How Users Decide Where to Go for Help.**

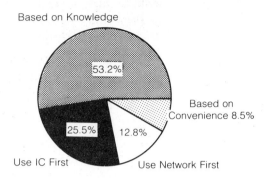

Based on Knowledge
53.2%

Based on Convenience 8.5%

25.5%

Use IC First

12.8%

Use Network First

Figure 15.5c **How Users Expect to Satisfy Need for Support in Future.**

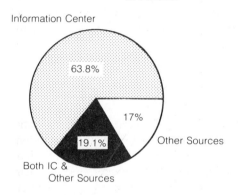

Information Center

63.8%

17%

19.1%

Other Sources

Both IC & Other Sources

(Source: Brancheau, J. C.; Vogel, D. R.; and Wetherbe, J. C. "An Investigation of the Information Center from the User's Perspective," *DATA BASE*, Volume 16, Number 1, Fall 1985, pp. 14–17.)

Figure 15.6 **End-Users' Perception of Information Centers**

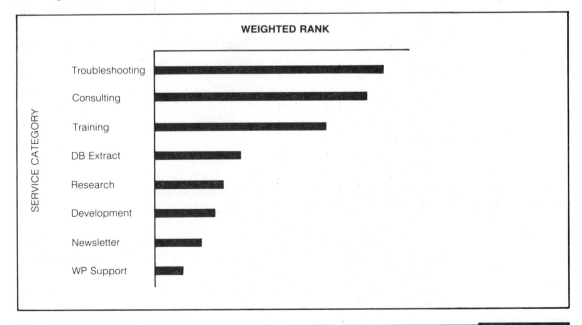

WEIGHTED RANK

SERVICE CATEGORY

Troubleshooting
Consulting
Training
DB Extract
Research
Development
Newsletter
WP Support

to an infrastructure of highways, the user of the personal computer needs access to infrastructure of databases and communication networks. Accordingly, the information architecture concept developed in Chapter 13 is crucial to truly supporting end-user computing. At the individual application level, the data file relationship analysis discussed in Chapter 7 is also crucial. The potential of end-user computing is greatly compromised if the MIS department fails in its role to develop an integrated database environment in which the users can "help themselves."

Summary

The growth of end-user computing offers MIS executives the opportunity to improve the performance of their firms. To make use of this opportunity, however, requires a new, more cooperative approach to working with end-users than has been generally demonstrated in the past. No longer can the MIS department simply dictate which application systems are to be built and how they are to be developed and used. It must now compete for computing business with services that meet the needs of users. Given the right support services, MIS executives can take the lead in increasing the amount of beneficial computing done in their firms.

The growth of end-user computing also creates new risks for MIS executives. Until the new cooperative arrangement is structured, the MIS department will be open to every kind of claim by end-users who run into trouble while computing. Without a clear specification of responsibilities, top management will, rightly or wrongly, hold the MIS department accountable for these computing claims. MIS executives must reduce the risks of end-user computing to themselves and to their firms.

Obtaining the benefits of end-user computing while minimizing its inherent risks requires a clear delineation of responsibilities and an effective service program. An organized approach for supporting end-user computing based on the establishment of service support levels is a good solution. With end-user computing becoming the dominant kind of computing in this decade, MIS executives need such an approach for dealing with the demands of their new computing environment.

Exercises

Exercises are based upon the following case study:

Case Study: High-Rise Engineering is an engineering construction firm specializing in large commercial buildings. It has an international reputation for designing modernistic office buildings that provide distinctive corporate headquarters. The emergence of information systems at High-Rise Engineering was similar to that of most organizations. High-Rise initially implemented its major accounting applications followed with personnel systems. Then it implemented a variety of mainstream applications, including inventory management and contracts, and eventually moved into engineering graphics.

Information systems have historically been a centralized function at High-Rise Engineering, with most processing done on a large-scale IBM computer system. In the mid-1970s, minicomputers began to emerge within the engineering groups of the company. Since these computers were used primarily for specialized engineering applications, they created very little concern for information systems management and were for the most part ignored by top management. In the early 1980s the advent of the personal computer resulted in a proliferation of personal computers in all functional areas of the organization.

Information systems management became concerned about the widespread purchasing of personal computers and, in some respects, viewed them as a threat to the centralized information systems function. In an attempt to address the situation, information systems management approached the information systems steering committee and suggested that all purchases of either minicomputers or personal computers be approved by the information systems function. This seemed reasonable to the steering committee, and the policy was implemented.

However, it quickly became clear that such authority was of little value. Whenever end-users wanted to buy a particular device, if they could not get it approved by information systems, they resorted to political tactics, arguing that information systems really shouldn't be deciding whether an end-user could make a $1,500 or $4,000 purchase. If that tactic didn't work, end-users simply went underground and made their purchases discreetly enough to be undetected by information systems. As a result, end-user computing continued to proliferate.

The managers at High-Rise, including the information systems managers, were busy enough with other issues to feel little concerned over the proliferation of computers throughout the organization. By the mid-1980s there was one computer for every one and one-half employees at High-Rise Engineering. Then the problems began.

A financial analyst, using a financial modeling language, made a recommendation on a building-procurement decision. The analyst was reasonably experienced at using the language; however, in this case he made a slight error in the way he constructed some computations. The result was that the analysis indicated a much more favorable financial profile for the purchase than was actually true. As a result, the company lost $250,000 a year for three years until it could sell the property. Needless to say, management was very upset, and some anti-computer sentiment emerged.

Several end-users bought a variety of personal computers to support such things as office automation, financial modeling, and engineering analysis. After using these computers for several months, they began to realize that they would need to connect to the central computer to get access to much of the data they needed to support their decision making. However, when they attempted to connect to the computer, they found that their hardware and software were incompatible with the processing environment used by the information systems department. The only way they could achieve interface was to replace the equipment they had purchased with new equipment at considerable expense. This upset a lot of budgets and a lot of managers.

The information systems department began to work on establishing a tele-communications network to support a variety of terminals and personal computing devices for direct access to corporate data. They constructed a financial forecast to determine how much it would cost to upgrade to the new processing environment. The original estimates indicated that the total upgrade, including a mainframe upgrade, would run approximately $4 million. Management approved the upgrade, and implementation began. However, as they began final plans for implementation, users began to "come out of the woodwork" raising concerns about how the PCs and terminals they had been using for the past several years would interface to the new processing environment. For the most part, information systems had been unaware of many of the users' devices and their access to the central computer and had made no provisions for a migration strategy to ensure that they would be compatible. The diverse variety of user interfaces would require the purchase of additional communications software and hardware. The additional total expenditure was a half a million dollars.

The accumulative effect of these three incidents caused top management to be very upset with the information systems department. When questioned as to how they had let this fiasco happen, Mr. Harry Burkstaller, vice-president of information systems, responded in a memo to Mr. Joe Childs, executive vice-president and chairman of the steering committee, that information systems didn't "just let this happen." He resurrected the memos and statements from several years earlier that indicated that no end-user was to purchase equipment without clearing it with information systems. Mr. Burkstaller explained that this policy had never been adhered to, that end-users overrode information systems' authority, or simply took their purchases underground and did whatever they pleased, and that now this violation of the policy was catching up with the organization.

This response was not satisfactory to either Mr. Childs or any member of the steering committee. The rather angry response was, "Don't use old memos to cover up your lack of leadership. We look to you for providing leadership in information systems, and this problem has been going on for some time now. The only constructive thing that information systems has done is sit on a five-year-old policy that obviously was not satisfactory to take care of the issue."

Mr. Burkstaller and his staff felt they had been unfairly treated by top management. In discussing the situation with their information systems colleagues in other organizations, they found that similar problems had emerged, but none as serious as the one Mr. Burkstaller was facing.

Realizing that management would need a scapegoat for this problem, and that information systems managers are very good scapegoats, Mr. Burkstaller concluded that arguing would be futile. The next week, he received a memo from Mr. Childs, directing him to come up with an action plan to correct the end-user computing dilemma. Specifically, he was instructed to resolve the problem of incompatible equipment and the quality problems with end-user software such as those experienced with the financial model used in the procurement decision. Further, he was told to have the action plan documented and available for the steering committee within a week.

1. Develop an action plan that includes policies and courses of action to resolve the problem that developed at High-Rise Engineering.

2. In hindsight, what positive initiatives could Mr. Burkstaller and the information systems staff have taken to have avoided the unpleasant situation that developed?

Selected References

1. Alavi, M., and Weiss, I. R. "Managing the Risks Associated with End-User Computing." *Journal of Management Information Systems* 2 (Winter 1985–86), pp. 5–20.

2. Alloway, R. M., and Quillard, J. A. "User Managers' Systems Needs." *MIS Quarterly* 7 (June 1983), pp. 27–41.

3. Benjamin, R. I. "Information Technology in the 1990's: A Long Range Planning Scenario." *MIS Quarterly* 6 (June 1982), pp. 11–31.

4. Benson, D. H. "A Study of End User Computing: Findings and Issues." *MIS Quarterly* 7 (December 1983), pp. 35–45.

5. Brancheau, J. C., Vogel, D. R., and Wetherbe, J. C. "An Investigation of the Information Center From the User's Perspective." *DATA BASE* 16 (Fall 1985), pp. 4–17.

6. "Info Centers Gaining." *Computerworld*, February 27, 1984, p. 6.

7. Davis, G. B. "Caution: User Developed Systems Can Be Dangerous to Your Organization." MISRC-WP-82-04, MIS Research Center, University of Minnesota, Minneapolis, Minnesota, 1984.

8. Dickson, G. W., Leitheiser, R. L., Wetherbe, J. C., and Nechis, M. "Key Information Systems Issues for the 1980's." *MIS Quarterly*, 8 (September 1984), pp. 135–54.

9. Hammond, L. W. "Management Considerations for an Information Center." *IBM Systems Journal* 21 (1982), pp. 131–61.

10. Leitheiser, R. L., and Wetherbe, J. C. "Approaches to End-User Computing: Service May Spell Success." *Journal of Information Systems Management* 3, (Winter 1986), pp. 9–14.

11. Rockart, J. F., and Flannery, L. S. "The Management of End User Computing." *Communications of the ACM* 26 (October 1983), pp. 776–84.

12. Summer, M. "Organizations and Management of the Information Center: Case Studies." *ACM Proceedings of the Computer Personnel and Business Data Processing Research Conference*, Minneapolis, Minnesota, May 2–3 1985, pp. 38–49.

13. Wetherbe, J. C., and Leitheiser, R. L. "Information Centers: A Survey of Services, Decisions, Problems, and Successes." *Journal of Information Systems Management* 2 (Summer 1985), pp. 3–10.

16 Future Considerations of Systems Analysis

Introduction

In this last chapter of the book, the future of systems analysis is discussed. Future organizations and computer technology are discussed in the first two sections of the chapter. The third section discusses the implications of future organizations and computer technology on systems analysis.

Future Organizations

It is difficult to attempt to forecast the future development of organizations. A multiplicity of unpredictable and unforeseen forces will shape future organizations. Because of the unpredictability of the future, it is presumptuous to attempt to describe future organizations without emphasizing the possibility of error in the forecasting. However, it is reasonable and worthwhile to make a few generalized forecasts about future organizations. Such forecasts are useful in preparing for the future. At least some, if not many, of the forecasts are likely to be valid.

Drawing from general systems theory, there are two basic factors to consider in discussing future organizations. One is the environment in which organizations operate; the second is the actual operation of organizations or the transformation process used to produce goods and services. These factors are discussed in the text following.

Environment of Future Organizations

Organizations do not operate in a vacuum. Rather, they are dependent upon, and must interact with, their environment (e.g., customers, competitors, governments, and natural resources). Accordingly, organizations must be cognizant of changes in their environment and make appropriate adjustments in order to survive.

Market Conditions
Traditionally, the primary environmental concern of organizations has been to recognize and respond to changing market conditions. For example, an organization producing buggy whips in the early twentieth century had to recognize that the proliferation of automobiles would displace the need for buggy whips. Consequently, the organization had to diversify its production efforts into other goods and/or services or face extinction.

Increasingly, the time in which an organization must recognize and adjust to changing market conditions is narrowing. Consider organizations in the fashion, ski equipment, video game, and personal computer industries. Popular products can quickly lose their market appeal as a result of new products released by competitors. For example, the NCR company, the leading producer of cash registers, temporarily lost its domination of the retail business machine market to the Singer company due to the advent of electronic retail terminals. In the intense competition that ensued, NCR regained its position in the market, and Singer withdrew from the market completely. Had NCR not recognized and responded to the changes in market conditions quickly enough, the organization might not have regained its market position; it might even have gone out of business. As another example, the American auto industries have been seriously threatened by international competition and increasing fuel costs. In the past decade, Chrysler Corporation was almost destroyed by these environmental factors.

It is reasonable to assume that future organizations will have to cope with increasingly unpredictable and less forgiving market conditions. Therefore, organizations not only are going to have to more quickly recognize changes in the marketplace, but also are going to have to be more flexible to respond to these changes.

Society and Natural Resources

There are other environmental concerns that present challenges to organizations in the future. These challenges can be broadly categorized as pertaining to society and natural resources. From a social perspective, organizations must respond increasingly to expanding social pressures. The news media in particular has created great social visibility for, and criticism of, organizations that are not socially responsive. If an organization fails to respond, these social pressures may manifest themselves in the form of rejection of the organization's products and/ or services, government legislative action, and even sabotage. Organizations of the future will be expected to assume (or will be legislated into assuming) greater social responsibility for such things as pollution, consumer safety, equality of employment, and retraining of workers displaced by new technologies. To keep social pressures from becoming unreasonable, organizations will have to increasingly influence public opinion and legislation in order to protect their interests.

The shortage of natural resources has become, and will continue to be, a problem. Early economic theory assumed an unlimited supply of natural resources. This assumption is obviously not true. Organizations are going to have to cope with these shortages and participate in, or at least support, efforts to discover and operationalize new forms of energy and materials.

Operation of Future Organizations

When organizations are viewed in terms of their internal operations, certain forecasts can be made.

Size and Complexity

Existing organizations are likely to continue to increase in size and complexity. Also, there will continue to be a proliferation of new entrepreneurs, particularly "high-tech" companies. Organizations are expanding both the market for their existing products and their product offerings. One of the most visible areas in which organizations are expanding existing product activities is in the international market. For example, a large portion of IBM's computer sales are transacted outside of the United States. The multinational organization has become a common phenomenon that will continue in the future.

Organizations are expanding the products and/or services they offer by diversifying their efforts into new and varied fields. Consider the diverse products and/or services offered by such organizations as General Electric, DuPont, General Motors, the federal government, and higher education. Universities, for example, have significantly expanded their academic programs into areas not even recognized a few years ago.

Technology

The technology used by organizations to produce goods and/or services has become, and will continue to be, more complex and less stable. Technology is used in this context to refer not only to the equipment used in organizations, but also to the skills of organizational participants (engineers, computer technicians, accountants, lawyers, etc.).

The field of computing and information systems provides a vivid example of the increasing complexity and instability of technology. Computer hardware has exhibited life cycles of from one to five years, and programming skills have been affected in a parallel fashion. Programming skills have been upgraded from board wiring in the 1950s to batch programming in the early 1960s, to on-line programming in the late 1960s, to database programming in the 1970s, and to distributed, 4th GL, CASE or artificial intelligence technology in the 1980s.

To cope with the dynamics of technology in all dimensions of the organization, it will be necessary to forecast future technologies and their impacts on the organization, just as it is necessary to forecast market conditions. If an organization fails to consider future technology in the purchasing of equipment and the selection and training of personnel, its survival may be seriously threatened.

Another vivid example of technological trends and their impact on organizations is the evolution of occupations during the past two hundred years. Though today, less than 3 percent of the work force is involved in agriculture, in 1800, 95 percent of all workers were involved in agriculture. By 1900, over 25 percent were involved in industrial labor. By 1950, that 25 percent had increased to over 65 percent, but this percentage dropped back in the 1980s to less than 15 percent. Today the dominant occupation is information- or knowledge-related. In other words, a simple history of the dominant careers in the United States would be farmer—laborer—knowledge worker. Accordingly, key productivity gains in the internal operations of organizations will be achieved not only in the production of goods and services, but also in the processing of information. Engineers, architects, lawyers, managers, salespeople all need to be made more productive.

Future Computer Technology

Predicting future computer technology is as risky as predicting future organizations. To illustrate, in 1950 computer and industry experts predicted the total number of computers that would be manufactured would be ten. This forecast was obviously in error. There are several million computers installed today. However, considering the size, cost, and difficulty of using early computers, the forecast was not unrealistic. What the forecasters were unaware of were the great strides forthcoming in the technology used to manufacture computers. Computers today can perform calculations, store data, and be programmed for a minute fraction of the cost and difficulty involved in the 1950s.

Improvements in the cost effectiveness of both hardware and software continue to expand the frontiers of possible computer applications. Some hardware and software developments likely to occur in the future are discussed next.

Hardware

Computer Processors
Computer processors of the future will be constructed of smaller, faster circuits. During the early 1990s computer circuits will perform at least ten times faster and be only one-tenth as expensive as those in use in the late 1980s. Superconductivity is the biggest technological breakthrough occurring in this area.

Lower-cost, higher-performance processing units will result that will not only proliferate the use of microcomputers, minicomputers, and large computers as stand-alone devices, but also increase the tendency to combine several processors into multicomputers. In the multicomputer environment, multiple processors (of varying sizes) perform different tasks traditionally performed by one processor. For example, processors are assigned to handle different input and output functions. This multicomputer architecture significantly increases the overall performance capabilities of a single computer system. It also allows a system to be totally backed up by a second set of processors that can be activated if one or more of the primary processors malfunctions.

File Storage
The ability to store massive databases will be significantly improved in future years due to improvements in magnetic disk technology. Compact disk read-only memory (CD ROM) provides a major breakthrough in disk storage for static files such as bibliographic data. It appears reasonable to expect a twenty-fold increase in the amount of data that can be stored on magnetic disk surfaces by the mid-1990s.

Besides improvements in disk technology, considerable research is being conducted into other file-storage technologies—magnetic bubbles, laser-holographic devices, charge-coupled devices, cryogenic devices, and others. This research will likely result in new types of file-storage devices that will augment, and possibly replace, magnetic disk devices.

Input/Output
Less dramatic improvement, however, is anticipated in the basic technologies of input/output devices. Electromechanically oriented devices such as card readers and line printers are relatively mature as far as their technologies go. However, higher-performance, nonimpact printers show real promise. Devices ranging from low-cost thermal printers to higher-cost electrostatic and ink-jet printers should eventually displace impact printers. The electrostatic and ink-jet technologies offer additional advantages and greater versatility—for example, multiple colors, multiple fonts, and graphic capabilities.

Computer output microfilm (COM) will be used increasingly to alleviate problems associated with distributing and storing computer printouts. Also, word-processing equipment will continue to increase in use as information systems expand from traditional data-processing activities into office environments. A novel input/output technology that will increase in use is voice recognition.

The most significant input/output technology of the future will continue to be on-line. In this area, it appears that the on-line personal computer will dominate because of its versatility and low cost. More applications-oriented terminals (e.g., banking, retailing devices, and graphics) can be expected as greater versatility in terminals evolves.

Software

Software is the area in which major computer technology accomplishments are likely to occur. Software capabilities have always lagged behind hardware capabilities. Hardware technology appears to be stabilizing; it is therefore reasonable to expect an increasing emphasis on software technology.

Major software advancements can be anticipated in the areas of database management, data communications, and programming languages.

Database Management

The various "housekeeping" functions of managing data, records, and files will be increasingly simplified by advancements in database technology. Such functions as disk track and cylinder management, secondary indexing, space allocation, address randomization, and staging of data volumes, are in the hierarchy of devices that will expand database management systems. These functions will become even more apparent to application programmers as the data manipulation languages coupled to DBMS are further perfected. Programmers will become more productive as they are relieved from these time-consuming tasks.

Data Communications

Data communications management functions such as adapting for varied line, terminal, and message characteristics will be increasingly accommodated by front-end controllers and systems management software. If industry standard communication architectures are forthcoming, as it is hoped, will simplify the interfacing required for on-line networks in which a wide variety of computers and terminals are linked together, including local area networks (LANs). The automatic handling of data communications protocols will save considerable programmer time and simplify systems design.

Easier Programming

It appears that a major revolution will involve computer programming, as COBOL, PL/1, and FORTRAN are already becoming the "languages of last resort" in sophisticated organizations.

Users will become programmers using new fourth- and fifth-generation languages, and system professionals will increasingly assume a consultant role for users. Programming by the systems groups will apply more to the large-scale, organization-wide information systems and leading-edge technology applications such as expert systems.

Implications for Systems Analysis

The future developments of organizations and computer technology will have its impact on what systems analysts do and how they do it. Several major implications are discussed next.

Organizational Implications

Environment
Future organizations are going to be faced with a more turbulent environment in terms of market conditions, society, and natural resources. To remain viable, organizations are going to require more information about their environment in order to make sound decisions. Computer-based information systems will continue to expand beyond the scope of processing information based on activities internal to the organization to processing information based on activities external to the organization.

Since it is not usually practical to collect and process data about the entire environment of an organization, statistics will be increasingly used to infer changes in the environment. For example, there will be increasing emphasis on marketing research and forecasting.

A single organization can incur considerable expense in obtaining information external to the organization. Therefore, the collection and maintenance of a multiplicity of databases about such things as market conditions and labor availability in particular industries is likely. For example, independent companies may develop databases of consumers and their preferences for the motel and hotel industry. Various motel and hotel chains would be able to purchase information extracted from these databases much more economically than they could collect and process it themselves.

A major concept that is emerging is interorganizational information systems. With this concept, organizations couple their information systems with those of other organizations to increase information availability and efficiency. The examples of Holden Business Forms and American Hospital Supply provided in Chapter 5 are illustrations of interorganizational information systems.

One of the critical environmental concerns for organizations is the increasing interest in, and pressure for, privacy legislation. Legislation to protect individuals from the abusive invasion of privacy possible through computer technology is continually being proposed and debated. Though individual privacy needs to be protected, the means by which it is protected can have serious implications for organizational databases. Organizations may be required to delete certain data from their databases, restrict the exchange of data among other organizations, and provide tighter security. They may also be subjected to more stringent audits. From a technical perspective, such legislation can severely impact both the effectiveness and the efficiency of computer-based information systems.

Internal Operations
The internal operations of future organizations can be characterized as expanding. This expansion is resulting in increasing organizational size and complexity. The technology used to produce goods and/or services is becoming less stable and less predictable. It is going to be both more difficult and more important to integrate the information about the more diversified and more volatile organizational activities in order to maintain management control and coordination.

Computer Technology Implications

Faced with the increasing need for information by organizations, it is encouraging to note that the future of computer technology looks very promising. The projected reductions in the cost of hardware will expand the number and scope of viable areas in which computer technology can be applied, including robotics, computer-assisted design and manufacturing, artificial intelligence, decision-support systems, and advanced office technology. The declining cost and increasing performance of computer hardware will make it possible for the hardware to cost-effectively absorb the overhead required to support the projected software enhancements (e.g., decision-support systems) without seriously increasing processing costs.

Clearly, more of the traditional technical details of computer technology will be handled by vendor-supplied hardware and software. Consequently, systems analysts will be able to concentrate more on the information requirements that must be satisfied by the information systems, and less on computer technology. Systems analysts and computer programmers will have to accept the obsolescence of many of their technologically elegant skills.

The systems analysts of the future will become even more involved in understanding organizations and organizational processes in order to assist organizations in determining how to logically integrate information and how to develop decision-support systems. Systems analysts will be less involved in the intricacies of the computer technology needed to support information systems.

Index

Developmental procedures, 83
Diagramming tools, 90
Dickson, Gary W., 242
Dickson and Wetherbe, 318
Digital Equipment Corporation, 352, 353
Digital patterns, 41
Direct access, 59, 70–72, 75
Direct-access storage device input/outputs (DASD I/
    Os), 318
Directing, 257
Directories (computer), 64
Discrete implementation, 14–15
Disk drives, 65, 70, 196
Distributed data processing, 62–64, 84, 348
Diversified conglomerates, 289
DO WHILE structure, 211
Documentation, 10, 12–13
    auditing for adequate, 252
    and CASE technology, 83
    centralization vs. decentralization and, 288
    of existing system, 118, 127–138, 141
    free-text, 156
    graphical, 128–138
    of inputs/output functions and access, 179–181
    narrative, 88, 127–128, 163
    and observation, 116–117, 141
    by procedures, 83
    programmers' workbench and, 225
    review of, 116–117, 141
    separate maintenance group and, 82
    standards for, 83
    tools for, 225
    user-developed systems and, 231
Dynamic system, 30
DYNAMO, 348
Dysfunctional behavioral reactions, 243, 244–246,
    253

E/M analysis. *See* Ends/means analysis
Economic evaluation, 237
Economies of scale, 283, 287, 288, 289
EDP audit, 100, 250–252, 253
EDP growth, 284–286, 291, 300, 304
Effectiveness, 13, 15, 108
    EDP audit for, 250
    in ends/means analysis, 315–316
    in future, 378
Efficiency, 13, 15, 108
    centralization vs. decentralization and, 287–288
    vs. economy, 101
    EDP audit for, 250
    in ends/means analysis, 315–316
    in future, 378
    managerial, vs. user service, 79–81
    problems and opportunities in, 98, 101

Electronic data processing (EDP). *See* EDP audit;
    EDP growth
Electronic mail systems, 64
Electrostatic printers, 376
Embezzlement, 100
End-user computing, 65–67, 357–368
    in future, 377
    to increase information system capacity, 249
    information architecture and, 366–368
    information centers and, 82
    opportunities and risks in, 357–360
    recommendations concerning, 359
    responses to, 359–360
    service support levels and, 360–366
    See also User training
Ends/means (E/A) analysis, 125–126
    MIS planning and, 312, 316–317, 320, 322
    OIRA and, 323, 325
Entities, 18, 19, 151
    attributes of, 28, 29, 151
    in closed/stable/mechanistic system, 20
    in data dictionary, 162
    data modeling to define, 149
    examples of, 152
    files and, 162
    in OIRA, 325–326
    in open/adaptive/organic system, 20–21
    in organization, 45
    in relational data model, 170–174
    system philosophy and, 27
    vs. subsystems, 25
Environment (of system), 18, 20, 21, 46
    future, 373–374, 379
    MIS planning and, 310
    of organization, 45
    systems approach and, 27
EPIC Realty Services Inc., 323, 326, 329
Equal Employment Opportunity Commission
    (EEOC), audits, 202, 203, 207
Error reports, 228
Evaluation, 15, 237, 246–253
Events
    in DFDs, 133
    on Gantt/PERT chart, 258, 263–265. *See also*
    Deliverables
Evolutionary development process, 30, 175
Excelerator, 115, 149
Exception reports, 38, 100, 119, 338
Excess capacity, 26, 100
Exclusion, 29
Executive-support systems, 64
Expected value, 350, 353
Expert systems, 39, 68, 351–353
    for constructing Gantt and PERT charts, 263
    vs. DSS, 351–352
    in future, 377
    systems analysts' skills required for, 334–335